540.3
H247c

DETROIT PUBLIC LIBRARY

CONELY BRANCH
4600 Martin
Detroit, MI 48210

DATE DUE

APR '06

JAN 2000

AGQ 2322dp CO

The CASSELL DICTIONARY of CHEMISTRY

PERCY HARRISON
and
GILLIAN WAITES

A CASSELL BOOK

This edition first published in the UK in 1998

by
Cassell
Wellington House
125 Strand
London WC2R 0BB

Copyright © Percy Harrison and Gillian Waites 1998

The right of Percy Harrison and Gillian Waites to be identified as the authors of the work has been asserted by them in accordance with the Copyright, Designs and Patents Act 1988.

All rights reserved. No part of this book may be reproduced or transmitted in any form or by any means, electronic or mechanical, including photocopying, recording or any information storage and retrieval system, without prior permission in writing from the publisher and copyright owner.

British Library Cataloguing-in-Publication Data
A catalogue record for this book is available from the British Library

ISBN 0-304-35038-9

Designed, edited and typeset by Book Creation Services, London

Printed and bound in Great Britain by
Mackays of Chatham PLC, Chatham, Kent

Contents

How to use *The Cassell Dictionary of Chemistry*

Arrangement of the dictionary

Entries are arranged alphabetically on a letter by letter basis, ignoring hyphens and spaces between words. Headwords – or main entries – are shown in **bold** type; ***bold italics*** are used to indicate an alternative form of the main headword.

Cross-references

Words that appear in SMALL CAPITALS in articles have their own entries elsewhere in the dictionary. Certain very common scientific words, such as 'element' or 'atom', are not automatically cross-referenced each time they are mentioned in the text.

See denotes a direct cross-reference to another article. *See also* indicates related articles or entries that contain more information about a particular subject.

Units

SI and metric units are used throughout the dictionary.

Abbreviations

In those cases where the part of speech of a headword is specified, the abbreviations used are as follows:

adj. adjective
n. noun
vb. verb

A

ablation The loss of material by melting or evaporation.

abrasive (*n.*, *adj.*) Any substance that is used to rub or wear away a softer material. Powders and grits based on diamond and aluminium oxide (CORUNDUM) are often used as abrasives.

ABS *See* ACRYLONITRILE-BUTADIENE-STYRENE.

absolute temperature A temperature scale in which the temperature is proportional to the energy of the random thermal motion of the molecules. This is the same as a temperature scale based on the PRESSURE LAW, which states that temperature is directly proportional to the pressure exerted by an IDEAL GAS held in a fixed volume. The SI UNIT of absolute temperature is the KELVIN. *See also* ABSOLUTE ZERO, INTERNATIONAL PRACTICAL TEMPERATURE SCALE.

absolute zero The lowest temperature that is theoretically attainable; zero on the ABSOLUTE TEMPERATURE scale (0 K). Absolute zero is the temperature at which molecules would stop moving and an IDEAL GAS would produce no pressure. It is equivalent to –273.15°C. It is impossible to achieve absolute zero, but temperatures as low as 2×10^{-9} K have been reached.

absorption 1. The taking up of a gas by a solid or a liquid, or of a liquid by a solid. The molecules of the absorbed substance penetrate throughout the whole of the absorbing substance. *Compare* ADSORPTION.

2. The removal of energy from a wave, or particles from a particle beam, as the wave or beam passes through a material. The energy of the wave or beam is usually converted into heat within the absorbing material. *See also* FILTER.

absorption coefficient The fraction of the light falling on a partially transparent material, especially a solution, that is absorbed per unit distance travelled through the material.

absorption spectrum A SPECTRUM formed when a sample absorbs certain wavelengths from a continuous background of ELECTROMAGNETIC RADIATION. *See also* ATOMIC ABSORPTION SPECTROSCOPY, SPECTROSCOPY.

absorption tower A column in which a gas, which rises up through the tower, is dissolved in a liquid, which trickles down the tower. To increase the rate of absorption, the liquid may be broken up into droplets, or passed over a suitable packing material to increase the surface area.

abundance 1. Of an element, the percentage of mass of the given element in the Earth's crust, or occasionally in a particular ore from which the element is to be extracted.

2. Of an ISOTOPE, the percentage of the atoms in a sample of a pure element that are of the specified isotope.

accelerator A catalyst that increases the rate of a chemical reaction. In particular, an accelerator is a catalyst that increases the rate of a POLYMERIZATION leading to the formation of a hard material, such as the setting of certain glues and RESINS.

acceptor atom An atom that receives a share in a pair of electrons in the formation of a CO-ORDINATE BOND.

accumulator An obsolete term for a BATTERY of LEAD-ACID CELLS.

acetaldehyde *See* ETHANAL.

acetals Organic compounds with the general formula

$$\mathrm{RHC(OR')_2}$$

(R and H bonded to C, which carries two OR′ groups)

Acetals are formed by the addition of an ALCOHOL to an ALDEHYDE. A hemiacetal is formed if one molecule of alcohol reacts with one molecule of aldehyde. A full acetal is formed by the further reaction of the hemiacetal with another alcohol molecule.

$$\underset{\text{aldehyde}}{\mathrm{R{-}CHO}} + \underset{\text{alcohol}}{\mathrm{R'OH}} \longrightarrow \underset{\text{hemiacetal}}{\mathrm{RHC(OR')(OH)}}$$

$$\begin{matrix} R \\ H \end{matrix} \!\!>\! C \!<\!\! \begin{matrix} OR' \\ OH \end{matrix} + R'OH \longrightarrow \begin{matrix} R \\ H \end{matrix} \!\!>\! C \!<\!\! \begin{matrix} OR' \\ OR' \end{matrix} + H_2O$$

acetal

The formation of acetals is reversible and under acidic conditions an aldehyde can again be formed. These reactions are useful in synthetic organic chemistry and some acetals are used as solvents. *See also* KETALS.

acetamide *See* ETHANAMIDE.

4-acetamidophenol Systematic name for PARACETAMOL.

acetanilide, N-*phenylethanamide* ($CH_3CONHC_6H_5$) A white crystalline solid, melting point 114°C; a primary AMIDE of ETHANOIC ACID. It is made by reacting PHENYLAMINE with excess ethanoic acid or ethanoic anhydride and is used in the manufacture of DYES and RUBBER.

acetate Common name for ETHANOATE.

acetic acid Common name for ETHANOIC ACID.

acetic ester *See* ETHYL ETHANOATE.

acetic ether *See* ETHYL ETHANOATE.

acetone Common name for PROPANONE.

acetylation *See* ETHANOYLATION.

acetylbenzoic acid *See* ASPIRIN.

acetyl chloride See ETHANOYL CHLORIDE.

acetylcholine A NEUROTRANSMITTER concerned with the transmission of nerve impulses from one nerve cell to another or from a nerve cell to a muscle cell, resulting in muscular contraction.

acetyl CoA *See* ACETYL COENZYME A.

acetyl coenzyme A, *acetyl CoA* COENZYME A that is carrying an ACYL GROUP.

acetylene Common name for ETHYNE.

acetylide *See* DICARBIDE.

acetylsalicylic acid *See* ASPIRIN.

acid (i) Any compound that contains hydrogen and liberates hydrogen ions when dissolved in water. Acids release hydrogen gas when they react with metals, and they react with BASES to form a SALT plus water. For example, sulphuric acid, H_2SO_4, is ionized in water:

$$H_2SO_4 \Leftrightarrow 2H^+ + SO_4^{2-}$$

Dilute sulphuric acid reacts with magnesium to release hydrogen

$$H_2SO_4 + Mg \rightarrow MgSO_4 + H_2$$

and with sodium hydroxide to form sodium sulphate plus water

$$H_2SO_4 + 2NaOH \rightarrow Na_2SO_4 + 2H_2O$$

(ii) In the Lowry–Brønsted theory of acids and bases, an acid is any compound that gives up a proton to some other substance. For example, when sodium chloride is dissolved in ammonia,

$$NH_3 + NaCl \Leftrightarrow NaNH_2 + HCl$$

the ammonia donates a proton and is thus regarded as an acid.

(iii) In the Lewis theory of acids and bases, an acid is defined as any substance that accepts electrons. This encompasses the traditional definition, as in reactions with metals and bases, hydrogen ions in the acid are neutralized to form hydrogen or water molecules. However, it also extends to REDOX REACTIONS and to the formation of CO-ORDINATE BONDS, for example:

$$H_3N: + BCl_3 \rightarrow H_3NBCl_3$$

where the ammonia is a LEWIS ACID because it donates a pair of electrons to form the bond with the boron.

See also ACIDIC, ACIDIC HYDROGEN, PH, PK, STRONG ACID, WEAK ACID, and under individual acids.

acid anhydride Any chemical compound obtained by the removal of water from another compound, usually an acid. The resulting compound is a dehydrated acid; for example sulphur trioxide, SO_3, is the anhydride of sulphuric acid, H_2SO_4.

In organic chemistry, an acid anhydride is formed by the removal of one molecule of water from two molecules of a CARBOXYLIC ACID, for example:

$$2CH_3COOH - H_2O \rightarrow CH_3COOCOCH_3$$

in which ethanoic acid is converted into ethanoic anhydride.

The general formula for an organic acid anhydride is RCOOCOR, where R is an ALKYL GROUP. The chemistry of acid anhydrides is similar to the ACID CHLORIDES except they are less reactive.

acid chloride, *acyl chloride* A CARBOXYLIC ACID in which the –OH HYDROXYL GROUP has been chlorinated, leaving an ACYL GROUP (RCO–, where R is an ALKYL GROUP) attached to chlorine. Thus the general formula for acid chlorides is RCOCl. An example is ethanoyl chloride, CH_3COCl, derived from ethanoic acid, CH_3COOH. Acid chlorides are named by replacing the suffix *-ic* of the carboxylic acid from which it is derived with *-yl*.

Acid chlorides are more reactive than acids since the chlorine atom is easily replaced by other NUCLEOPHILES. They are readily hydrolysed (*see* HYDROLYSIS) in cold water to form hydrochloric acid and induce tears if the vapour is close to the eyes. Acid chlorides are termed acylating agents since they are able to add an acyl group to a molecule.

acid dye *See* DYE.

acidic (*adj.*) Describing an ACID, a solution with a high concentration of hydrogen ions, or a material that produces an acid when dissolved in water. Thus sulphur dioxide, SO_2, which forms sulphurous acid, H_2SO_3, when dissolved in water, may be termed acidic. *See also* AMPHOTERIC, BASIC.

acidic hydrogen The portion of hydrogen in an ACID that will be replaced by a metal to form a SALT. The acidic hydrogen will form positive ions if the acid DISSOCIATES in water. *See also* ACIDIC SALT.

acidic salt Any salt in which not all of the ACIDIC HYDROGEN has been replaced by a metal. The hydrogen will be released as positive ions, forming an acidic solution, if the salt is soluble. An example is the HYDROGENCARBONATES, which contain the HCO_3^- ion. These are formed when only one of the hydrogen atoms in carbonic acid, H_2CO_3, a DIBASIC ACID, are replaced by a metal ion. *See also* BASIC SALT, NORMAL SALT.

acid rain Rain with a high acidity, caused mainly by sulphur dioxide (from volcanic emissions and the burning of FOSSIL FUELS) dissolving in water to form sulphuric and sulphurous acids. Nitrogen oxides (from industry and car exhausts) also contribute to acid rain. Acid rain causes damage in particular to coniferous forest species and some aquatic species, either directly due to the acidity or indirectly when the rain leaches toxic aluminium from soils (*see* LEACHING). *See also* POLLUTION.

acridine ($C_{13}H_9N$) An organic compound used to make dyes; melting point 110°C. Its structure consists of three fused rings with a nitrogen HETEROATOM in the centre. Derivatives of acridine, such as acridine orange, are used as dyes or biological stains.

acrylic A synthetic wool-like fabric made from the COPOLYMERIZATION of PROPENOIC ACID derivatives.

acrylic acid *See* PROPENOIC ACID.

acrylic resins Synthetic RESINS made by polymerizing derivatives of PROPENOIC ACID (acrylic acid). An example is PERSPEX, made from polymethyl methacrylate.

acrylonitrile *See* PROPENONITRILE.

acrylonitrile-butadiene-styrene (ABS) A synthetic RUBBER consisting of a copolymer (*see* COPOLYMERIZATION) of BUTADIENE, PHENYLETHENE and PROPENONITRILE (acrylonitrile). ABS is rigid and tough and used in telephone receivers and suitcases.

actin A protein that is a major constituent of muscle fibres. GLOBULAR PROTEIN monomers of actin (G-actin) polymerize to form long fibrous molecules of filamentous actin (F-actin), two of which then twist around one another to form the so-called thin filaments characteristic of muscle fibres. During muscular contraction, filaments of actin and MYOSIN contract together as ACTOMYOSIN.

actinide, *actinoid* Any one of the series of elements with atomic numbers from 89 (ACTINIUM) to 103 (LAWRENCIUM). The actinides all have two electrons in the 7s ORBITAL. Increasing atomic number corresponds to filling the 5f (and sometimes 6d) orbitals. All are radioactive and have similar chemical properties, which differ only slightly with atomic number.

actinium (Ac) The element with atomic number 89; melting point 1,050°C (approx.). It is a white metal, the first in the ACTINIDE series of elements. All known ISOTOPES of the element are radioactive with HALF-LIVES of 21.7 years (actinium–227) or less. Despite this, actinium occurs in nature in very small quantities, being produced in the decay of heavier nuclei with longer half-lives. Actinium is used as a source of high-energy ALPHA PARTICLES.

actinoid *See* ACTINIDE.

activation analysis A technique for detecting small quantities of an element present in a sample. The sample is bombarded with neutrons, forming unstable ISOTOPES of the elements present. These decay, emitting GAMMA RADIATION with energies characteristic of the elements present.

activation energy (E_a) The minimum amount of energy required before a particular process can take place. The term is usually applied to chemical reactions, where the activation energy is the ENERGY BARRIER that must be overcome for the reaction to occur. *See also* ACTIVATION PROCESS, CATALYST.

activation process Any process in which the particles involved can only take part in the process if they have more than a specified amount of energy, known as the ACTIVATION ENERGY. This acts as an ENERGY BARRIER, which must be overcome for the process to take place. To do this, energy may be supplied externally (for example by light in the case of some chemical reactions). The energy barrier may also be overcome by the action of a catalyst that allows the reaction to proceed via some intermediate state that requires less energy. Another possibility is to raise the temperature of the material so that more molecules have sufficient energy. Many chemical reactions are activation processes. *See also* MAXWELL–BOLTZMANN DISTRIBUTION, REACTION PROFILE.

active anode An ANODE that is chemically involved in an ELECTROLYSIS process.

active site In biochemistry, the part of an ENZYME molecule to which the SUBSTRATE binds. The active site is formed by the three-dimensional structure of the enzyme and the distribution of electric charge in the molecule. The substrate specificity of an enzyme is determined by the active site. An enzyme can have more than one active site. Enzyme inhibitors can also reversibly bind to the active site, thereby blocking the action of the enzyme. *See also* ENZYME INHIBITION.

active transport In biochemistry, an energy-requiring process (usually involving ATP) where substances, usually molecules or ions, are moved across a membrane against a concentration gradient – that is, from a region of low concentration to one of higher concentration. The process involves 'pumps' of protein molecules in the membrane that carry specific ions across, such as sodium. *Compare* DIFFUSION.

activity 1. (α) A thermodynamic function used to calculate the EQUILIBRIUM CONSTANT for a REAL GAS. It is a correction factor that allows for the effect that the INTERMOLECULAR FORCES between gas molecules have on the equilibrium concentrations of reacting gases. In a reaction

$$A \Leftrightarrow B + C$$

the equilibrium constant (K) is given by

$$K = \alpha_B \alpha_C / \alpha_A$$

where α_A is the activity for A, etc. The activity coefficient, γ, for a gas of pressure p is defined as

$$\gamma = \alpha / p$$

For a solution,

$$\gamma = \alpha X$$

where X is the MOLE FRACTION.

2. (A) The level of IONIZING RADIATION emitted by a radioactive material. Activity is usually measured in BECQUEREL, or becquerel per litre, although other units, such as the CURIE, are sometimes used. *See also* RADIOACTIVITY.

activity coefficient *See* ACTIVITY.

actomyosin A complex formed by the interaction of the proteins ACTIN and MYOSIN. This occurs in muscle in the presence of calcium ions and is the basis of muscular contraction. Actomyosin dissociates in the presence of ATP.

acyclic (*adj.*) Describing a compound that has no rings in its structure, only chains, sometimes branched. *Compare* CYCLIC. *See also* ALIPHATIC.

acylating agent Any organic compound capable of adding an ACYL GROUP to a molecule. *See also* ACID CHLORIDE.

acylation The substitution of an ACYL GROUP (RCO–, where R is an ALKYL GROUP) into a molecule, usually in exchange for a hydrogen atom of a HYDROXYL GROUP. For example:

$$CH_3COCl + NH_3 \rightarrow CH_3CONH_2 + HCl$$

acyl chloride *See* ACID CHLORIDE.

acyl group The RCO– group of a CARBOXYLIC ACID (RCOOH) that remains when the –OH HYDROXYL GROUP has been removed. R is an ALKYL GROUP. Acyl groups are named after the carboxylic acid from which they are derived; for example, ethanoyl, CH_3CO–, is derived from ethanoic acid, CH_3COOH.

acyl halide A CARBOXYLIC ACID in which the –OH HYDROXYL GROUP has been replaced by a HALOGEN, leaving an ACYL GROUP (RCO–, where R is an ALKYL GROUP) attached to a halogen. An example is ethanoyl chloride, CH_3COCl. *See also* ACID CHLORIDES.

addition polymerization *See* POLYMERIZATION.

addition reaction A chemical reaction in which a molecule is added to an UNSATURATED COMPOUND across a double or triple COVALENT BOND. An example is the addition of bromine to ethene to give 1,2-dibromoethane:

$$CH_2{=}CH_2 + Br_2 \rightarrow BrCH_2CH_2Br$$

See also ELECTROPHILIC ADDITION.

additive Any substance that is added, usually in small amounts, to confer specific, desirable properties on a bulk material. An example in chemistry is the use of anti-foaming additives in lubricating oils. Additives are widely used in the food industry to improve the colour, flavour or nutritional value of food, or to prolong shelf-life. These additives can be natural or artificial and their use is regulated since some can cause side-effects, such as asthma, hyperactivity and cancer in certain people. A group of additives approved by the European community are termed E NUMBERS.

adduct A compound formed by an ADDITION REACTION. In particular, an adduct is a compound formed between a LEWIS ACID and a LEWIS BASE.

adenine, *6-aminopurine* ($C_5H_5N_5$) An organic base called a PURINE that occurs in NUCLEOTIDES. *See also* BASE PAIR, DNA, RNA.

adenosine A PURINE NUCLEOSIDE, consisting of the organic base ADENINE and the sugar RIBOSE. In its phosphorylated forms (*see* PHOSPHORYLATION) adenosine is AMP, ADP and ATP.

adenosine diphosphate *See* ADP.

adenosine monophosphate *See* AMP.

adenosine triphosphate *See* ATP.

adhesive Any substance that is used to join solid surfaces together. Some adhesives are based on gums dissolved in a SOLVENT; as the solvent evaporates the gum holds the solids together. Others, such as EPOXY RESINS, contain a substance that reacts under the action of a catalyst (called a hardener), or ultraviolet light to form a solid bond. *See also* CEMENT.

adiabatic (*adj.*) Describing a change in which there is no exchange of energy between a system and its surroundings.

ADP (adenosine diphosphate) The product formed by the PHOSPHORYLATION of AMP during energy-yielding biochemical reactions, or produced from the HYDROLYSIS of ATP. ADP can be phosphorylated to form ATP.

adrenaline, *epinephrine* ($C_9H_{13}NO_3$) A HORMONE, derived from the AMINO ACID tyrosine, secreted by the adrenal gland in response to external stress, such as fear, anger or pain. It prepares the body for action by increasing blood flow to the heart and muscles, causing the heart rate to quicken and dilating airways in the lungs to enable more oxygen to be delivered to cells of the body, whilst constricting blood vessels in the skin and gut. Adrenaline also increases the amount of sweat, causes hair to stand up, eye pupils to dilate and increases breakdown of GLYCOGEN to GLUCOSE in the liver. It can be used in the treatment of allergic reactions or circulatory collapse.

adsorption The taking up of a gas by the surface of a solid. Unlike ABSORPTION, the gas does not penetrate the solid material, but is held on the surface either by the formation of chemical bonds (called chemisorption) or by VAN DER WAALS' FORCES (called physisorption).

adsorption chromatography *See* CHROMATOGRAPHY.

adsorption indicator An INDICATOR used in reactions that involve the formation of PRECIPITATES. Many DYES, such as fluorescein, have a different colour when adsorbed onto the surface of precipitate particles, but other ions will be adsorbed in preference (*see* ADSORPTION). When such ions are no longer present, at the end point of a TITRATION for example, the precipitate starts to adsorb the dye molecules and a colour change occurs.

aerosol A COLLOID in which liquid particles are suspended in a gas. The term aerosol is also used to refer to the mechanism for producing an aerosol, in which a propellant gas forces liquid out of a tube through a fine nozzle. CHLOROFLUOROCARBONS (CFCs) have been used as the propellant, but alternatives such as BUTANE are now more common since concerns have emerged over damage to the ozone layer from CFCs in the atmosphere.

affinity chromatography A CHROMATOGRAPHY technique that depends on the affinity between specific molecules. The stationary phase, usually packed into a column, contains a substance that the molecule under investigation will attach to, thus separating it from the mixture. The bound molecule can then be eluted from the stationary phase using a different solvent. This technique is useful, for example, in ANTIBODY purification, where a specific ANTIGEN is attached to the matrix. It can also be used to separate groups of substances, such as sugars, from a mixture. Sometimes it is easier to bind unwanted molecules to the stationary phase.

agar An extract of red seaweeds that forms a gel at room temperature. Agar is a mixture of two POLYSACCHARIDES, agarose and agaropectin. It is widely used as a gelling agent in foodstuffs, medicine, cosmetics and as a culture media in microbiology.

agarose A POLYSACCHARIDE containing 3,6-anhydro-L-galactose and D-galactose as the repetitive units. Agarose is one of the constituents of AGAR and is used in CHROMATOGRAPHY and ELECTROPHORESIS.

agate A form of QUARTZ found in some igneous rocks. It is usually red-brown in colour due to the presence of iron oxide and often shows banded patterns of differing depths of colour.

air The mixture of gases forming the Earth's atmosphere. Dry air contains 78 per cent nitrogen, 21 per cent oxygen, 0.9 per cent argon, 0.03 per cent carbon dioxide and traces of other NOBLE GASES. In addition, air usually contains a few per cent water vapour, though the concentration varies widely.

alanine ($C_3H_7NO_2$) An AMINO ACID found in most proteins.

alcohol Any one of a group of organic chemicals with the structure of an ALKANE but with one or more of the hydrogen atoms replaced by HYDROXYL GROUPS (–OH). An alcohol thus has the general formula $C_nH_{2n+1}OH$. The oxygen atom is added between the carbon–hydrogen bond. *Compare* ETHER.

Alcohols are classified by the position of their ALKYL GROUPS; primary alcohols contain CH_2OH, secondary alcohols contain CHOH and tertiary alcohols contain COH.

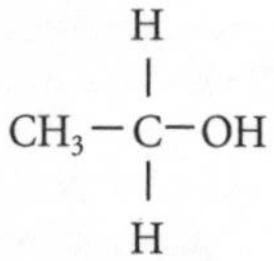

```
       H                 H
       |                 |
 CH3 - C - OH      CH3 - C - OH
       |                 |
       H                 CH3

   Primary           Secondary

            CH3
             |
       CH3 - C - OH
             |
            CH3

          Tertiary
```

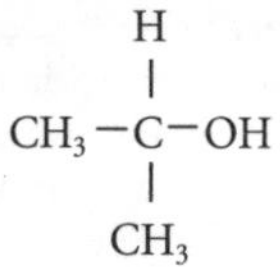

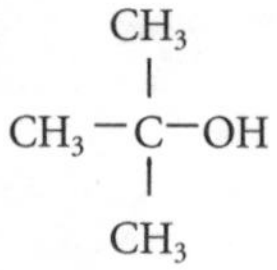

The nomenclature used for alcohols is based on that of the alkane forming the carbon skeleton, replacing the *-ane* suffix with *-anol*, for example propanol. The position of the hydroxyl group is indicated by a number placed before the *-ol*, for example propan-1-ol. Where there is more than one hydroxyl group this is also indicated in the name, for example ethane-1,2-diol has two hydroxyl groups and is called a dihydric alcohol. An alcohol with three hydroxyl groups is called trihydric and so on.

OXIDATION of primary alcohols yields ALDEHYDES, which can in turn be oxidized to CARBOXYLIC ACIDS.

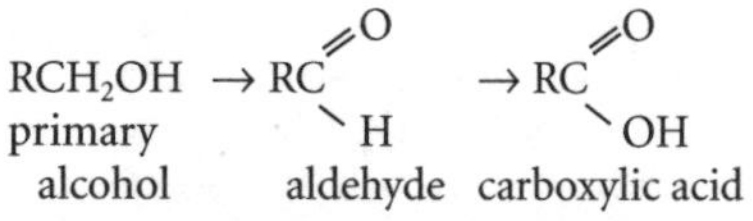

```
                    //O               //O
RCH2OH    →  RC            →  RC
                    \ H               \ OH
primary
 alcohol       aldehyde     carboxylic acid
```

Oxidation of secondary alcohols yields KETONES, which are not easily oxidized further. Alcohols react with acids to form ESTERS. HALOGENOALKANES can be manufactured by reacting alcohols with phosphorus HALIDES. The lower alcohols – methanol, ethanol, propanol, butanol – are liquids that mix with water. The higher members are oily liquids and the highest are waxy solids.

Ethanol is produced naturally during FERMENTATION and is used to manufacture alcoholic beverages. Other uses of alcohols are as solvents, in dye manufacture, in the cosmetics and medical industries, in DETERGENTS and in the manufacture of POLYESTER. Ethane-1,2-diol is used as an ANTIFREEZE. A common laboratory test for an alcohol is the evolution of hydrogen gas when sodium is added (*see* ALKOXIDE).

alcoholic fermentation *See* FERMENTATION.

aldehyde, *alkanal* Any of a group of organic compounds containing the group

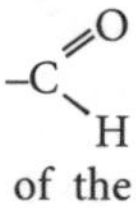

```
    //O
 –C
    \ H
```

The carbon of the CARBONYL GROUP (C=O) can be attached to another hydrogen atom as H-CO-H (written as HCHO), or to an ALKYL GROUP (R) as R-CO-H (written as RCHO). The general formula of an aldehyde is $C_nH_{2n}O$, the same as a KETONE, except that in the latter this comprises two alkyl groups. The nomenclature of aldehydes follows that of the ALKANE with the same carbon skeleton, with the ending *-ane* being replaced by *-anal.* The carbonyl carbon atom is given the positional number 1.

Aldehydes can be ALIPHATIC (which are usually colourless liquids) or AROMATIC (the higher members are solids). Examples of aldehydes include methanal, ethanal and benzenecarbaldehyde.

O
||
C
/ \
H H
methanal

O
||
C
/ \
CH_3 H
ethanal

CHO
|
benzene-
carbaldehyde

Aldehydes are formed by the oxidation of primary ALCOHOLS, hence their name, since the alcohol loses a hydrogen atom (*al*cohol *dehyd*rogenation). Aldehydes can be reduced back to primary alcohols and are themselves readily oxidized to the corresponding CARBOXYLIC ACID. FEHLING'S TEST and TOLLEN'S REAGENT are used to test for aldehydes.

See also ALDOL CONDENSATION, CANNIZZARO REACTION, SCHIFF'S REAGENT.

aldol *See* HYDROXYALDEHYDE.

aldol condensation The reaction between two ALIPHATIC ALDEHYDES to yield a HYDROXYALDEHYDE (formerly known as aldol, hence the name). The hydroxyaldehydes easily lose water to yield unsaturated aldehydes. An example is the reaction of ethanal in dilute alkaline solution to give 3-hydroxybutanol, which is then dehydrated to give but-2-enal.

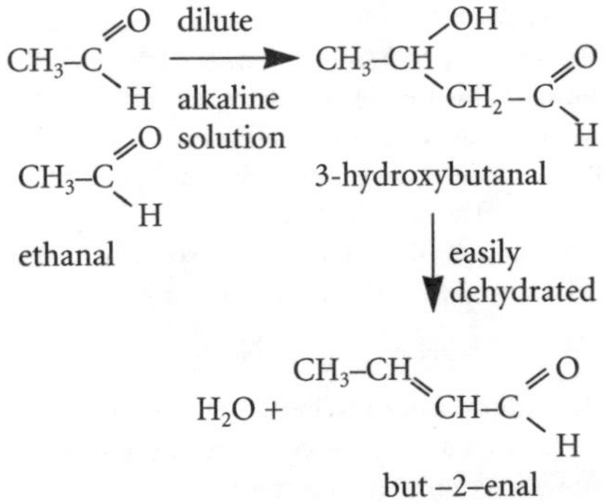

$$2\,CH_3{-}CHO \xrightarrow{\text{dilute alkaline solution}} CH_3{-}CH(OH){-}CH_2{-}CHO$$

ethanal → 3-hydroxybutanal

easily dehydrated

$$H_2O + CH_3{-}CH{=}CH{-}CHO$$

but –2–enal

Only those aldehydes containing the $-CH_2CHO$ group will react in this way; other aldehydes undergo the CANNIZZARO REACTION. KETONES can also react in this way to produce unsaturated ketones but they do so less readily. These type of reactions where new carbon–carbon bonds are formed are useful in synthesizing large carbon skeletons from smaller molecules.

aldose, *aldo-sugar* A sugar containing an ALDEHYDE group (CHO). *See* MONOSACCHARIDE.

aldosterone ($C_{21}H_{28}O_5$) A MINERALOCORTICOID hormone that regulates water retention in the kidney by controlling the distribution of sodium in the body tissue. It also affects centres in the brain, creating a sensation of thirst to stimulate the animal to seek water.

aldo-sugar *See* ALDOSE.

alicyclic (*adj.*) Describing any organic compound that is CYCLIC but does not possess an AROMATIC ring. These compounds are therefore cyclic with ALIPHATIC characteristics, for example CYCLOHEXANE. *See also* HETEROCYCLIC.

aliphatic (*adj.*) Describing an organic chemical in which the carbon atoms are linked by COVALENT BONDS in straight chains, such as pentane, C_5H_{12}, or in branched chains, such as methylpropane, $CH_3CH(CH_3)CH_3$. This is in contrast to CYCLIC compounds.

alizarin ($C_{14}H_8O_4$) An orange/red compound that is a derivative of ANTHRAQUINONE. It is a DYE used in the production of red/purple LAKES, when dissolved in alkaline solutions and precipitated with heavy metal salts. Alizarin occurs naturally or can be made synthetically.

alkali A base that dissolves in, or reacts with, water to produce HYDROXIDE ions, OH^-. Examples are sodium hydroxide, NaOH, which dissolves to form Na^+ and OH^- ions, and ammonia, NH_3, which produces NH_4^+ and OH^- ions.

alkali metal Any element, except hydrogen, from GROUP 1 (formerly Group IA) of the PERIODIC TABLE. The alkali metals are LITHIUM, SODIUM, POTASSIUM, RUBIDIUM, CAESIUM and FRANCIUM. They are classified by their electronic configuration, which is a NOBLE GAS electron structure plus a single electron in an outer S-ORBITAL. They are all highly reactive and readily lose an electron to form a positive ion with a single charge (M^+). Alkali metals form IONIC compounds and react with water to form HYDROXIDES plus hydrogen. Their reactivity increases with increasing ATOMIC NUMBER.

alkaline (*adj.*) Having the properties of an alkali.

alkaline earth Any element from GROUP 2 (formerly Group IIA) of the PERIODIC TABLE. The alkaline earth metals are BERYLLIUM, CALCIUM, MAGNESIUM, STRONTIUM, BARIUM and RADIUM. They are all chemically reactive, though less so than the ALKALI METALS. Reactivity increases with increasing ATOMIC NUMBER, though beryllium is more reactive than its position would suggest.

All compounds formed by alkaline earths are IONIC in character, with the metal atoms losing two electrons to form M^{2+} ions. This is due to their electron structure, which is that of a NOBLE GAS plus two electrons in an outer S-ORBITAL. Many, but not all, of the compounds are soluble in water, but the solubilities are generally lower than for compounds of the alkali metals.

alkaloid One of a group of organic substances that occur in many plants and are usually poisonous. Many drugs used in medicine owe their properties to the presence of alkaloids. Examples include morphine, caffeine and nicotine. Alkaloids vary in their constitution but all are basic and combine with acids to form salts that are usually water-soluble.

alkanal *See* ALDEHYDE.

alkane, *paraffin* The general name for an ALIPHATIC HYDROCARBON having the general formula C_nH_{2n+2}. Alkanes have only single COVALENT BONDS and are therefore said to be SATURATED, for example:

CH_3CH_3 ethane or

```
   H   H
   |   |
H— C — C —H
   |   |
   H   H
```

The first four alkanes of the group – methane, ethane, propane, butane – are gases, whilst those with higher relative molecular mass are liquids or solids (the latter being those larger than $C_{16}H_{34}$). The liquids form the basis of petrol, kerosene and lubricating oil, whilst the solids form paraffin WAXES used in cosmetics and ointments. The names of all the alkanes end in *-ane* with the prefix referring to the number of carbon atoms present, for example the alkane with five carbon atoms is called pentane, the one with six is hexane.

Alkanes are found in PETROLEUM and NATURAL GAS. They are insoluble in water but soluble in benzene and chloroform. They are relatively chemically unreactive, compared to hydrocarbons with attached FUNCTIONAL GROUPS such as the HALOGENOALKANES. They are all flammable.

alkanone *See* KETONE.

alkene, *olefin* The general name for any ALIPHATIC HYDROCARBON that possesses one or more carbon-carbon double COVALENT BONDS and has the general formula C_nH_{2n}. Alkenes are therefore said to be UNSATURATED compounds, for example:

$CH_2{=}CH_2$ ethene or

```
H\     /H
  C = C
H/     \H
```

The names of all the alkenes end in *-ene* with the prefix referring to the number of carbon atoms present, for example the alkene with two carbon atoms is called ethene, the one with five is pentene.

The lower alkenes, such as ethene and propene, are gases obtained from PETROLEUM by CRACKING and provide the raw materials for most of the organic chemical industry. Products such as poly(ethene), polystyrene, polyvinyl chloride (PVC) and a vast range of detergents, paints and pharmaceuticals are all derived from the alkenes.

Alkenes can be reduced to ALKANES and they burn in air to give carbon dioxide and water. The alkenes are more reactive than the alkanes and react by ELECTROPHILIC ADDITION reactions, adding groups across the DOUBLE BOND. Their reactivity is due to the nature of the carbon-carbon double bond, which consists of a SIGMA-BOND and a PI-BOND and is weaker than two single (sigma) carbon-carbon bonds. The double bond is therefore the site of most of the reactions of alkenes.

alkoxide Metal derivatives formed by the reduction of ALCOHOLS by ALKALI METALS. For example, methanol is reduced by sodium to give sodium methoxide and hydrogen gas:

$$2CH_3OH + 2Na \rightarrow 2NaOCH_3 + H_2$$

This and similar reactions with other alcohols form the basis of the common laboratory test for an alcohol, which is the evolution of hydrogen gas from a liquid when sodium is added.

Alkoxides can also be formed as a result of the partial IONIZATION of alcohols in the presence of sodium hydroxide or potassium hydroxide:

$$ROH + HO^- \Leftrightarrow RO^- + H_2O$$

where RO^- indicates an alkoxide.

Alkoxides are used as catalysts and reagents in organic chemistry.

alkyd resin A type of RESIN used in coatings, particularly paints. It is a copolymer (*see* COPOLYMERIZATION) of PHTHALIC ACID and GLYCEROL, and is brittle.

alkylation The introduction of an ALKYL GROUP into a HYDROCARBON chain or ring structure.

alkyl group (C_nH_{2n+1}) The group that remains after removing a hydrogen atom from an ALKANE. They are named after the alkane from which they are derived by replacing the *-ane* ending with *-yl*, for example methyl, ethyl. The symbol R is used to denote an unspecified alkyl group. The extent of chain branching in an alkyl group determines its classification as primary, secondary or tertiary.

```
    H          R′         R′
    |          |          |
  R—C—       R—C—       R—C—
    |          |          |
    H          H          R″

 primary   secondary   tertiary
```

alkyl halide *See* HALOGENOALKANE.

alkyne The general name for UNSATURATED HYDROCARBONS that possess one or more carbon-carbon triple COVALENT BONDS with the general formula C_nH_{2n-2}, for example:

CH≡CH ethyne or H–C≡C–H

Lower alkynes such as ethyne are gases, whilst those with higher relative molecular masses are liquids or solids.

allenes DIENES with the general formula C_nH_{2n-2} and containing the group

```
\         /
 C = C = C
/         \
```

There are two adjacent double bonds linking three carbon atoms and each outer carbon is linked to another group by a single bond. Allenes are isomeric with ALKYNES and under basic conditions convert to alkynes. They undergo typical reactions of ALKENES but are less stable. Allenes are usually colourless liquids with a garlic-like odour; the higher members are solids. 'Allene' is also the common name of the simplest member of the group, whose systematic name is 1,2-propadiene (CH_2=C=CH_2).

allosteric enzyme An ENZYME that has a binding site on its surface in addition to the ACTIVE SITE. The second site is termed an allosteric site. This gives the enzyme two forms: an active form and an inactive form. In the active form, the SUBSTRATE is able to bind to the active site and the enzyme is functional. In the inactive form, a regulatory molecule can bind to the allosteric site and alter the overall shape of the enzyme, thereby preventing the substrate binding to the active site. In the inactive form the enzyme is not functional.

In reactions catalysed by allosteric enzymes, the products often act as the regulatory molecules, binding to the allosteric site and causing feedback inhibition (*see* NEGATIVE FEEDBACK). In non-competitive inhibition (*see* ENZYME INHIBITION) an inhibitor can bind to an allosteric site on an enzyme and prevent it functioning.

allotrope Any one of two or more forms of an element or compound that have different physical properties but exist in the same state of matter (solid, liquid or gas). *See* ALLOTROPY.

allotropy The existence of an element or compound in more than one form (called allotropes). Allotropes have different physical properties, but exist in the same state of matter (solid, liquid or gas). For example, the solid allotropes of carbon are diamond, graphite and AMORPHOUS carbon, such as coal, charcoal and soot. Both diamond and graphite are CRYSTALLINE and all three allotropes are stable at all temperatures.

Some substances have allotropes that are stable at different temperatures. An example is the allotropes of sulphur. Red sulphur is stable at higher temperatures, whilst yellow sulphur is stable at lower temperatures. A third, non-crystalline allotrope, amorphous sulphur also exists. This is unstable at all temperatures, but a high ACTIVATION ENERGY means that it is converted only extremely slowly to the other forms. Allotropy that involves only crystalline solids is called POLYMORPHISM.

Gaseous oxygen exists in two allotropes: O_2 ('normal' oxygen) and O_3 (OZONE). These differ in their molecular configurations. *See also* ENANTIOTROPY, MONOTROPY.

alloy A material with metallic properties consisting of two or more metals or a metal with a nonmetal. An alloy may be a SOLID SOLUTION, a compound or a mixture of two or more crystalline solids. Alloys are very often used in engineering applications. They are often stronger, harder or more resistant to corrosion than their constituent metals.

The most common example of an alloy is steel, which consists of a few per cent of carbon in iron. Many steels also contain other elements, such as chromium and manganese. The carbon atoms are much smaller than the iron atoms, and become INTERSTITIAL atoms, occupying the gaps between the roughly spherical iron atoms. The carbon atoms have relatively little effect on the ELASTIC properties of the iron, which depend on the interatomic forces between the iron atoms. However, they do prevent the onset of PLASTIC behaviour, caused by imperfections or gaps in the lattice, called DISLOCATIONS. These dislocations can move through a metal: as one atom moves to fill a gap in the lattice it leaves a gap in its original location, so the dislocation effectively moves through the lattice in the opposite direction. This makes the metal far softer than it would have been without any dislocations. In small quantities, the carbon in steel 'pins' the dislocations, increasing the strength of the metal without destroying its ductility (*see* DUCTILE). Larger quantities of carbon in steel produce a very hard but brittle material called cast iron. Similar properties apply to other metal alloys: for example, aluminium-magnesium alloy has a high strength for its density. (This high strength with low weight is exploited in the manufacture of aircraft.)

Although they may have enhanced mechanical properties, alloys are generally poorer conductors of electricity and heat than pure metals. This is because the lattice structure is less highly ordered, making it harder for any FREE ELECTRONS to pass through the metal.

See also INTERMETALLIC COMPOUND.

Alnico (*Trade name*) A family of magnetic alloys containing aluminium, nickel and cobalt, used in the manufacture of permanent magnets.

alpha decay The spontaneous disintegration of an unstable atomic nucleus with the emission of an ALPHA PARTICLE. *See* ALPHA RADIATION.

alpha helix, *α-helix* A common type of secondary structure of PROTEINS in which the POLYPEPTIDE chain is coiled into a corkscrew shape. HYDROGEN BONDS form between successive turns of the helix, stabilizing the structure.

alpha particle A helium–4 nucleus (a stable particle consisting of two protons and two neutrons), emitted during ALPHA RADIATION.

alpha radiation The emission of alpha particles, which are helium–4 nuclei (two protons and two neutrons bound together in a stable entity). Alpha radiation occurs when large nuclei, which are unstable due to the ELECTROSTATIC repulsion of the protons in the nucleus for one another, spontaneously disintegrate.

Alpha particles are highly ionizing (*see* IONIZING RADIATION), and hence lose their energy very quickly. They have a range of only a few centimetres in air and can be stopped by a thin sheet of paper. When a nucleus emits an alpha-particle, it changes into a new nucleus with an ATOMIC NUMBER smaller by 2, and a MASS NUMBER smaller by 4. For example, the metal radium–226 decays to the gas radon–224 with the emission of an alpha particle.

alum Any double SULPHATE containing a MONOVALENT metal A and a TRIVALENT metal B in the crystalline form $A_2SO_4.B_2(SO_4)_3.24H_2O$. The term particularly refers to aluminium potassium sulphate, $Al_2(SO_4)_3.K_2SO_4.24H_2O$, commonly known as potash alum. Potash alum occurs naturally and is important as a MORDANT for dyes and in the processing of leather.

alumina *See* ALUMINIUM OXIDE.

aluminate Any salt containing the aluminate ion $[Al(OH)_4]^-$, formed by the reaction of aluminium hydroxide, $Al(OH)_3$, with strong bases.

aluminium (Al) The element with atomic number 13; relative atomic mass 26.98; melting point 660°C; boiling point 1,800°C; relative density 2.7. It is a chemically reactive TRIVALENT metal, forming compounds containing the Al^{3+} ion.

Aluminium is the most abundant metal in the Earth's crust. It is extracted from its main ore, BAUXITE, by the process of ELECTROLYSIS. In the main industrial method, the bauxite is first purified to obtain the OXIDE Al_2O_3, which is mixed with CRYOLITE, Na_3AlF_6, to lower its melting point. The molten mixture is then electrolysed with graphite electrodes. Molten aluminium is produced at the CATHODE (generally the lining of the CELL) and tapped off.

Aluminium reacts with atmospheric oxygen to form a hard surface layer of aluminium oxide, which prevents further chemical attack. Aluminium is thus far more useful as an engineering material than its reactivity would suggest and is highly corrosion resistant. In its pure form it is very soft, but when alloyed with other metals, such as magnesium, it forms light low-density alloys that are used in the aerospace industry. Aluminium is also used for the manufacture of cooking utensils, overhead power cables and other applications where its lightness and/or high electrical conductivity may be exploited.

aluminium chloride ($AlCl_3$) A white solid; sublimes at 178°C (the ANHYDROUS salt); relative density 2.44. Aluminium chloride reacts violently with water and fumes in moist air. Aluminium chloride can be formed by passing hydrogen chloride over hot aluminium:

$$6HCl + 2Al \rightarrow 2AlCl_3 + 3H_2$$

Commercially, it is manufactured by passing chlorine over heated aluminium oxide and carbon:

$$6Cl_2 + 2Al_2O_3 + 3C \rightarrow 4AlCl_3 + 3CO_2$$

Aluminium chloride contains POLAR BONDS and acts as a LEWIS ACID. It is also a powerful drying agent. In industry, it is used as a catalyst, particularly in the CRACKING of oil.

aluminium hydroxide ($Al(OH)_3$) A white crystalline compound; relative density 2.4–2.5. Aluminium hydroxide is formed as a characteristic gelatinous white precipitate by the reaction of alkalis with aqueous aluminium salts:

$$Al^{3+} + 3OH^- \rightarrow Al(OH)_3$$

Aluminium hydroxide is AMPHOTERIC, and the precipitate will dissolve in an excess of alkali to form the ALUMINATE ion, $[Al(OH)_4]^-$. Aluminium hydroxide decomposes on heating, losing water to form aluminium oxide:

$$2Al(OH)_3 \rightarrow Al_2O_3 + 3H_2O$$

aluminium oxide, *alumina* (Al_2O_3) A white crystalline solid; melting point 2,020°C; boiling point 2,980°C. Aluminium oxide occurs naturally as the ore BAUXITE and in a more pure form in the gemstones ruby and sapphire. Aluminium oxide is insoluble and unreactive. It is a very hard material, widely used as an abrasive and in the manufacture of REFRACTORY materials.

aluminium potassium sulphate *See* ALUM.

aluminium sulphate ($Al_2(SO_4)_3$) A white crystalline compound, commonly occurring as the HYDRATE, $Al_2(SO_4)_3.18H_2O$. It loses water at 86°C and decomposes at 770°C. The hydrate is soluble in water and may be formed by the reaction of aluminium hydroxide with sulphuric acid:

$$Al_2(OH)_3 + 3H_2SO_4 \rightarrow Al_2(SO_4)_3 + 3H_2O$$

Aluminium sulphate is important in the treatment of sewage and drinking water and is also used as a fireproofing agent.

amalgam An alloy of a metal with mercury. Amalgams are soft and have low melting points. Silver and gold amalgams were traditionally used in dentistry to fill decayed teeth, but fears about the toxicity of mercury have led to a decline in this use.

americium (Am) The element with atomic number 95; melting point 994°C; boiling point 2,607°C; relative density 13.7. Americium occurs in minute quantities in uranium ores, and is made in commercial quantities in nuclear reactors. It is used as an ALPHA PARTICLE source in smoke detectors and other devices that require a strong source of IONIZING RADIATION. Ten ISOTOPES with HALF-LIVES up to 7,700 years (americium–243) are known.

amide An organic compound derived from ammonia, NH_3, in which one or more of the hydrogen atoms has been replaced by an organic acid group. A primary amide has one hydrogen of ammonia replaced ($RCONH_2$, where R is an ALKYL GROUP), a secondary amide has two hydrogens replaced ($(RCO)_2NH$) and a tertiary amide has all three replaced ($(RCO)_3N$). Primary amides are formed by the reaction of ammonia or AMINES with an ACID CHLORIDE, ANHYDRIDE or an ESTER. For example, ethanamide is formed in the reaction between ethanoyl chloride and ammonia:

$$CH_3COCl + NH_3 \rightarrow CH_3CONH_2 + HCl$$

or between ethyl ethanoate and ammonia:

$$CH_3COOC_2H_5 + NH_3 \rightarrow CH_3CONH_2 + C_2H_5OH$$

Ethanamide (acetamide) is an example of a primary amide. Amides are weakly basic and react with nitrous acid to form CARBOXYLIC ACIDS.

$$RCONH_2 + HNO_2 \rightarrow RCOOH + N_2 + H_2O$$

Secondary and tertiary amides are formed by treating primary amides or NITRILES with organic acids or their anhydrides.

See also HOFMANN DEGRADATION.

amination The addition of the AMINO GROUP, $-NH_2$ to a molecule. An example is the reaction of halogenated hydrocarbons with ammonia under conditions of high pressure and temperature.

amine An organic compound derived from ammonia, NH_3, in which one or more of the hydrogen atoms has been replaced by an ALKYL GROUP. Amines are classified as primary, secondary or tertiary depending on whether one, two or three alkyl groups are present respectively. The general formula for an amine is $C_nH_{2n+3}N$. The names of amines are derived from the alkyl groups attached to the nitrogen followed by the ending *-amine*, such as methylamine, CH_3NH_2, ethylmethylamine, $C_2H_5NHCH_3$.

Amines can be ALIPHATIC, AROMATIC or a mixture. They are weak bases and easily form complexes with LEWIS ACIDS (the complexing agent EDTA is an amine). Many amines have distinctive odours, for example ethylamine smells of rotting fish.

Amines may be produced by the reduction of NITRILES,

$$RCN \rightarrow RCH_2NH_2$$

where R is an ALKYL GROUP. They may also be produced by the reduction of of AMIDES,

$$RCONH_2 \rightarrow RNH_2$$

The latter reaction is known as the HOFMANN DEGRADATION.

Primary and secondary amines react with ACID CHLORIDES and ANHYDRIDES to give amides. Primary amines react with nitrous acid to yield molecular nitrogen, which effervesces, providing a useful test for primary amines.

$$RNH_2 + HNO_2 \rightarrow N_2 + ROH + H_2O + \text{other organic products}$$

Artificial sweeteners are derived from amines and the aromatic amines are used in the dyeing industry.

amino acid One of a group of water-soluble molecules mainly composed of carbon, oxygen, hydrogen and nitrogen containing a basic AMINO GROUP, $-NH_2$, and an acidic CARBOXYL GROUP, $-COOH$. There are 20 amino acids that make up all the different proteins known. Some of these 20 amino acids are called the ESSENTIAL AMINO ACIDS.

All amino acids have the same core structure (two carbon atoms, two oxygen atoms, a nitrogen and four hydrogen atoms) with a variable ALKYL GROUP (denoted R) attached to this. This can be as simple as another hydrogen atom, as in GLYCINE, NH_2CH_2COOH, or more complex, as in TYROSINE, $C_6H_4OH.CH_2CH.(NH_2).COOH$. Some amino acids contain sulphur groups (for example CYSTEINE, CYSTINE and METHIONINE). Many amino acids are neutral because they have one acidic and one basic group, for example VALINE. Others have more basic, $-NH_2$, groups, for example ARGININE and LYSINE, and some have more acidic (COOH) groups, for example ASPARTIC ACID and GLUTAMIC ACID.

All amino acids except glycerine form ISOMERS; all naturally occurring amino acids are of the L-form (*see* LAEVOROTATORY, OPTICAL ISOMERISM). Amino acids can join together to form a PEPTIDE or POLYPEPTIDE.

See also ASPARAGINE, GLUTAMINE, HISTIDINE, ISOLEUCINE, LEUCINE, PHENYLALANINE, SERINE, THREONINE, TRYPTOPHAN.

2-aminobenzoic acid, *anthranilic acid* ($C_7H_7NO_2$) An important DYESTUFFS intermediate that is manufactured by HOFMANN DEGRADATION. It is used in the synthesis of indigo dye and it can be diazotized (*see* DIAZOTIZATION) and used as a first component in AZO DYES.

4-(2-aminoethyl)-imidizole *See* HISTAMINE.

amino group The NH_2 group. *See* AMINO ACIDS.

α-aminoisocaproic acid *See* LEUCINE.

***β*-aminopropylbenzene** *See* AMPHETAMINE.

ammine Any COMPLEX ION containing ammonia molecules, NH_3, as LIGANDS, for example copper tetrammine $[Cu(NH_3)_4]^{2+}$.

ammonia, *nitrogen hydride* (NH_3) A colourless gas; melting point –74°C; boiling point –31°C. Ammonia is an irritant gas, with a characteristic smell and is highly soluble in water, where it forms ammonium ions, NH_4^+. The molecule has a TRIGONAL PYRAMIDAL shape: the nitrogen atom forms the apex of a triangle-based pyramid, with a hydrogen atoms at each of the three corners of the base. The nitrogen atom has a LONE PAIR of electrons, which has a profound effect on the properties of ammonia. HYDROGEN BONDS may be formed with other ammonia molecules and with water molecules.

Ammonia is manufactured commercially in the HABER PROCESS and is an important material in the manufacture of fertilizers, explosives and dyes. In nature, the formation of ammonia is vitally important in the NITROGEN CYCLE. Nitrogen-fixing bacteria (*see* NITROGEN FIXATION) convert atmospheric nitrogen into ammonia, which is used by nitrifying bacteria (*see* NITRIFICATION) to produce NITRITES and NITRATES.

ammonification The breakdown of PROTEINS and AMINO ACIDS by bacteria to produce AMMONIA. *See* NITROGEN CYCLE.

ammonium carbonate ($(NH_4)_2CO_3$) A white crystalline solid that decomposes slowly at room temperature and more rapidly on heating:

$$(NH_4)_2CO_3 \rightarrow 2NH_3 + CO_2 + H_2O$$

Ammonium carbonate can be formed by heating ammonium chloride and calcium carbonate:

$$2NH_4Cl + CaCO_3 \rightarrow (NH_4)_2CO_3 + CaCl_2$$

The ammonium carbonate is more volatile than the calcium chloride and SUBLIMES.

ammonium chloride (NH_4Cl) A white crystalline solid that sublimes at 340°C; relative density 1.5. Ammonium chloride is soluble in water. It can be prepared by the reaction between ammonia and hydrogen chloride, either in aqueous solution or directly between the two gases:

$$NH_3 + HCl \rightarrow NH_4Cl$$

The gas phase reaction, in which the product appears as white fumes, is used commercially. Ammonium chloride is widely used as the ELECTROLYTE in ZINC-CARBON CELLS.

ammonium nitrate (NH_4NO_3) A white crystalline solid; melting point 197°C; boiling point 210°C; relative density 1.7. Ammonium nitrate is soluble in water and can be prepared by the reaction between ammonia and dilute nitric acid:

$$NH_3 + HNO_3 \rightarrow NH_4NO_3$$

Commercially, the same reaction is used to make large quantities of ammonium nitrate, with ammonia gas being bubbled through nitric acid. The compound is very important commercially, with large amounts used in the manufacture of fertilizers as it is a good source of nitrogen. It is also used in the manufacture of explosives.

ammonium sulphate ($(NH_4)_2SO_4$) A white crystalline solid; decomposes at 235°C; relative density 1.7. Ammonium sulphate is manufactured by the reaction between ammonia and dilute sulphuric acid:

$$2NH_3 + H_2SO_4 \rightarrow (NH_4)_2SO_4$$

Ammonium sulphate is used commercially as a fertilizer.

amorphous (*adj.*) Having no particular shape. The term especially refers to solids such as glass, where the molecules have no regular LATTICE arrangement. *See also* DISORDERED SOLID.

AMP (adenosine monophosphate) A NUCLEOTIDE component of DNA and RNA. It is formed by the HYDROLYSIS of ATP and ADP. PHOSPHORYLATION of AMP yields ADP. AMP can be converted to CYCLIC AMP by the enzyme adenylate cyclase in response to the appropriate extracellular signals. Cyclic AMP is important in many biochemical pathways.

ampere (*abbrev.* amp; *symbol* A) The SI UNIT of electric current. Since a current produces a magnetic field and a current in a magnetic field experiences a force, two currents flowing close to one another will produce a force on one another. The size of this force is used to define the ampere: one ampere is equal to that current which, when flowing in two infinitely long parallel wires one metre apart in a vacuum, will produce a force of 2×10^{-7} N on each metre of their length. This force will be attractive if the currents are in the same direction, repulsive if they are in opposite directions. This definition of the ampere determines the strength of the magnetic field produced by a given current.

amperometric titration Any TITRATION in which the END POINT is determined by measuring the current generated by a vessel in which the titration is performed as a HALF-CELL, along with a second half-cell of known ELECTRODE POTENTIAL.

amphetamine, *β-aminopropylbenzene* ($C_9H_{13}N$) A drug that stimulates the release of certain NEUROTRANSMITTERS and inhibits sleep, reduces appetite and affects mood. It is addictive and so its use is restricted.

amphiboles The members of a group of minerals comprsing the SILICATES of sodium, calcium magnesium, iron and aluminium. They are important constituents of IGNEOUS rocks.

ampholyte A solution containing an AMPHOTERIC substance that will act as an ALKALI in the presence of a STRONG ACID, or as an ACID in the presence of a strong base.

amphoteric (*adj.*) Displaying the properties of both an ACID and a BASE. The OXIDES and HYDROXIDES of some TRANSITION METALS are amphoteric. For instance, zinc oxide is basic in that it will react with an acid to produce a salt plus water, for example:

$$ZnO + 2HCl\,(aq) \rightarrow ZnCl_2 + H_2O$$

but is acidic in its reaction with alkalis, forming the complex ANION $Zn(OH)_4^{2-}$, for example:

$$ZnO + H_2O + 2NaOH \rightarrow Na_2\,Zn(OH)_4$$

amyl alcohol *See* PENTANOL.

amylase One of a group of enzymes that breaks down STARCH into its constituent sugars. It occurs in saliva and other digestive juices.

amylopectin A POLYSACCHARIDE made up of about 20 GLUCOSE molecules in a branched structure cross-linked by GLYCOSIDIC BONDS. Amylopectin is a component of STARCH. *See also* AMYLOSE.

amylose A straight-chained POLYSACCHARIDE made up of hundreds of GLUCOSE molecules. Amylose is a component of STARCH. *See also* AMYLOPECTIN.

anabolic steroid A STEROID HORMONE that has anabolic (*see* ANABOLISM) effects; that is, to speed up tissue growth, particularly muscle. The male sex hormone TESTOSTERONE is a natural anabolic steroid and many synthetic variants exist. Anabolic steroids are useful in medicine but they have also been widely abused by athletes, who use them to build up body muscles.

anabolism The synthesis of components of living tissue (for example, proteins and fats) from simpler precursors. This requires energy in the form of ATP. *Compare* CATABOLISM. *See also* METABOLISM.

anaesthetic A drug that is used to render a person insensitive to pain. Local anaesthetics act only at site of application, whereas general anaesthetics cause loss of consciousness. General anaesthesia has been induced by a number of different agents in the past, including ETHER, CHLOROFORM and DINITROGEN OXIDE. HALOTHANE, $CF_3CHBrCl$, is now commonly used since it has less side-effects.

analgesic A pain-relieving agent. OPIATE analgesics are the strongest of these drugs, whilst non-opiates such as ASPIRIN and PARACETAMOL are useful for less severe pain.

analysis The determination of the elements present in a compound or mixture. QUALITATIVE ANALYSIS establishes the components of a chemical sample and the way they are combined in the case of a compound. QUANTITATIVE ANALYSIS measures the proportions of known components of a mixture.

A wide range of techniques are used for analysis, from simple chemical tests that give characteristic results in the presence of certain materials to more sophisticated physical techniques, including MASS SPECTROSCOPY, CHROMATOGRAPHY and spectroscopic techniques, such as ATOMIC ABSORPTION SPECTROSCOPY. Chemical techniques of quantitative analysis may be classified as VOLUMETRIC ANALYSIS, such as TITRATION, or GRAVIMETRIC ANALYSIS.

See also ACTIVATION ANALYSIS, ELECTROPHORESIS, FLAME TEST, INFRARED SPECTROSCOPY, MASS SPECTROMETER, MICROWAVE SPECTROSCOPY, SPECTROSCOPY, THERMAL ANALYSIS.

Andrews titration A titration in which a REDUCING AGENT under investigation is dissolved in concentrated hydrochloric acid and titrated with potassium iodate solution. Tetrachloromethane is used as an INDICATOR, the END POINT being indicated by the disappearance of the purple colour of dissolved iodine from the indicator. *See also* TITRATION.

angstrom (Å) A non-SI UNIT of length, equal to 10^{-10} m. It is still sometimes used to specify the wavelengths and intermolecular distances, but has largely been superseded by the nanometre (nm), which is 10^{-9} m.

anhydride A compound that reacts with water to produce a new compound, rather than simply dissolving to form an aqueous solution. For example, sulphur trioxide, SO_3, is the anhydride of sulphuric acid, H_2SO_4:

$$SO_3 + H_2O \rightarrow H_2SO_4$$

anhydrite A mineral form of ANHYDROUS calcium sulphate, $CaSO_4$. It is used in the chemical industry as a raw material for the manufacture of cement.

anhydrous (*adj.*) Containing no water, in particular no WATER OF CRYSTALLIZATION.

aniline The common name for PHENYLAMINE.

aniline dyes *See* AZO DYES.

anion A negatively charged ion. So called because it will be attracted to the ANODE in ELECTROLYSIS. *Compare* CATION.

anisotropic (*adj.*) Describing a medium, usually a crystalline solid, in which certain physical properties, such as conductivity, are different in different directions. Graphite (a crystalline ALLOTROPE of carbon) is an example: electric conduction can take place relatively easily along the planes of carbon atoms, but with much more difficulty across the planes.

annealing A process in which metal is heated and then allowed to cool slowly. The result is that DISLOCATIONS are formed, under the influence of thermal vibrations. Some of these disappear as the material cools, but sufficient remain for the material to be soft and easily worked. If the material then undergoes sufficient PLASTIC deformations, it may become hard and brittle (WORK HARDENING), as the dislocations become tangled with one another or run up against the edges of the individual crystals in the POLYCRYSTALLINE metal. If the material is again annealed, the effects of work hardening are reversed and the material again becomes soft and easily worked. *See also* QUENCHING.

annulenes Simple conjugated (*see* CONJUGATION) polyalkenes consisting of rings of an even number of carbon atoms linked by alternating single and double bonds. The number of carbons in the structure is denoted by the prefix n, for example [18]-annulene, $C_{18}H_{18}$. The lower members are not stable because of interactions of hydrogen atoms inside the ring. [18]-annulene is a stable, brown/red reactive solid. Some annulenes are AROMATIC in character.

anode A positively charged ELECTRODE.

anodic oxidation The process that takes place at the ANODE of an ELECTROLYSIS, regarded as an OXIDATION. Since the anode is positively charged, electrons are removed from the ELECTROLYTE at this point. For example, in the electrolysis of copper sulphate with copper ELECTRODES, copper metal is oxidized to form Cu^{2+} ions:

$$Cu \rightarrow Cu^{2+} + 2e^-$$

See also CATHODIC REDUCTION.

anodize (*vb.*) To apply a protective coating of ALUMINIUM OXIDE to a piece of aluminium by making it the ANODE in an ELECTROLYSIS process. A solution of sulphuric acid or chromic acid is normally used as the ELECTROLYTE and the oxygen released at the anode reacts with the aluminium to produce a thin oxide layer.

anthocyanin A group of soluble, coloured pigments in plants, particularly flowers, that are responsible for the red, blue and purple coloration. They are GLYCOSIDES.

anthracene ($C_{14}H_{16}$) An AROMATIC HYDROCARBON with three fused rings; melting point 216°C; boiling point 351°C. Anthracene is separated from the high boiling point fractions of COAL TAR by FRACTIONAL DISTILLATION. It is a white crystalline substance with a slight blue FLUORESCENCE.

anthracite A type of hard, shiny coal consisting of more than 90 per cent carbon with a low percentage of impurities. It is therefore a clean fuel that burns slowly without flame, smoke or smell. Anthracite gives out intense heat but is not suitable for open fires as it is hard to light and slow to burn.

anthranilic acid *See* 2-AMINOBENZOIC ACID.

anthraquinone ($C_{14}H_8O_2$) A colourless crystalline solid; a QUINONE. It is used as the basis of a number of DYES, such as ALIZARIN. Anthraquinone can be made by reacting benzene with phthalic anhydride.

antibiotic A chemical substance produced by a micro-organism that prevents the growth of other micro-organisms (but not viruses) and is used to combat many animal and human illnesses. Many antibiotics have been isolated, mainly from bacteria and fungi, but only a few are medically useful and commercially viable. Their use may be restricted because of side-effects, for example toxicity and allergy. Overuse of antibiotics can lead to resistance of a micro-organism to a particular antibiotic. The common antibiotics include PENICILLIN, STREPTOMYCIN, TETRACYCLINE and chloramphenicol. The action of antibiotics varies – streptomycin affects DNA, RNA and PROTEIN SYNTHESIS, while penicillin prevents formation of bacterial cell walls.

antibody A protein secreted by certain cells in the body, called lymphocytes, in response to the presence of a foreign substance (called the ANTIGEN), for example a viral or bacterial infection. Each antibody produced is specific for a particular antigen and they bind non-covalently forming an antigen-antibody complex. Antibodies can act in number of ways to remove or destroy the foreign substance, for example by precipitation of the antigen-antibody complex, agglutination of antigens or neutralization of toxins produced by micro-organisms. Antibodies can remain in the blood after an infection and protect against infection by the same organism.

antibonding orbital The higher energy of the two MOLECULAR ORBITALS formed when two atomic ORBITALS overlap. This orbital tends to push atoms apart, preventing closer bonding. The high energy of this orbital means that atoms only form COVALENT BONDS if this orbital is not filled. *See also* BONDING ORBITAL.

anti-bumping granules Small pieces of porous material that provide an irregular surface on which bubbles can form more easily, preventing the problems of bumping in chemistry experiments. *See* SUPERHEATED.

antifreeze An ALCOHOL added to water in the cooling system of internal combustion engines to lower the point at which the water freezes. The most common antifreeze used is GLYCOL, which can be mixed with water in any concentration and lowers the point at which it freezes accordingly, thereby preventing freezing in cold weather.

antigen A substance (usually a protein or glycoprotein) that, when introduced into the body of an animal, induces the production of ANTIBODY. Antigens may be, for example, proteins on the surface of bacteria, viruses, pollen grains or foodstuffs. Body tissues and blood cells can also act as antigens and matching of these is necessary in blood transfusions or organ transplants.

anti-knock agent A substance added to PETROL to reduce KNOCKING, in which the fuel burns explosively rather than evenly. Lead tetraethyl, $Pb(C_2H_5)_4$ has long been used in this context, but concerns about lead levels in the environment have lead to a reduction in its use. Modern engines are designed to operate with lower levels of anti-knock protection in the fuel. 1,2, dibromoethane $(CH_2Br)_2$ is also used as an anti-knock agent.

antimony (Sb) The element with atomic number 51; relative atomic mass 121.8; melting point 630°C; boiling point 1,750°C; relative density 6.7. It is a METALLOID occurring mainly in the ore STIBNITE.

Antimony has a VALENCY of 5 and is used as a source of electrons in some semiconductors. The main use of the metal is as an alloying agent, while its compounds are used in fireproofing, pigments and rubber technology.

antioxidants Substances that slow down the rate of oxidation reactions. Antioxidants can be substituted PHENOLS, AROMATIC AMINES or sulphur compounds. They are used as preservatives in foods, to prevent ageing in plastics or rubber and in a number of other products. Some antioxidants sequester metal ions that would catalyse oxidation reactions; others remove oxygen free radicals. Some natural antioxidants, such as VITAMIN E, limit damage caused to body tissues by, for example, toxins.

antiseptic A substance that prevents or inhibits the growth of micro-organisms. The first antiseptic used was CARBOLIC ACID and many substances used today are based on it.

apatite A common mineral PHOSPHATE, calcium fluorophosphate or calcium chlorophosphate. It is mined commercially as a source of phosphates for fertilizers.

aquaion A hydrated ion, particularly one that is found even in non-aqueous solutions. An example is $[Fe(H_2O)_6]^{3+}$, which is responsible for the purple colour of the gemstone amethyst.

aqua regia A mixture of concentrated hydrochloric and nitric acids in the ratio 1:3. It will dissolve all metals except silver, including the NOBLE METALS gold and platinum.

aqueous (*adj.*) Relating to water, particularly a solution in water.

arachidonic acid, *cis,cis,cis,cis-5,8,11,14-eicosatetraenoic acid* $(C_2OH_36O_2)$ An ESSENTIAL FATTY ACID that is an important precursor of many biological compounds, particularly PROSTAGLANDINS. Arachidonic acid is a polyunsaturated fatty acid and plays a role in fat metabolism and membrane production. It can be easily synthesized from LINOLEIC ACID.

aragonite An ANHYDROUS mineral form of CALCIUM CARBONATE. It is found it some limestone areas, particularly around hot springs, and in METAMORPHIC rocks. It slowly recrystallizes to CALCITE.

Araldite Trade name for an EPOXY RESIN used as a household adhesive and to mount specimens to be viewed in an ELECTRON MICROSCOPE.

arene A HYDROCARBON containing a BENZENE RING. The term is derived from AROMATIC and ALKENE.

argentic (*adj.*) Describing silver compounds containing the Ag^{2+} ion (for example argentic oxide, AgO) rather than the more common ARGENOUS compounds.

argenous (*adj.*) Describing silver compounds containing the Ag^+ ion, for example argenous chloride, AgCl. *See also* ARGENTIC.

arginine An AMINO ACID essential for nutrition and the production of UREA.

argon (Ar) The element with atomic number 18; relative atomic mass 39.95; melting point –189°C; boiling point –185°C. Argon is the most common inert gas in the atmosphere, making up about 1 per cent. It is separated from liquid air by FRACTIONAL DISTILLATION. It is used to provide an inert atmosphere in some welding processes and in light bulbs.

aromatic (*adj.*) Describing any organic compound containing a BENZENE RING and some non-benzene compounds that are HETEROCYCLIC. Aromatic compounds undergo SUBSTITUTION REACTIONS. The ring usually contains nitrogen, sulphur or oxygen if it does not contain all carbon atoms. These compounds were originally named because of their fragrance, although not all of them are pleasant.

Aromatic compounds are more stable than other UNSATURATED COMPOUNDS (such as ALKENES) and thus less reactive, usually undergoing substitution rather than ADDITION REACTIONS. Examples include benzene, C_6H_6, and pyridine, C_6H_5N. Where an aromatic compound contains a FUNCTIONAL GROUP the reactivity of both the benzene ring and the functional group is altered. This is because the delocalized electrons of the ring and those of functional groups containing DOUBLE BONDS interact. This interaction controls the chemistry of aromatic compounds.

See also BENZENE, DELOCALIZED ORBITAL.

Arrhenius' equation An equation that describes the rate of any reaction involving an ACTIVATION ENERGY. Arrhenius' equation is

$$k = A\exp(-E/RT)$$

where k is the rate constant for the reaction, A is a constant for a given reaction, E is the activation energy per mole, R is the MOLAR GAS CONSTANT and T the ABSOLUTE TEMPERATURE. *See also* RATE OF REACTION.

arsenic (As) The element with atomic number 33; relative atomic mass 74.9; sublimes at 613°C; relative density 3.9. It is a powdery METALLOID that occurs quite widely in nature and is best known for its toxic properties. Arsenic is a cumulative poison, and although its use in certain pigments was once quite common, it is little used nowadays due to fears about its toxicity. It is produced as a by-product of the extraction of many metals from their ores.

arsenic hydride *See* ARSINE.

arsenic oxide Either of the two oxides arsenic(III) oxide, As_4O_6 (otherwise known as arsenic trioxide, arsenous oxide or white arsenic) and arsenic(V) oxide, As_2O_5.

Arsenic(V) oxide is a white solid that decomposes to arsenic(III) oxide at 315°C and has a relative density of 3.5. It is formed by the dehydration of arsenous acid:

$$2HAsO_3 \rightarrow As_2O_5 + H_2O$$

Arsenic(III) oxide is a white compound with three solid ALLOTROPES. It sublimes at 193°C and the OCTAHEDRAL form has a relative density of 3.9. It is formed by burning arsenic in air:

$$4As + 5O_2 \rightarrow 2As_2O_5$$

It is extremely toxic, and its former use as a poison for vermin has been vastly reduced on account of concerns over its accumulation in the food chain.

arsine, *arsenic hydride* (AsH_3) A colourless gas; melting point –116°C; boiling point –55°C. Arsine is manufactured by the reaction of strong acids with metal arsenides, for example,

$$3H_2SO_4 + 2K_3As \rightarrow 3K_2SO_4 + 2AsH_3$$

Arsine is used commercially in small quantities as a vehicle for depositing arsenic in the manufacture of integrated circuits.

arsenic trioxide *See* ARSENIC OXIDE.

arsenide Any BINARY COMPOUND containing arsenic and a metal.

arsenous oxide *See* ARSENIC OXIDE.

aryl group A group obtained by removing a hydrogen atom from a hydrocarbon of the BENZENE series.

aryne Transient intermediates in certain organic reactions. They do not exist in isolation. Arynes are derived from ARENES by the removal of two adjacent hydrogen atoms to convert a double bond into a triple bond. An example is BENZYNE.

asbestos A naturally occurring form of AMPHIBOLE, containing small needle-shaped fibres that can be spun into a thread and woven into cloth. Asbestos cloth was widely used for its fireproof and chemical resisting qualities, as an insulation for pipes, and in other high temperature components. Exposure to asbestos fibres has been found to cause a severe respiratory disease called asbestosis and its use is now far more strictly controlled. The removal of old asbestos insulation from buildings is a major problem for the construction industry.

ascorbic acid *See* VITAMIN C.

asparagine ($C_4H_8N_2O_3$) A non-essential AMINO ACID that is an uncharged derivative of ASPARTIC ACID. It is found mainly in asparagus, hence its name, potatoes and beetroot.

aspartame An artificial sweetener (trade name Nutrasweet) consisting of two amino acids (ASPARTIC ACID and PHENYLALANINE) linked by a methylene ($-CH_2^-$) group. It is 200 times sweeter than sugar and does not have the aftertaste that SACCHARIN has. It cannot be used for cooking as it breaks down on heating.

aspartate *See* ASPARTIC ACID.

aspartic acid ($C_4H_7NO_4$) A non-essential acidic AMINO ACID, commonly called aspartate to indicate its negative charge at physiological pH. ASPARAGINE is the uncharged derivative. Aspartic acid is involved in the production of UREA through the UREA CYCLE and also acts as a NEUROTRANSMITTER. It is found in young sugar cane and sugar beet.

aspirin, *acetylsalicylic acid, acetylbenzoic acid* ($C_9H_8O_4$) A widely used pain-killing drug that acts by inhibiting PROSTAGLANDINS. It also reduces fever and inflammation, for example in arthritis. Side-effects of long-term usage are stomach bleeding and kidney damage, although more recently it has been suggested that an aspirin a day can reduce the risk of heart attacks and thrombosis.

astatine (At) The element with atomic number 85; melting point 302°C; boiling point 377°C. Only unstable ISOTOPES are known, with the longest-lived (astatine–210) having a HALF-LIFE of 8 hours. Astatine is present in nature in minute amounts, being formed by the decay of longer lived radioactive elements.

atactic polymer A POLYMER in which the substituted carbons have a random arrangement with respect to the carbon chain, if it is considered that the carbon atoms all lie in the same plane. This results in a sticky polymer unsuitable for manufacture. *Compare* ISOTACTIC POLYMER, SYNDIOTACTIC POLYMER.

atmosphere, *standard atmosphere* (atm.) A unit of pressure equivalent to 101,325 Pa. This is equal to the pressure that will support a column of mercury 760 mm high, at 0°C, at sea level and latitude 45°.

atmospheric pressure The pressure produced by gravity acting on the atmosphere. At the surface of the Earth this produces a pressure of about 1×10^5 Pa (1 ATMOSPHERE), though the exact value varies from day to day.

atom The smallest particle into which an ELEMENT can be divided without losing its chemical identity. Atoms were originally believed to be indivisible objects, but it is now known that they comprise a small dense positive NUCLEUS of PROTONS and NEUTRONS, surrounded by ELECTRONS.

Almost all the mass of an atom is contained in its nucleus: protons and neutrons have similar masses, while electrons are about 1/1,836 the mass of protons. In a neutral atom, the number of electrons is equal to the number of protons in the nucleus (called the atomic number). The number and arrangement of the electrons in an atom is what gives each element its distinct chemical properties.

In a simple model of the atom, electrons are regarded as orbiting the nucleus in shells. However, a more sophisticated treatment recognizes the influence of quantum mechanics on the atom, and places the electrons in ORBITALS, each having a certain defined energy. The sequence in which these orbitals are filled, and the energies involved are crucial in explaining the chemical properties of each element.

Chemistry generally concerns itself with the reactions of neutral atoms or of ions that contain a few electrons more or less than the number needed to make the atom neutral.

See also ATOMIC THEORY, BOHR THEORY, CHEMICAL COMBINATION, DALTON'S ATOMIC THEORY, LIQUID DROP MODEL, MOLECULE, QUANTUM THEORY, RUTHERFORD–BOHR ATOM, RYDBERG EQUATION, SHELL MODEL.

atomic absorption spectroscopy A technique of chemical ANALYSIS based on the ABSORPTION SPECTRUM formed when white light is shone through a sample in the form of a vapour. The wavelengths of light absorbed are characteristic of the elements present.

atomic emission spectroscopy A technique of chemical analysis in which a sample is IONIZED and the wavelengths in the EMISSION SPECTRUM so produced are measured. These wavelengths are characteristic of the elements present.

atomic force microscope A microscope that produces an image using a diamond-tipped probe, which is moved over the surface of a sample and responds to the interatomic forces between the probe and the sample. The probe in effect 'feels' its way over the contours of the surface, and its up-and-down movements are transmitted to a computer that produces a profile of the sample. The atomic force microscope can resolve single molecules, and is useful for biological specimens as the sample does not have to be electrically conducting.

atomicity The number of atoms present in a single molecule of a given compound. For example, ethanol, C_2H_5OH, has an atomicity of 9.

atomic mass unit (amu) The unit in which nuclear masses are usually measured. The mass of one atom of carbon–12 (the isotope of carbon with MASS NUMBER 12) is defined to be 12 amu exactly. Nuclear masses can be measured using a MASS SPECTROMETER.

atomic number The number of protons in a nucleus.

atomic theory The idea that all materials are made up of small particles called atoms. The motion of these particles leads to KINETIC THEORY in physics, whilst the way in which they combine to form molecules is the foundation of DALTON'S ATOMIC THEORY in chemistry.

atomic volume The RELATIVE ATOMIC MASS of an element divided by its volume, usually expressed in $cm^3 mol^{-1}$.

atomic weight *See* RELATIVE ATOMIC MASS.

ATP (adenosine triphosphate) The short-term energy storage and carrier molecule found in all living cells. It transfers energy from where there is plenty to where it is needed for cellular reactions. Energy is released when one of the three PHOSPHATE groups of ATP is removed (catalysed by a number of enzymes) by a process called HYDROLYSIS. This yields ADP (adenosine diphosphate). Hydrolysis of ADP then yields AMP (adenosine monophosphate). The phosphate molecules can be added back to AMP to reconvert to ATP by a process called PHOSPHORYLATION.

atropine ($C_{17}H_{23}NO_3$) A poisonous ALKALOID extracted from the plant deadly nightshade. It is used in medicine to dilate the pupil of the eye, to reduce secretion of saliva and in the treatment of colic.

aufbau principle (German = build up) The idea that the electron configuration of any element can be determined from the sequence of ENERGY LEVELS of the ORBITALS, which are filled up in the sequence 1s, 2s, 2p, 3s, 3p, 4s, 3d, 4p, 5s, 4d, 5p, 6s, 4f, 5d, 6p, 7s, 5f, 6d. Although the principle usually works well, there are some anomalies in the TRANSITION METALS and LANTHANIDES in which the last S-ORBITAL contains only a single electron rather than the two predicted.

Auger effect The ejection of an electron from a highly excited atom rather than an X-ray PHOTON. The kinetic energy of the electron is equal to the energy of the photon that would have been emitted less the binding energy of the electron. The measurement of the electron energy can be used as a form of SPECTROSCOPY. *See also* AUGER SPECTROSCOPY.

Auger spectroscopy A technique for obtaining information about the ENERGY LEVELS of an atom from the electrons ejected in the AUGER EFFECT. This can also be used as an analytical technique to determine the chemical composition of a sample.

auric (*adj.*) Describing compounds containing the gold ion Au^{3+}, for example auric chloride, $AuCl_3$. *See also* AUROUS.

aurous (*adj.*) Describing compounds containing the gold ion Au^+, for example aurous chloride, AuCl. *See also* AURIC.

autocatalysis Any chemical reaction in which one of the products of the reaction is also a CATALYST for the reaction. Such a reaction will start slowly then accelerate as the products are produced and catalyse further reactions, finally slowing down as the reagents are used up.

autoclave A vessel designed to contain liquids and gases under high pressure, used for some chemical reactions and more often with boiling water for the sterilization of small pieces of equipment. The high pressure means that a higher temperature can be reached than with water boiling at atmospheric pressure, so killing a wider range of bacteria.

autoradiography A technique used in the laboratory for visualizing a substance that has been radioactively labelled (i.e. attached to a radioactive ISOTOPE). The substance is placed in contact with a photographic film in a light-tight cassette for a period of time. The radioactivity causes the film to darken and thus the substance to be identified.

Autoradiography can be used to analyse polyacrylamide gels (*see* POLYACRYLAMIDE GEL ELECTROPHORESIS) containing radiolabelled proteins. It can be also used to examine the incorporation of a radiolabelled substance into living tissues and cells.

auxin A group of chemical substances (natural or artificial) that modify plant growth. They are termed plant growth substances or plant hormones. The most common naturally occurring auxin is indolacetic acid. At high concentrations auxins can inhibit growth and cause death of a plant. Many synthetic auxins have been developed and used, for example, to aid root development and to prevent fruit drop. Others are used in weedkillers, where they can cause such rapid growth that the plant dies. Some synthetic auxins have different effects on different plants, which is useful in selective weedkillers to kill only the unwanted plants.

Avogadro constant, *Avogadro number* (L or N_A) The number of atoms in one MOLE of atoms, molecules in one mole of molecules, etc. It is equal to 6.022×10^{23}. The mass of this number of carbon–12 atoms is 12 g.

Avogadro number *See* AVOGADRO CONSTANT.

Avogadro's hypothesis A given number of molecules of any gas at a given temperature and pressure will occupy the same volume, regardless of the nature of the gas. In particular, one MOLE of any gas occupies a volume of 22.4 dm^3 under conditions of STANDARD TEMPERATURE AND PRESSURE (atmospheric pressure and a temperature of 0°C). This result was originally based on EMPIRICAL observations but is now seen to be a consequence of KINETIC THEORY.

azeotropic Describing a liquid mixture, such as that of ethanol in water, that produces a minimum in the graph of boiling point against composition. Such a minimum will be the lowest temperature at which the liquid can boil and is lower than the boiling point of either material on its own. This mixture will be produced if an attempt is made to separate the mixture by FRACTIONAL DISTILLATION. For example, in the fractional distillation of ethanol from water, an azeotropic mixture is produced, with a composition of 96 per cent ethanol and a boiling point of 78.3°C, compared to a boiling point of 78.5°C for pure ethanol.

azides Compounds containing the group $-N_3$ or the ion N^{3-}.

azine Organic compounds containing a six-membered ring of carbon and nitrogen atoms. PYRIDINE is an example: it contains one nitrogen atom (C_5H_5N). Diazines have two nitrogen atoms in the ring ($C_4H_4N_2$) and triazines have three nitrogens.

azo compound One of a group of organic compounds formed by reacting DIAZONIUM SALTS with AROMATIC AMINES or a PHENOL. The reaction is a diazo-coupling reaction in which a second BENZENE RING is added to the diazonium salt to give two benzene rings joined by the –N=N– group. The products formed are brightly coloured and form the basis of AZO DYES. The compounds are usually very stable. An example is METHYL ORANGE:

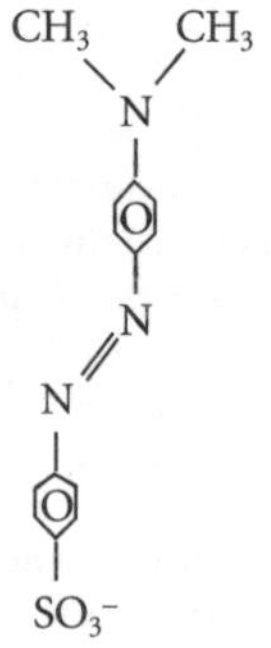

CH_3 CH_3
N
N
N
SO_3^-

azo dyes, *aniline dyes* A large group of DYES, constituting about half of all dyes made. Azo dyes are formed by the reaction of a DIAZONIUM SALT with an AROMATIC AMINE or a PHENOL. They consist of two BENZENE RINGS joined by the –N=N– group. The simplest azo dyes derive from PHENYLAMINE (aniline) and they are therefore also known as aniline dyes. They are cheap to make and since they are stable it is easy to add groups in order to change their colour and fabric bonding properties. Azo dyes are usually red, brown or yellow and can be used on most types of fabrics. They are also used as pigments in the photographic industry.

azulene ($C_{10}H_8$) A blue crystalline solid; melting point 99°C. It is an AROMATIC compound with an odour similar to that of NAPHTHALENE. Azulene consists of a five-membered ring joined to a seven-membered ring. It is converted to naphthalene on heating. The term azulene is also generally applied to the blue oils that are produced on heating many essential oils from plants.

azurite A MINERAL form of BASIC copper carbonate, $Cu_3(OH)_2(CO_3)_2$. It has a bright blue colour and is occasionally used as a pigment and as a gemstone.

B

back bonding A form of DATIVE BOND in which a LIGAND forms a SIGMA-BOND with an atom or ion by donating a pair of electrons in the usual way, whilst the central atom also donates electrons from its D-ORBITALS to empty p- or d-orbitals in the ligand.

background radiation The collective name for the many sources of IONIZING RADIATION. The most important of these are naturally occurring radioactive materials in rocks, soil and atmosphere. The other main source of background radiation is cosmic radiation – high-energy charged particles, mostly protons, that enter the atmosphere from space.

Compared to these two sources of radiation, the radiation present from nuclear reactors and nuclear weapon tests represents only 1 or 2 per cent of the total exposure to ionizing radiation for the average human. In addition, individuals often experience significant doses of ionizing radiation from medical sources, mainly X-rays. The level of medical exposure can vary widely from one individual to another, though in the West it typically accounts for about 13 per cent of the lifetime dose.

See also RADIOACTIVITY.

back titration A TITRATION in which the substance being analysed is first reacted with a measured excess of some other reagent. A titration is then performed to measure the quantity of this second reagent that remains unreacted. Thus the quantity of the unknown reagent initially present can be deduced.

Bakelite Trade name for certain PHENOL-FORMALDEHYDE RESINS. Bakelite was the first synthetic plastic to be made and is tough, hard, heat resistant and an electrical insulator. It has many uses, for example, to make telephone receivers, electric plugs and sockets.

balance A device for weighing. A beam balance consists of a lever balanced on a central pivot, with known and unknown masses being suspended on opposite sides of the pivot, usually at equal distances. The two masses are compared by using the pull of gravity on the masses to produce turning moments about the pivot: when the masses are equal, the beam is exactly horizontal. A top-pan balance does not compare one mass with another, but uses the pull of gravity on an unknown mass to deform a strain gauge, with the mass being presented directly as a digital reading.

Balmer series A series of lines in the HYDROGEN SPECTRUM. The wavelengths are mostly in the visible part of the ELECTROMAGNETIC SPECTRUM, though the series extends slightly into the ultraviolet. Each line corresponds to a transition between the second ENERGY LEVEL and some higher level. *See also* BOHR THEORY.

band gap In the BAND THEORY of solids, the gap between one energy band and the next, particularly between a full VALENCE BAND and an empty CONDUCTION BAND.

band theory The branch of quantum mechanics that explains the properties of solids in terms of ENERGY LEVELS.

The electrons in a single atom exist in discrete, sharply defined energy levels. When the atoms come together to form a solid, these sharply defined energy levels become bands of allowed energies. These bands may be filled with electrons, or may be partially full or empty. Between the allowed bands are 'forbidden' bands.

The VALENCE ELECTRONS, those involved in chemical bonding, form the VALENCE BAND of a solid. In an ideal crystal, the valence band is fully occupied. The only electrons free to move through the solid, carrying heat and electricity, are those in partially full bands. The PAULI EXCLUSION PRINCIPLE forbids an electron from gaining energy unless there is an empty energy level for it to move to. If the valence band is full, the electrons must move to an unfilled band, called the CONDUCTION BAND. Conductors are those materials whose valence band and conduction band are unfilled, or whose conduction and valence bands overlap – in either case there are vacant energy levels.

In insulators, the valence band and conduction band are separated by a wide forbidden band. Such materials do not conduct because the electrons do not have enough energy to cross from one band to another. In semiconductors, the energy difference between the valence band and the next band is sufficiently small that thermal vibrations may give electrons enough energy to enter the empty conduction band. The energy required is called the BAND GAP. If the band gap has an energy corresponding to a PHOTON of visible light, light will be absorbed or emitted as electrons cross the band gap. This effect is exploited in electronic devices.

bar A unit of pressure in the C.G.S. SYSTEM. One bar is equal to 10^5 PASCAL, or approximately one ATMOSPHERE. The millibar (100 Pa) is used as the unit of pressure in meteorology.

barbiturate Any of a group of drugs derived from barbituric acid, $C_4H_4N_2O_3$, that depress the central nervous system. Their original use as sleeping pills and sedatives is now limited since prolonged use can be addictive. They are still used in the treatment of epilepsy and as anaesthetics.

barium (Ba) The element with atomic number 56; relative atomic mass 137.3; melting point 725°C, boiling point 1,640°C; relative density 3.5. Barium is an ALKALINE EARTH metal that occurs in the minerals barytes, $BaSO_4$, and witherite, $BaCO_3$.

Barium forms an insoluble SULPHATE, which is used as a contrast medium in medical X-rays. Barium is used in alloys, and its compounds are used in pigments, matches and fireworks.

barium carbonate ($BaCO_3$) A white powder; relative density 4.4; decomposes on heating. Barium carbonate occurs naturally, and being insoluble it can be formed as a precipitate by adding a carbonate to a dissolved barium salt:

$$Na_2CO_3 + BaCl_2 \rightarrow BaCO_3 + 2HCl$$

Barium carbonate decomposes on heating to give barium oxide, BaO:

$$BaCO_3 \rightarrow BaO + CO_2$$

barium hydroxide ($Ba(OH)_2$) A white solid; melting point 408°C; decomposes on further heating; relative density 4.5. Barium hydroxide occurs naturally and is slightly soluble in water. It can be produced as a precipitate in the reaction between sodium hydroxide and barium chloride:

$$2NaOH + BaCl_2 \rightarrow Ba(OH)_2 + 2HCl$$

barium oxide (BaO) A white solid; melting point 1,920°C; boiling point 2,000°C; relative density 5.7. Barium oxide is insoluble in water and can be manufactured by heating BARIUM CARBONATE:

$$BaCO_3 \rightarrow BaO + CO_2$$

barium peroxide (BaO_2) A creamy white solid; melting point 450°C; decomposes on further heating; relative density 5.0. Barium peroxide can be manufactured by heating barium oxide in oxygen:

$$2BaO + O_2 \rightarrow 2BaO_2$$

On heating in air it decomposes to barium oxide:

$$2BaO_2 \rightarrow 2BaO + O_2$$

Barium peroxide reacts with acids to form hydrogen peroxide, for example:

$$BaO_2 + 2HCl \rightarrow H_2O_2 + 2HCl$$

barium sulphate ($BaSO_4$) A white powder; melting point 1,580°C; relative density 4.5. Barium sulphate occurs naturally and can be made as an insoluble precipitate in the reaction between sulphuric acid and barium chloride solution:

$$BaCl_2 + H_2SO_4 \rightarrow BaSO_4 + 2HCl$$

Barium sulphate is opaque to X-rays and a suspension of the sulphate can be ingested to provide a contrast medium for X-rays of the digestive system. The sulphate is very insoluble and is not absorbed into the body, so is non-toxic. Barium sulphate is also used as a white pigment in some paints.

barytes The chief ore of barium, the mineral form of BARIUM SULPHATE.

basalt A fine-grained basic IGNEOUS rock, the commonest type of LAVA.

base (i) Any compound that will react with an ACID to produce a SALT plus water. An example is sodium hydroxide:

$$NaOH + HCl \rightarrow NaCl + H_2O$$

Most bases are the OXIDES or HYDROXIDES of

metals – ammonia is an important exception. A base that is soluble in water is termed an alkali.

(ii) In the Lowry–Brønsted theory of acids and bases, a base is any compound that accepts a proton. For example, sodium chloride behaves as a base when it dissolves in ammonia:

$$NH_3 + NaCl \rightarrow NaNH_2 + HCl$$

(iii) In the Lewis theory of acids and bases, a base is defined as any substance that donates electrons. This encompasses not only the traditional base reactions, but also REDOX REACTIONS and the formation of CO-ORDINATE BONDS. Thus boron trichloride acts as a base when it forms a bond with ammonia:

$$H_3N{:} + BCl_3 \rightarrow H_3NBCl_3$$

See also ALKALI, BASICITY.

base exchange A chemical reaction in which one ANION changes places with another. An example is the precipitation of copper(II) carbonate in a solution of sodium sulphate from the reaction between solutions of copper sulphate and sodium carbonate:

$$CuSO_4 + Na_2CO_3 \rightarrow CuCO_3 + Na_2SO_4$$

base metal A common inexpensive metal, especially one used as a substitute for a NOBLE METAL, such as gold. Base metals are those which are chemically more reactive, so corrode rapidly.

base pair Two NUCLEOTIDE BASES linked by HYDROGEN BONDS in DNA that link the two strands of the double helix. The pairing is always between a PURINE and a PYRIMIDINE, so that the bases adenine and thymine always link and cytosine and guanine always link. Base pairing also occurs between RNA and DNA during PROTEIN SYNTHESIS. Uracil in RNA pairs with adenine.

basic (*adj.*) Having the properties of a base.

basicity The number of hydrogen atoms in an acid that can be replaced by metal ions in the formation of a salt. Thus hydrochloric acid, HCl, has a basicity of 1 (monobasic), whilst sulphuric acid, H_2SO_4, has a basicity of 2 (dibasic).

basic lead carbonate *See* LEAD CARBONATE HYDROXIDE.

basic salt Any SALT formed by replacing some of the OXIDE or HYDROXIDE of a base with some other negative ion. The compound consists of a NORMAL SALT combined with a base in a simple molecular ratio. An example is basic lead carbonate, $2PbCO_3.Pb(OH)_2$, produced by the reaction of excess lead oxide with carbonic acid. *See also* ACIDIC SALT.

battery Several electrochemical CELLS connected together, usually in series to produce a larger voltage.

bauxite An aluminium ore containing mainly aluminium oxide, Al_2O_3. Bauxite is the most important source of aluminium in the Earth's crust.

beam balance *See* BALANCE.

Beckmann thermometer A thermometer for measuring temperature changes over a narrow range, for example when measuring the change in melting or boiling point of a substance as a result of the presence of a second, dissolved, material (*see* COLLIGATIVE PROPERTIES). A Beckmann thermometer comprises a bulb filled with mercury and a short length of narrow capillary tube at the top of which is a second, smaller mercury reservoir. The range of temperatures over which the thermometer operates can be varied by running mercury from the smaller reservoir into the lower bulb.

becquerel (Bq) The SI UNIT of radioactive ACTIVITY, one becquerel being an activity of one ionizing particle per second.

beehive shelf A cylindrical ceramic device with a hole in the top and the side, used to support a GAS JAR when collecting gas over water.

Benedict's test A test for REDUCING SUGARS based on a modification of FEHLING'S TEST. Only one solution is used, containing sodium citrate, sodium carbonate and copper sulphate in water. This is added to the test sample and if a reducing sugar is present a rust-brown precipitate forms on boiling.

benzaldehyde *See* BENZENECARBALDEHYDE.

benzene (C_6H_6) A colourless, liquid HYDROCARBON derived from COAL TAR, with a characteristic smell; melting point 5°C; boiling point 80°C. Benzene is the simplest AROMATIC compound, consisting of six carbon atoms arranged so that they form a regular hexagon (with 120° angles and carbon–carbon bonds of equal length, intermediate between single and DOUBLE BOND lengths).

The molecule is represented by a hexagon either with three double bonds or with a circle drawn inside the hexagon.

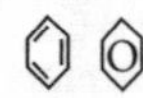

The PI ELECTRONS represented by the double bonds are not localized but are uniformly distributed around the ring. These are said to be delocalized and provide stability to the benzene ring. Reactions of benzene are usually ELECTROPHILIC SUBSTITUTION REACTIONS, which retain this delocalization of electrons, rather than ADDITION REACTIONS.

Benzene derivatives are named by prefixing 'benzene' with the name of the group attached, for example, methyl benzene, $C_6H_5CH_3$, 1,2-dimethylbenzene, $C_6H_4(CH_3)_2$. In the latter example, the numbers are chosen to be as low as possible. In addition, the terms *ORTHO*, *META* and *PARA* are used to describe 1,2-, 1,3- and 1,4- disubstituted benzenes respectively, for example, *ortho*-dimethylbenzene (or *o*-dimethylbenzene). If there are different groups in the same ring they are named alphabetically, for example, methyl before nitro.

Benzene is used in the synthesis of many chemicals and as a solvent. Recently, it has become known that benzene vapours are dangerous, causing respiratory problems and cancer, and so derivatives such as METHYLBENZENE are more often used.

See also DELOCALIZED ORBITAL.

benzenecarbaldehyde, *benzaldehyde* (C_6H_5CHO) An ALDEHYDE that is a colourless liquid with a smell of almonds; boiling point 180°C. It occurs in nature in certain leaves and nuts and can be extracted from these or from toluene. It is readily oxidized to benzenecarboxylic acid. It is used in the manufacture of perfumes and dyes and as a solvent.

benzenecarboxylic acid, *benzoic acid* (C_6H_5-COOH) A white solid occurring in some natural resins and oils; melting point 122°C; boiling point 249°C. It is a CARBOXYLIC ACID and is manufactured from the oxidation of TOLUENE or BENZENECARBALDEHYDE. It is used as a food preservative and an antiseptic.

1,4-benzenedicarboxylic acid, *terephthalic acid* ($C_8H_6O_4$) A DIBASIC ACID; melting point 300°C. It is used in the manufacture of terylene and is produced by the OXIDATION of dimethylbenzene.

benzene-1,2-dicarboxylic acid *See* PHTHALIC ACID.

benzene hexachloride *See* BHC.

benzene ring A term referring to the six carbon ring structure of benzene and found in many other AROMATIC COMPOUNDS.

benzfuran, *coumarone* (C_8H_6O) An AROMATIC HYDROCARBON with a benzene ring fused to a five-membered FURAN ring.

benzoquinone *See* CYCLOHEXADIENE-1,4-DIONE.

benzoic acid The common name for BENZENECARBOXYLIC ACID.

benzyne (C_6H_4) An ARYNE, thought to be a very short-lived and highly reactive intermediate in certain reactions. Benzyne consists of a hexagonal ring with two double bonds and one triple bond.

berkelium (Bk) The element with atomic number 97. It does not occur in nature, and is named for Berkeley, California, where it was first synthesized. The longest-lived ISOTOPE (berkelium–247) has a HALF-LIFE of 1,400 years; all the others have half-lives of a few days or less.

beryl The mineral form of beryllium aluminium silicate, $Be_3Al_2Si_6O_{18}$. It occurs widely in granite rocks and is the chief source of BERYLLIUM.

beryllate Any compound containing the beryllate ion, BeO_2^{2-}, formed by the reaction of beryllium with strong alkalis, for example:

$$Be + 2NaOH \rightarrow Na_2BeO_2 + H_2$$

beryllium (Be) The element with atomic number 4; relative atomic mass 9.0; melting point 1,285°C; boiling point 2,970°C; relative density 1.9. Beryllium is a hard, white ALKALINE EARTH metal with a low density. It is used to manufacture low density alloys and as a moderator in some nuclear reactors. Beryllium is highly toxic.

beryllium hydroxide ($Be(OH)_2$) A white crystalline solid that decomposes on heating. Beryllium hydroxide can be precipitated by the addition of ALKALIS to soluble beryllium salts, for example:

$$BeCl_2 + 2NaOH \rightarrow Be(OH)_2 + 2NaCl$$

Beryllium hydroxide is AMPHOTERIC and will dissolve in strong bases to form the BERYLLATE ion.

beryllium oxide (BeO) A white solid; melting point 2,550°C; boiling point 4,120°C; relative density 3.0. Beryllium oxide is insoluble in water. It is AMPHOTERIC, forming salts with dilute acids, and reacting with strong alkalis to form BERYLLATES.

Bessemer converter A device, now largely obsolete, for the manufacture of steel from PIG IRON. The converter consists of a vessel with a REFRACTORY lining into which molten pig-iron is poured. Air is blown through the base of the furnace to oxidize any impurities, in particular carbon. Additional materials, including some extra carbon, may then be added to control the properties of the steel. The converter is then tipped onto its side and the steel poured into moulds. The Bessemer converter has now largely been replaced by the OXYGEN FURNACE and the ELECTRIC ARC FURNACE. *See also* OPEN HEARTH FURNACE.

beta decay The emission of BETA PARTICLES, which may be either positive or negative, from an unstable atomic nucleus. Negative beta particles are fast moving electrons, produced by atomic nuclei with too many neutrons; a neutron turns into a proton, and an electron is emitted together with an antineutrino (*see* NEUTRINO). Positive beta particles are POSITRONS, produced when a nucleus with too few neutrons converts a proton into a neutron emitting a positron and a neutrino.

Beta radiation is less ionizing than ALPHA RADIATION, but more ionizing than GAMMA RADIATION. Beta particles have a range of several metres in air but can be stopped by a layer of aluminium a few millimetres thick. In beta decay, the MASS NUMBER is unchanged, but the ATOMIC NUMBER is increased by 1 in negative beta decay and decreased by 1 in positive beta decay.

beta particle An electron or POSITRON produced in an atomic nucleus in BETA DECAY.

beta pleated sheet, *β-sheet* A type of PROTEIN secondary structure resulting from HYDROGEN BONDING. Hydrogen atoms from one side of the protein molecule link with the oxygen atoms of the parallel side, causing anti-parallel folding of the molecule.

beta radiation A stream of BETA PARTICLES, emitted from unstable nuclei by BETA DECAY.

BHC (benzene hexachloride), *1,2,3,4,5,6-hexachlorocyclohexane* ($C_6H_6Cl_6$) A powerful INSECTICIDE formed by the reaction of benzene with chlorine in sunlight or ultraviolet irradiation. BHC actually exists in several isomeric forms (*see* ISOMER), only one of which, gamma-BHC, is active against insects. The isomers have differing melting points, gamma-BHC is 113°C. Despite its success in the combat of a number of disease-carrying insects gamma-BHC now has restricted use due to its dangerous accumulation in the food chain.

binary compound A compound containing just two elements.

biochemistry The study of the chemistry of living organisms. This includes the structure and functioning of, for example, ENZYMES, PROTEINS and NUCLEIC ACIDS, and molecular biology. The study of biochemistry is fundamental to the understanding of life processes.

bioluminescence The emission of visible light by living organisms such as the firefly, glow-worms, fungi, bacteria and various fish. The light is produced as a result of the oxidation of a substance called luciferin and can serve as a means of protection or as a mating signal or for species identification.

bioreactor A large stainless steel tank used in the large-scale production of chemicals such as enzymes. The tank is used to grow the micro-organisms needed for the production process and consists of innoculated culture medium maintained at the optimum conditions for enzyme production.

biosynthesis The processes by which living organisms build up molecules, compounds and structures needed for their growth and reproduction.

biotin A VITAMIN of the VITAMIN B COMPLEX, also called vitamin H. Biotin is found in yeast, liver, milk and vegetables. It is also made by intestinal bacteria. Biotin is a coenzyme in carboxylation reactions, the incorporation of carbon dioxide into compounds, and lack of it causes dermatitis.

bismuth (Bi) The element with atomic number 83; relative atomic mass 208.98; melting point 271°C; boiling point 1,560°C; relative density 9.8. It is the highest atomic number known to have a stable ISOTOPE. Bismuth is a white crystalline metal with a reddish tint. It is used in certain alloys to lower the melting point.

bituminous coal *See* COAL.

biuret test A test for the presence of PROTEINS in solution. The test solution is warmed with sodium hydroxide solution and then a few drops of copper(II) sulphate solution are added. The formation of a violet ring indicates the presence of protein as a result of the formation of a complex involving copper ions and PEPTIDE bonds.

blast furnace A device for extracting iron from its ores. Iron ore, COKE and LIMESTONE are fed into the top of the furnace and heated by the oxidation of the coke by oxygen blown upwards from the base of the furnace. The detailed chain of reactions is complex, but the carbon in the coke is oxidized to carbon dioxide whilst the iron ore is reduced to metallic iron. Molten iron, with dissolved carbon, and molten SLAG formed from the impurities fall to the bottom of the furnace, where they form separate layers. The hot gases leaving the top of the furnace are used to pre-heat the incoming air. The heated air is blown into the furnace through a ring of pipes called tuyers. The molten metal, called PIG IRON, and the slag are regularly tapped off from the base of the furnace and fresh raw materials are added at the top, so the furnace can operate continuously.

blende A mineral SULPHIDE of certain metals. In particular, zinc blende, which is ZINC SULPHIDE, ZnS, the chief ore of ZINC.

body centred cubic A crystalline structure in which the UNIT CELL is a cube with an atom at the centre surrounded by one eighth of an atom at each corner. This structure has a CO-ORDINATION NUMBER of 8.

All the ALKALI METALS form crystals with this structure, which accounts for their low density compared to other metals of comparable atomic mass. The structure is also found in IONIC compounds where the two ions have equal charges and similar radii. If the ions differ much in radius, such as in sodium chloride, a different cubic structure is favoured with a co-ordination number of 6.

Bohr atom *See* BOHR THEORY.

Bohr theory, *Bohr atom* A simple model of ATOMS put forward in 1913 by Niels Bohr (1885–1962) to explain the LINE SPECTRUM of hydrogen. The Bohr theory has three basic postulates: (i) electrons revolve around the nucleus in fixed orbits without the emission of ELECTROMAGNETIC RADIATION; (ii) electrons can only occupy orbits in which the angular momentum of the electron is a whole number times $h/2\pi$, where h is PLANCK'S CONSTANT – in other words, within each orbit an electron has a fixed amount of energy; and (iii) when an electron jumps from one orbit to another, energy is emitted or absorbed as a PHOTON of electromagnetic radiation.

Whilst this model explains the broad features of the HYDROGEN SPECTRUM, it does not account for the fine detail and has been superseded by more complete quantum mechanical descriptions. It was however an important early step in the development of QUANTUM THEORY.

boil (*vb.*) To turn from a liquid to a vapour, with bubbles of gas being formed in the liquid.

boiling point The temperature at which a liquid boils, usually quoted at a pressure of one ATMOSPHERE. More technically, the one temperature for a given VAPOUR PRESSURE at which the liquid and its vapour can exist in equilibrium together. Boiling points increase with pressure, as it is harder for a bubble to form if the liquid is at high pressure. All materials expand on boiling and have boiling points that increase with increasing pressure. *See also* SATURATED VAPOUR PRESSURE.

boiling point diagram A diagram showing how the boiling point of a mixture of two MISCIBLE liquids, for example water and ethanol, varies with the composition of the mixture. If the graph has a minimum at some intermediate composition, this will be the composition of an AZEOTROPIC mixture of the two liquids.

boiling tube A large thick-walled glass tube, used to hold samples being strongly heated. *See also* TEST TUBE.

Boltzmann constant The constant k in the IDEAL GAS EQUATION

$$pV = NkT$$

where N is the number of molecules, p the pressure, V the volume and T the absolute temperature. The Boltzmann constant is equal to $1.38 \times 10^{-23}\,\mathrm{J\,K^{-1}}$.

Boltzmann factor The factor $e^{-E/kT}$ in the MAXWELL–BOLTZMANN DISTRIBUTION. It describes the number of particles with an energy greater than E when the ABSOLUTE TEMPERATURE is T, where k is the BOLTZMANN CONSTANT.

bomb calorimeter A strong metal container used to measure the CALORIFIC VALUE of a fuel or food. A sample of known mass is burnt inside the container and the calorific value is calculated from the quantity of heat produced.

bond *See* CHEMICAL BOND.

bond energy The energy released when a CHEMICAL BOND is formed, or, alternatively, the energy needed to break a chemical bond. Typical values are of the order of a few hundred kilojoules per mole. Bond energies can be deduced from the HEAT OF FORMATION of the compound and the HEAT OF ATOMIZATION of the elements.

The energy of a bond may be slightly altered by its chemical environment, but bond energies generally give a fair indication of the strength of a particular bond, regardless of the molecule in which it occurs.

See also HESS'S LAW.

bonding orbital The lower energy of the two MOLECULAR ORBITALS formed when two atomic ORBITALS overlap. It is the structure that holds covalently bonded atoms together. For the bond to be stable, the bonding orbital must be full and the ANTIBONDING ORBITAL empty. Since each orbital can hold two electrons, this explains why COVALENT BONDS involve the sharing of pairs of electrons in an orbital.

borane, *boron hydride* Any one of a family of BINARY COMPOUNDS of boron with hydrogen. All are readily oxidized by atmospheric oxygen. The simplest is diborane, B_2H_6, formed by the action of an acid on magnesium boride, for example:

$$MgB_2 + 6HCl \rightarrow B_2H_6 + MgCl_2 + Cl_2$$

borate Any compound containing a CATION that contains boron. Lithium borate, $LiB(OH)_4$, is amongst the most highly ionic of these compounds and also contains the most highly HYDRATED ion. At the other extreme, many borates have an ANHYDROUS form based on the BO_3^{2-} ion. This ion does not usually occur on its own but in covalently bonded combinations, such as the triborate $B_3O_6^{3-}$ ion, which contains alternate boron and oxygen atoms in a ring.

borax Hydrated sodium borate, $Na_2B4O_7.10H_2O$. Borax is a white solid that loses water on heating; relative density 1.7. Borax occurs naturally and is an important source of boron. It is used directly in the manufacture of glass and ceramics and its ability to dissolve metal oxides makes it an important flux for soldering.

boric acid Any acid containing boron and oxygen, in particular, H_3BO_3, a white solid; melting point 169°C; relative density 1.4. It occurs naturally in small quantities, but can be manufactured by the reaction between sulphuric acid and borax:

$$Na_2B_4O_7 + H_2SO_4 + 5H_2O \rightarrow 4H_3BO_3 + Na_2SO_4$$

On heating, H_3BO_3 loses water to form a polymeric form of boric acid, $(HBO_2)_n$.

boride Any BINARY COMPOUND containing a metal and boron, such as zinc boride, ZnB_2. Borides are unusual in that they combine the characteristic high melting points and hardness of ceramic materials with the good electrical and thermal conductivities of a metal.

Born–Haber cycle A cycle of reactions used to calculate the LATTICE ENERGY of an IONIC crystal lattice. The lattice energy (ΔH_L) of the compound NaCl, for example, is the HEAT OF FORMATION of the solid lattice from its ions; that is:

$$Na^+ (g) + Cl^- (g) \rightarrow NaCl (s)\ \Delta H_L$$

The STANDARD HEAT OF FORMATION (ΔH_f) of the solid is the enthalpy of the reaction

$$Na (s) + \tfrac{1}{2}Cl_2 (g) \rightarrow NaCl (s)\ \Delta H_f$$

which can be determined experimentally. The Born–Haber cycle involves equating this enthalpy to the sum of the enthalpies of the steps needed to form the solid from the elements. These are: (i) the atomization of sodium ΔH_1; (ii) the atomization of chlorine, ΔH_2; (iii) the IONIZATION of sodium (the first IONIZATION ENERGY) ΔH_3; (iv) the ionization of chlorine (the first ELECTRON AFFINITY) ΔH_4; and (v) the formation of the lattice ΔH_L. The lattice energy can therefore be calculated from

$$\Delta H_L = \Delta H_f - (\Delta H_1 + \Delta H_2 + \Delta H_3 + \Delta H_4)$$

bornite An ore of copper, sometimes called peacock ore for its iridescent reddish-purple sheen. It is a mixed iron-copper sulphide, Cu_5FeS_4.

boron (B) The element with atomic number 5; relative atomic mass 10.8; melting point 2,079°C; boiling point 2,550°C; relative density 2.4. It is a non-metal, occurring as a brown powder or as clear crystals. In nature it occurs in the mineral BORAX. Boron is widely used in the ceramics and glass industries and its neutron-absorbing properties led to its use in control rods in some nuclear reactors.

boron carbide (B_4C) A hard black solid; melting point 2,350°C; boiling point 3,500°C (approx.); relative density 2.5. Boron carbide is made by the reduction of boron oxide with carbon:

$$2B_2O_3 + 4C \rightarrow B_4C + 3CO_2$$

Boron carbide is extremely hard and is used as an abrasive.

boron hydride *See* BORANE.

boron nitride A black solid; sublimes at 3,200°C; relative density 2.3. Boron nitride is formed by the reaction of boron oxide with ammonia:

$$B_2O_3 + 2NH_3 \rightarrow 2BN + 3H_2O$$

It has an unusually high thermal conductivity for an electrical insulator and therefore has some uses in the electronics industry, for example in the manufacture of supports for semiconductor devices.

borosilicate Any material containing both BORATE and SILICATE ions. Borosilicate glasses such as Pyrex have particularly low thermal expansions and are easier to work than pure silica glasses since they have a lower melting point.

Bosch process An industrial process, now largely obsolete, for extracting hydrogen from WATER GAS, a mixture of carbon monoxide and hydrogen. In the Bosch process, water gas and steam are passed over a catalyst at a temperature of around 450°C. Iron, with chromium(III) oxide as a PROMOTER is often used as the catalyst. The carbon dioxide in the water gas is oxidized, increasing the hydrogen content at the expense of the poisonous carbon monoxide:

$$CO + H_2O \rightarrow CO_2 + H_2$$

The carbon dioxide is removed by POTASSIUM CARBONATE solution in a SCRUBBER. The Bosch process was formerly used to produce hydrogen for the HABER PROCESS in the manufacture of ammonia. However, the necessary hydrogen is now largely produced from natural gas.

Boyle's law For a fixed mass of an IDEAL GAS at constant temperature, the pressure, p, is inversely proportional to the volume, V, i.e. the pressure multiplied by the volume is a constant:

$$pV = \text{constant}$$

See also CHARLES' LAW, GAS CONSTANT, IDEAL GAS EQUATION, PRESSURE LAW.

Bragg's law In X-RAY DIFFRACTION, the relationship between the angles at which maximum intensity of the scattered waves will occur (measured from the surface of a crystal) and the separation of the planes of atoms in the crystal. For a beam of X-rays of wavelength λ, striking a crystal at an angle θ to the planes of the crystal, which have a separation d between one plane and the next, maximum intensity will occur if

$$2d \sin\theta = n\lambda$$

where n is a whole number.

brass An alloy of copper with around 20 per cent zinc. Harder than copper but still soft enough to be worked easily, it is used for small metal parts but is too expensive and too weak to find widespread use.

brighteners PHOSPHATES and other fluorescent compounds included in some DETERGENTS, particularly domestic washing powders, to make white clothes appear brighter due to FLUORESCENCE.

brine A concentrated solution of SODIUM CHLORIDE (common salt) in water.

British thermal unit (Btu) An obsolete unit of energy, originally defined as the energy required to heat one pound of water by one degree Fahrenheit. One Btu is equivalent to 1,055 JOULES.

bromic acid The name given to bromic(I) acid (hypobromous acid), HBrO, or, more usually, bromic(V) acid, $HBrO_3$.

Bromic(I) acid is a WEAK ACID, but a strong OXIDIZING AGENT. It is formed by the action of hydrogen peroxide on bromine:

$$H_2O_2 + Br_2 \rightarrow 2HBrO$$

Bromic(V) acid, stable only in aqueous solution, is a REDUCING AGENT and a STRONG ACID. It can be produced by the reaction between barium bromate(V) and sulphuric acid:

$$Ba(BrO_3)_2 + H_2SO_4 \rightarrow 2HBrO_3 + BaSO_4$$

bromide Any BINARY COMPOUND containing BROMINE.

bromine (Br) The element with atomic number 35; relative atomic mass 79.91; melting point –7°C; boiling point 58°C. It is a HALOGEN and is a deep brown, volatile liquid, producing brown fumes of bromine vapour. It is extracted as sodium bromide from the salts present in sea water. Bromine is extracted from sodium bromide by ELECTROLYSIS. Silver bromide is an important component of many photographic materials. Large quantities of bromine are also used in the manufacture of halogenated HYDROCARBONS (*see* HALOGENATION).

bromine water A red-brown solution of bromine in water. It is used in organic chemistry to test for UNSATURATED COMPOUNDS that decolourize the water when they react with bromine, for example:

$$CH_2{=}CH_2 + Br_2 \rightarrow CH_2BrCH_2Br$$

2-bromo-2-chloro-1,1,1-trifluoroethane *See* HALOTHANE.

bromoethane, *ethyl bromide* (C_2H_5Br) A colourless flammable liquid HALOGENOALKANE; melting point –7°C; boiling point 59°C. Bromoethane undergoes SUBSTITUTION REACTIONS more readily than ELIMINATION reactions. Examples of such reactions are: the reaction with aqueous sodium hydroxide to give ethanol,

$$OH^- + C_2H_5Br \rightarrow C_2H_5OH + Br^-$$

the reaction with a CYANIDE to give propanenitrile,

$$NC^- + C_2H_5Br \rightarrow C_2H_5CN + Br^-$$

the reaction with sodium methoxide, $NaOCH_3$, in methanol, to give methoxyethane,

$$CH_3O^- + C_2H_5Br \rightarrow C_2H_5OCH_3 + Br^-$$

which is an example of WILLIAMSON ETHER SYNTHESIS.

bromoform *See* TRIBROMOMETHANE.

bromomethane, *methyl bromide* (CH_3Br) A colourless, nonflammable, volatile liquid HALOGENOALKANE; melting point –93°C; boiling point 5°C. It is made from sodium bromide, methanol and concentrated sulphuric acid. It is widely used as a fumigant and is very toxic to humans.

bronze An alloy of copper with up to about 30 per cent tin. Bronze is far harder than tin and was important historically as the first alloy to be made by man, since it melts at a far lower temperature than iron.

Brownian motion The rapid, random motion of small particles suspended in a fluid, which is seen when such particles are viewed through a microscope. Brownian motion provides evidence for KINETIC THEORY, as it is caused by the bombardment of the particles by far smaller, but fast moving, molecules in the fluid. Brownian motion was first observed in pollen grains suspended in water, but is now usually demonstrated with smoke particles in air.

brown-ring test A test for dissolved NITRATE ions. Iron(II) sulphate solution is mixed with the solution under test and concentrated sulphuric acid is added slowly to form a separate layer. The presence of nitrates is confirmed by the formation of a brown ring. The ring results from the formation of iron sulphate with nitrogen monoxide as a LIGAND,

$$Fe^{2+} + NO^{3-} + H_2SO_4 \rightarrow Fe(NO)SO_4 + 2OH^-$$

buckminsterfullerene A recently discovered ALLOTROPE of carbon. Each molecule contains 60 carbon atoms at the vertices of a roughly spherical structure, rather like a soccer ball. Several similar allotropes have since been discovered, known collectively as fullerenes or as buckyballs if roughly spherical and buckytubes if elongated.

buckyball *See* BUCKMINSTERFULLERENE.

buckytube *See* BUCKMINSTERFULLERENE.

buffer A solution designed to maintain a fairly uniform pH level despite other changes that may take place. Buffer solutions are made by mixing a WEAK ACID or a weak BASE with a salt of the same acid or base. For example, an acidic buffer solution may be made from ethanoic acid and sodium ethanoate. The sodium ethanoate is fully IONIZED in solution:

$$CH_3COONa \rightarrow CH_3COO^- + Na^+$$

while the ethanoic acid is only weakly ionized:

$$CH_3COOH + H_2O \Leftrightarrow CH_3COO^- + H_3O^-$$

The resulting mixture contains large numbers of ethanoate ions, mostly coming from the sodium ethanoate. The free ethanoate ions will tend to associate with any excess hydrogen ions, thus reducing excess acidity, whilst any HYDROXIDE ions will be mopped up by the ethanoic acid. Many commercially made buffer solutions are available.

Natural buffers exist in many biological systems where enzymes are highly sensitive to changes in pH, and similar solutions are widely used in medicine. The most common natural buffers involve HYDROGENCARBONATE (HCO_3^-) and hydrogenphosphate (HPO_4^{2-}) ions.

bumping The violent boiling of a SUPERHEATED liquid. *See also* ANTI-BUMPING GRANULES.

bunsen burner A simple gas burner widely used as a source of heat for chemistry experiments. Gas, usually METHANE, emerges from a small hole and is mixed with air, which enters via an adjustable hole. The gas and air mixture burn at the end of a short metal tube. A large amount of air produces a hot roaring blue flame with a paler cone of unburnt gas. Less air produces a quieter, cooler flame. If the air hole is closed, a highly luminous yellow flame is produced, which will leave a deposit of unburnt carbon on a cool surface.

burette A glass tube with a calibrated scale and a tap, used to measure the amount of material involved in certain reactions, such as TITRATIONS.

butadiene, *buta-1,3-diene* ($CH_2{=}CHCH{=}CH_2$) A colourless gas; boiling point –5°C. It is made from PETROLEUM by CRACKING. It is used in the preparation of artificial RUBBER and RESINS. Butadiene is often combined with PHENYLETHENE (styrene) to give SBR (styrene-butadiene rubber) or with phenylethene and PROPENONITRILE (acrylonitrile) to make ABS (ACRYLONITRILE-BUTADIENE-STYRENE).

butaldehyde *See* BUTANAL.

butanal, *butaldehyde, butyraldehydes* (C_3H_7CHO) A colourless liquid ALDEHYDE with a strong odour; melting point –99°C; boiling point 75°C. It is used in the manufacture of RUBBER.

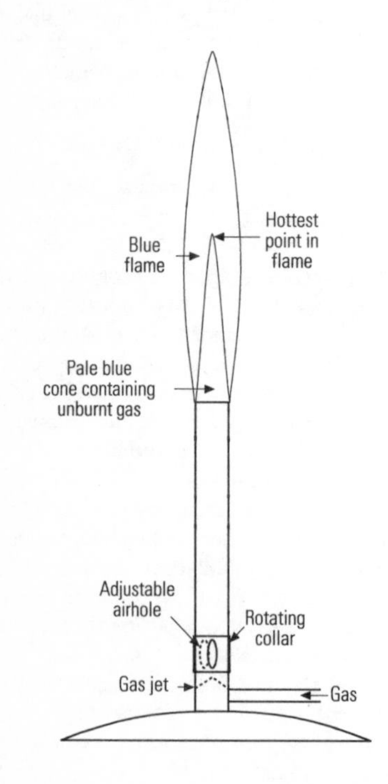

Bunsen burner.

butane (C_4H_{10}) An ALKANE, the first in the series to exhibit isomerism (*see* ISOMER). The isomers are normal or *n*-butane ($CH_3CH_2CH_2CH_3$) and isobutane ($CH_3CH(CH_3)CH_3$). Both are colourless gases found in NATURAL GAS which may be separated by FRACTIONAL DISTILLATION. The boiling point of *n*-butane is –0.3°C and of isobutane –10.3°C. Normal butane is obtained from natural gas and used as a domestic fuel for heating or portable cookers. Isobutane is a by-product of PETROLEUM manufacture.

butanedioic acid, *succinic acid* ($(CH_2)_2(COOH)_2$) A weak CARBOXYLIC ACID; melting point 185°C; boiling point 235°C. It is an intermediate in metabolism of living organisms, and has an important role in the KREBS CYCLE. It is also produced by the fermentation of sugars and is used in dye manufacture.

butanoic acid, butyric acid (C_3H_7COOH) A colourless liquid weak CARBOXYLIC ACID with a smell of rancid butter; boiling point 163°C. It occurs as an ESTER in butter and is used to make esters for use as flavourings and in perfumery.

butanol, *butyl alcohol* (C_4H_9OH) The fourth member in the series of ALCOHOLS, a colourless liquid important as a solvent for RESINS and lacquers. It has four structural ISOMERS.

butanone, *methyl ethyl ketone (M.E.K.)* (C_4H_8O, $CH_3COCH_2CH_3$) A colourless liquid with a pleasant odour; boiling point 80°C. It is one of the products of the DESTRUCTIVE DISTILLATION of wood. It is used as a solvent in the manufacture of RESINS and ACETATE film.

butene (C_4H_8) The fourth member of the hydrocarbon series of ALKENES. It is a colourless gas with an unpleasant odour. Three ISOMERS are obtained from PETROLEUM by CRACKING, which have some industrial use as various POLYMERS.

butenedioic acid ($C_4H_4O_4$) A compound exhibiting GEOMETRIC ISOMERISM and therefore existing in two forms. In the cis form it is MALEIC ACID; in the trans form it is FUMARIC ACID. Both are derivatives of ETHENE obtained by replacing a hydrogen atom on both carbons by a –COOH group.

butyl alcohol *See* BUTANOL.

butyraldehyde *See* BUTANAL.

butyric acid *See* BUTANOIC ACID.

C

cadmium (Cd) The element with atomic number 48; relative atomic mass 112.4; melting point 321°C; boiling point 765°C; relative density 8.7. Cadmium is a soft silvery white metal. Chemically, it is similar to zinc and the two elements often occur together. Cadmium is widely used in batteries and for ELECTROPLATING. Its neutron-absorbing properties allow it to be used in control rods in some nuclear reactors. Cadmium is a cumulative poison and its widespread use in industry has led to concerns about the pollution of water supplies by the dumping of cadmium compounds.

cadmium sulphide (CdS) A brown or yellow solid; decomposes on heating; relative density 4.8. Cadmium sulphide occurs naturally as the mineral GREENOCKITE, which can be roasted to provide cadmium. In its AMORPHOUS form, it is PHOTOCONDUCTIVE and is used in the manufacture of electronic deivices.

caesium, *cesium* (Cs) The element with atomic number 55; relative atomic mass 132.9; melting point 28°C; boiling point 678°C; relative density 1.9. Caesium is the most reactive of the ALKALI METALS, exploding violently on contact with water. The natural ISOTOPE, caesium–133, is stable and occurs in a number of minerals. There are 15 RADIOISOTOPES. Caesium–137 is a common product of nuclear fission, and causes particular concern in the handling of nuclear waste. It has a half-life of 20 years, and, being chemically similar to potassium, it is easily incorporated into the food chain. Caesium is used in photoelectric cells and as a catalyst.

calcite, *calcspar* The mineral calcium carbonate, $CaCO_3$. It is the main constituent of LIMESTONE and MARBLE.

calcium (Ca) The element with atomic number 20; relative atomic mass 40.1; melting point 840°C; boiling point 1,484°C; relative density 1.6. Calcium is a soft white ALKALINE EARTH metal that tarnishes rapidly in air. It is the fifth most abundant element in the Earth's crust, occurring mostly as the carbonate, $CaCO_3$, in limestone, coral, marble and CALCITE. It is extracted by reacting the carbonate with hydrochloric acid to form calcium chloride:

$$CaCO_3 + 2HCl \rightarrow CaCl_2 + CO_2 + H_2O$$

The fused chloride is then electrolysed.

Calcium is an essential element for living organisms. In animals it is an important constituent of bones and teeth.

calcium acetylide *See* CALCIUM DICARBIDE.

calcium carbide *See* CALCIUM DICARBIDE.

calcium carbonate ($CaCO_3$) A white solid; relative density 2.8; decomposes on heating. Calcium carbonate occurs naturally in limestone, chalk and marble, and as a purer crystalline form in the mineral CALCITE. The passage of rainwater through rocks containing calcium carbonate, which is very slightly soluble, is responsible for TEMPORARY HARDNESS in water. Industrially, calcium carbonate is roasted to produce calcium oxide (quicklime),

$$CaCO_3 \rightarrow CaO + CO_2$$

and is the main raw material in the SOLVAY PROCESS for the manufacture of sodium carbonate.

calcium chloride ($CaCl_2$) A white crystalline solid; melting point 772°C, boiling point 7,600°C; relative density 2.2 (ANHYDROUS). Calcium chloride is DELIQUESCENT and a number of HYDRATED crystalline forms exist. Anhydrous calcium chloride is used as a drying agent. Calcium chloride is produced as a by-product of the SOLVAY PROCESS. In the laboratory it can be prepared by the action of hydrochloric acid on calcium carbonate:

$$CaCO_3 + 2HCl \rightarrow CaCl_2 + H_2O$$

Fused calcium chloride is electrolysed as a source of metallic calcium.

calcium cyanamide ($CaCN_2$) A white solid that sublimes at 1,150°C. Calcium cyanamide can be prepared by heating calcium acetylide in nitrogen:

$$CaC_2 + N_2 \rightarrow CaCN_2 + C$$

Calcium cyanamide decomposes on contact with water to give calcium carbonate and ammonia:

$$CaCN_2 + 3H_2O \rightarrow CaCO_3 + 2NH_3$$

calcium dicarbide, *calcium acetylide, calcium carbide* (CaC_2) A white or grey solid; melting point 450°C; boiling point 2,300°C. Calcium dicarbide is an ionic salt that can be manufactured by reducing calcium oxide with coke at high temperatures:

$$CaO + 3C \rightarrow CaC_2 + CO$$

Calcium dicarbide reacts with water to produce ETHYNE:

$$CaC_2 + 2H_2O \rightarrow Ca(OH)_2 + C_2H_2$$

In addition to being an important source of ethyne for the synthesis of other organic reagents, this reaction is also used to produce a simple portable source of combustible gas in remote locations.

calcium fluoride (CaF_2) A white crystalline solid; melting point 1,360°C; boiling point 2,500°C; relative density 3.2. Calcium fluoride occurs naturally as the mineral fluorite and is an important source of fluorine, which is extracted by electrolysing the fused salt.

calcium hydroxide, *slaked lime* ($Ca(OH)_2$) A white solid that decomposes on heating; relative density 2.2. It is formed by the addition of water to calcium oxide:

$$CaO + H_2O \rightarrow Ca(OH)_2$$

Calcium hydroxide is slightly soluble in water, the solution being known as lime water. Calcium hydroxide is widely used in agriculture as an alkali to neutralize excess acidity in some soils.

calcium nitrate ($Ca(NO_3)_2$) A white solid; melting point 561°C; decomposes on further heating; relative density 2.5. Calcium nitrate is a DELIQUESCENT solid and several hydrated crystalline forms exist. It can be prepared by the reaction of nitric acid on calcium carbonate:

$$CaCO_3 + 2HNO_3 \rightarrow Ca(NO_3)_2 + CO_2 + H_2O$$

Calcium nitrate decomposes on strong heating:

$$2Ca(NO_3)_2 \rightarrow 2CaO + 4NO_2 + O_2$$

calcium oxide, *quicklime* (CaO) A white solid; melting point 2,600°C; boiling point 2,850°C; relative density 3.4. Calcium oxide can be prepared by burning calcium in air or industrially by heating calcium carbonate:

$$CaCO_3 \rightarrow CaO + CO_2$$

Calcium oxide is used in the manufacture of calcium hydroxide and in many metal extraction processes to form a SLAG with SILICA impurities in metal ores.

calcium phosphate ($Ca_3(PO_4)_2$) A white solid; decomposes on heating; relative density 3.1. Calcium phosphate is the major mineral constituent of animal bones, and is widely used as a fertilizer.

calcium sulphate ($CaSO_4$) A white solid; melting point 1,450°C; decomposes on further heating; relative density 3.0. Calcium sulphate occurs naturally as the mineral gypsum, which is comprised mostly of the DIHYDRATE $CaSO_4.2H_2O$. On heating gently, this loses water to form another HYDRATED form, $2CaSO_4.H_2O$, known as plaster of Paris. When water is added to plaster of Paris it forms a paste, which sets hard.

Calcium sulphate is an important raw material in the manufacture of sulphuric acid. It is only slightly soluble, and is a source of PERMANENT HARDNESS in water. In the laboratory, it can be prepared as a PRECIPITATE in the reaction between aqueous calcium chloride and sodium sulphate:

$$CaCl_2 + Na_2SO_4 \rightarrow CaSO_4 + 2NaCl$$

calcspar *See* CALCITE.

Calgon *See* WATER SOFTENING.

caliche A mixture of soluble salts that occurs naturally as crystalline deposits where inland seas have evaporated, particularly in Chile. Caliche is rich in SODIUM NITRATE and is mined commercially as a source of NITRATES.

californium (Cf) The element with ATOMIC NUMBER 98. It does not occur in nature, but nine isotopes have been synthesized. The longest-lived ISOTOPE has a HALF-LIFE of 800 years. Some of its isotopes decay by spontaneous nuclear fission, which makes it useful as a neutron source.

calomel half-cell A HALF CELL consisting of a mercury electrode coated with mercury(I) chloride (calomel) with an ELECTROLYTE of potassium chloride and mercury chloride. It is used as a standard against which to measure the ELECTRODE POTENTIAL of other materials since it is easier to use than a HYDROGEN ELECTRODE. It has a STANDARD ELECTRODE POTENTIAL of –0.2415 V.

calorie (cal) A unit of quantity of heat in the C.G.S. SYSTEM. One calorie is the amount of heat needed to raise the temperature of one gram of water by 1°C. The calorie has been largely replaced by the JOULE; it is roughly equivalent to 4.2 J.

calorific value The energy content of a fuel or food. It is the amount of heat generated by completely burning a given mass of fuel (which can be food) in a piece of apparatus called a BOMB CALORIMETER. Thus it is equal to the HEAT OF COMBUSTION. Calorific value is measured in JOULES per kilogram.

calorimeter A container for performing experiments related to heat transfer and temperature changes, such as the measurement of SPECIFIC HEAT CAPACITY and LATENT HEAT. Calorimeters are generally made of a metal (a good conductor of heat, so the entire vessel reaches the same temperature), of known heat capacity. *See also* CALORIFIC VALUE.

Calvin cycle Another term for the light-independent stage of PHOTOSYNTHESIS, named for Melvin Calvin (1911–), who established the mechanisms of the reactions in the cycle. The Calvin cycle involves the conversion of carbon dioxide from the air into a form of carbohydrate that plants can store as an energy source.

Carbon dioxide from the air diffuses into a leaf and in most plants combines with a 5-carbon compound called ribulose biphosphate to form a six-carbon intermediate, which is unstable and breaks down into two molecules of the 3-carbon compound glycerate-3-phosphate (GP). This is then converted, in the presence of ATP and reduced NADP, into triose phosphate which combines to form hexose sugars. These POLYMERIZE to yield STARCH for the plant to store. Some of the triose phosphate is used to regenerate ribulose biphosphate.

In some plants, carbon dioxide combines with phosphoenol pyruvate (PEP) instead of ribulose biphosphate and thus produces the 4-carbon oxaloacetate instead of GP.

camphor ($C_{10}H_{16}O$) A white, solid cyclic KETONE; melting point 179°C; boiling point 204°C. It has a characteristic odour of mothballs, in which it is used. Camphor occurs naturally in the wood of the camphor tree but can also be synthesized. It is used in the manufacture of CELLULOID and explosives. It is used in medicine as a cold remedy or stimulant in the form of camphor oil, which is a 20 per cent solution in olive oil.

candela (Cd) The SI UNIT of LUMINOUS INTENSITY. One candela is equal to the luminous intensity in a given direction of a source of monochromatic radiation of frequency 5.4×10^{14} Hz and has a radiant intensity in that direction of 1/683 watt per steradian.

Cannizzaro reaction The interaction of two ALDEHYDES in the presence of dilute alkalis where one is reduced to the corresponding ALCOHOL and the other is oxidized to the corresponding acid. Many but not all aldehydes react in this way. *See also* ALDOL CONDENSATION, OXIDATION, REDUCTION.

canonical form Any one of the possible structures of a molecule that coexist when the molecule is a RESONANCE HYBRID.

caprolactam ($C_6H_{11}NO$) A white, crystalline solid; melting point 68–70°C; boiling point 139°C. On heating, caprolactum gives POLYAMIDES and is therefore used in the manufacture of NYLON. It is a LACTAM, consisting of a seven-membered ring containing a –NH.CO– group.

carbaminohaemoglobin The product formed when carbon dioxide combines with HAEMOGLOBIN. A small proportion, about 10 per cent, of carbon dioxide is carried from the tissues to the respiratory system in this form. The remainder is carried in the form of hydrogen carbonate. *See also* CARBONIC ANHYDRASE.

carbanion An organic ion of the type R_3C^-, with a negative charge on the carbon atom. Carbanions are intermediates in certain organic reactions.

carbazole ($C_{12}H_9N$) A white solid; melting point 238°C; boiling point 335°C. It occurs with ANTHRACENE. Carbazole is used in the manufacture of DYESTUFFS.

carbene A highly reactive species of the type R_2C:, where the carbon atom has two spare electrons. Carbenes are intermediates formed during certain organic reactions. They can insert between the carbon and hydrogen atoms of a C–H bond or attack a double bond. METHYLENE, $:CH_2$, is the simplest example.

carbenium ion *See* CARBONIUM ION.

carbide A BINARY COMPOUND containing a metal and carbon. An example is silicon carbide (carborundum), SiC, which has a macromolecular structure (*see* MACROMOLECULE) with a high boiling point and is extremely hard. Silicon carbide is commonly used as an abrasive. The carbides of the ALKALI METALS and ALKALINE EARTHS are IONIC compounds, containing the C_2^{2-} ion. They react with water to produce ETHYNE (acetylene) and are known as DICARBIDES.

carbocation *See* CARBONIUM ION.

carbohydrate One of a large group of organic compounds with the general formula $C_x(H_2O)_y$. They are the main energy-providing components of the human diet. There are three main groups of carbohydrates: MONOSACCHARIDES, DISACCHARIDES and POLYSACCHARIDES. Monosaccharides, such as GLUCOSE and FRUCTOSE, are single sugars with the general formula $(CH_2O)_n$ that cannot be split into smaller carbohydrate units. Disaccharides, such as SUCROSE, MALTOSE and LACTOSE, are double sugars, where two monosaccharides are combined, which can be split into their single sugar components. Polysaccharides, such as STARCH and GLYCOGEN, consist of variable numbers of monosaccharides joined together in chains that can be branched or not and can fold for easy storage, and be broken down into their constituent disaccharides or monosaccharides for use. Some polysaccharides are structural, for example, CELLULOSE and CHITIN, and some are food reserves, for example, starch and glycogen.

Most carbohydrates can form ISOMERS, each one having a different functional property. ENZYMES usually only react with one form. Naturally occurring carbohydrates have a D(+) form (*see* DEXTROROTATORY and OPTICAL ISOMERISM).

There are two tests that can be used to identify sugar in a solution: BENEDICT'S TEST and FEHLING'S TEST. Both rely on the ability of monosaccharides and some disaccharides to reduce copper(II) sulphate to copper oxide, causing a colour change on boiling. These sugars can also be classified as reducing or non-reducing. The test for starch is the addition of iodine in potassium iodide, which integrates into the starch polymer causing a colour change from yellow/orange to blue/black.

carbolic acid The common name for PHENOL.

carbon (C) The element with atomic number 6; relative atomic mass 12.00; sublimes at 3,500°C; relative density 2.3 (graphite). Carbon occurs naturally in AMORPHOUS forms, such as coal, charcoal and soot and as the two crystalline ALLOTROPES graphite and diamond. A third allotrope, BUCKMINSTERFULLERENE, was discovered in 1995. Carbon has two stable isotopes (carbon–12 and –13) and four radioactive ones. The radioactive isotope carbon–14 is used in RADIOCARBON DATING.

Carbon's most important chemical property is its ability to form long chain molecules. This is the starting point for all organic compounds, including the complex molecules responsible for all known living organisms.

See also CARBON CYCLE.

carbonaceous (*adj.*) Describing any mineral containing carbon other than in CARBONATES.

carbonate Any compound containing a metal CATION and the carbonate ANION, CO_3^{2-}. Carbonates are formed by the reaction of the many metals with carbonic acid, for example:

$$Mg + H_2CO_3 \rightarrow MgCO_3 + H_2$$

All carbonates except those of the ALKALI METALS decompose on heating (lithium carbonate also decomposes), for example:

$$MgCO_3 \rightarrow MgO + CO_2$$

They also react with acids to release carbon dioxide, this reaction being the classic test for the presence of the carbonate ion, for example:

$$MgCO_3 + 2HCl \rightarrow MgCl_2 + CO_2 + H_2O$$

Only the carbonates of the alkali metals (except lithium) are soluble. These form alkaline solutions, due to the formation of the HYDROGENCARBONATE (or bicarbonate) ion:

$$CO_3^{2-} + H_2O \rightarrow HCO_3^- + OH^-$$

carbonation The addition of carbon dioxide, either to form a CARBONATE, or in aqueous solution as in the manufacture of fizzy drinks. In the latter case, the fizz is produced by carbon dioxide coming out of solution when the pressure is reduced.

carbon black, *lamp black* An AMORPHOUS form of CARBON, formed by burning a HYDROCARBON in a limited supply of oxygen. Carbon black is used as a black pigment in some inks and paints and is added to rubber to improve wear resistance in tyres.

carbon cycle The constant circulation of carbon between organic and inorganic sources in nature. This recycling maintains the balance between carbon dioxide in the atmosphere and carbon in organisms, and is vital for all forms of life.

Carbon dioxide in the atmosphere provides a major source of the carbon used by photosynthetic plants and organisms to make CARBOHYDRATES (*see* PHOTOSYNTHESIS). Carbon passes into the bodies of animals when they eat the plants (*see* FOOD CHAIN) and is returned, as carbon dioxide, back to the atmosphere through their respiration.

The natural processes of photosynthesis and respiration balance each other out, but in recent years human intervention has disturbed this balance. The burning of FOSSIL FUELS and the destruction of large areas of tropical forest causes the atmospheric levels of carbon dioxide to rise, contributing to the GREENHOUSE EFFECT.

Part of any natural cycle includes not only the biological (living) component but also a geological (non-living) component. The geological components of the carbon cycle include rocks and other deposits in the oceans and atmosphere, for example in the form of coal and oil under the ocean, peat in wetlands and limestone rocks. They provide the largest reservoir of carbon.

carbon dioxide (CO_2) A colourless gas; sublimes at –78°C. It is present in small quantities in the atmosphere, playing an important part in biological processes as it is produced by RESPIRATION and taken up in PHOTOSYNTHESIS. In the laboratory carbon dioxide is prepared by the action of acids on carbonates, for example:

$$CaCO_3 + 2HCl \rightarrow CO_2 + CaCl_2 + H_2O$$

Industrially, carbon dioxide is produced as a by-product of the production of calcium oxide (quicklime) from calcium carbonate (in the form of limestone). It is stored as a liquid under pressure and used for a wide variety of purposes – as a fire-extinguishing gas, for instance, and in its solid form for refrigeration.

The increase of carbon dioxide in the atmosphere, from the burning of fossil fuels and the destruction of vast areas of forest, is thought to be the greatest cause of global warming.

See also CARBON CYCLE.

carbon disulphide (CS_2) A colourless liquid; melting point –110°C; boiling point 46°C; relative density 1.3. Carbon disulphide can be produced by the action of methane on sulphur:

$$CH_4 + 4S \rightarrow CS_2 + 2H_2S$$

It is used as a solvent for many organic materials, including rubber, and will also dissolve sulphur and phosphorus.

carbon fibre A COMPOSITE MATERIAL made in a similar way to GLASS-FIBRE REINFORCED PLASTICS, but the carbon fibres are even stronger. However, like glass fibre, the material does not lend itself to mass-production techniques, so is found only in relatively expensive high-technology products.

carbonic acid (H_2CO_3) A weak DIBASIC ACID formed by dissolving carbon dioxide in water. The acid exists only in equilibrium with dissolved carbon dioxide, and cannot be isolated from aqueous solution. It dissociates strongly into HYDROGENCARBONATE ions, HCO_3^-, which in turn dissociate weakly into CARBONATE ions, CO_3^{2-}. It therefore forms two series of salts: NORMAL SALTS (carbonates) and ACIDIC SALTS (hydrogencarbonates). Soda water is a solution of carbonic acid.

carbonic anhydrase A zinc-containing enzyme that catalyses the reaction between carbon dioxide and water to form carbonic acid, H_2CO_3, in an organism. Carbonic acid subsequently dissociates into hydrogen, H^+, and hydrogen carbonate, HCO^{3-}, ions. It is in the form of hydrogen carbonate that the majority (85 per cent) of carbon dioxide produced by body tissues is transported to the lungs to be removed as waste. Carbonic anhydrase is found in red blood cells, where this reaction takes place. It is also found in the kidney, where it controls the pH of urine.

carboniferous (*adj.*) Describing a material containing NATIVE carbon, particularly coal.

carbonium ion, *carbenium ion, carbocation* Any positively charged ion containing a TRIVALENT carbon atom, R_3C^+, where R is an alkyl group. Carbonium ions have a strong affinity for NUCLEOPHILES. The most stable is the tertiary carbonium ion where all three bonds lead to other carbon atoms. The stability decreases in secondary and more so in primary carbonium ions where only two and one bonds respectively lead to other carbon atoms.

Carbonium ions are thought to be intermediates in many reactions, for example:

$$CH_3-C(CH_3)_2-Br \underset{\text{in water and ethanol}}{\overset{\text{dissolved}}{\rightleftharpoons}} CH_3-C^+(CH_3)_2 + Br^-$$

2, bromo-2-methyl propane → dimethylethyl cation (carbonium ion)

$$\downarrow$$

$$CH_3-C(CH_3)_2-OH$$

2-methylpropan–2–ol

Overall this reaction is a NUCLEOPHILIC SUBSTITUTION.

carbonization The DESTRUCTIVE DISTILLATION of COAL by heating, in the absence of air, to produce COKE, COAL GAS and COAL TAR.

carbon monoxide (CO) A colourless gas; melting point –199°C; boiling point –192°C. Carbon monoxide is produced by the combustion of carbon in limited supplies of oxygen:

$$2C + O_2 \rightarrow 2CO$$

It burns readily in air to give carbon dioxide:

$$2CO + O_2 \rightarrow 2CO_2$$

Carbon monoxide is toxic, as it replaces oxygen in HAEMOGLOBIN in the blood, and so prevents oxygen being transported to body tissues. The gas is very dangerous, as it is colourless and odourless and therefore difficult to detect. Its traditional use as a fuel gas has largely been replaced by methane, though it is still used as a REDUCING AGENT in some industrial processes.

carbon tetrachloride, *tetrachloromethane* (CCl_4) An organic chemical; boiling point 76°C; melting point –23°C. It is manufactured from the CHLORINATION of methane or other hydrocarbons. It was used in the past as a dry cleaning agent but is now thought to be carcinogenic (*see* CARCINOGEN). It is used as a solvent and in fire extinguishers.

carbonyl chloride, *phosgene* ($COCl_2$) A colourless gas with a distinctive odour, produced by the reaction between carbon monoxide and chlorine:

$$CO + Cl_2 \rightarrow COCl_2$$

Carbonyl chloride is used as a chlorinating compound in the manufacture of some chlorinated hydrocarbons, and is highly toxic. It was used as a chemical weapon in the First World War.

carbonyl group In organic compounds, a carbon atom attached by a DOUBLE BOND to an oxygen atom. The group is found in ALDEHYDES, KETONES and the related CARBOHYDRATES. *See also* HALOFORM REACTION, TRIIODOMETHANE TEST.

carboxyhaemoglobin The product formed when HAEMOGLOBIN combines irreversibly with carbon monoxide. Haemoglobin has a far grater affinity for carbon monoxide than for oxygen and carboxyhaemoglobin is therefore very stable. In the presence of carbon monoxide there is less haemoglobin available for the transport of oxygen, which accounts for the toxic effects of carbon monoxide on the respiratory system.

carboxyl group The –COOH FUNCTIONAL GROUP found in CARBOXYLIC ACIDS and their derivatives.

carboxylic acid Any one of a group of organic compounds that contain the CARBOXYL GROUP (–COOH) attached to another group (R). The other group can be a hydrogen atom (HCOOH, methanoic acid) or a larger molecule, with up to 24 carbon atoms. When R is a straight chain ALKYL GROUP (such as CH_3, C_2H_5 or $CH_3(CH_2)_{16}COOH$) the compound is termed a FATTY ACID.

Although carboxylic acids contain the C=O CARBONYL GROUP and the OH HYDROXYL GROUP, these groups behave differently when together as the –COOH carboxyl group. The hydroxyl group can be replaced by another FUNCTIONAL GROUP giving rise to compounds termed carboxylic acid derivatives. For example, RCOCl is an ACID CHLORIDE, RCOOR′ is an ESTER and $RCONH_2$ is an AMIDE. NITRILES contain a carbon-nitrogen TRIPLE BOND, RC≡N, and are also related to carboxylic acids.

Carboxylic acids are named after the ALKANE from which they are derived, replacing the ending *-ane* by *-anoic acid.* Examples include CH_3COOH, ethanoic acid and propanoic acid, CH_3CH_2. The carbonyl carbon atom is numbered one.

Some derivatives of carboxylic acids

General formula	Class name	Example	
R–CO–Cl	acid chloride or acyl chloride	CH_3COCl	ethanoyl chloride
R–CO–O–CO–R	acid anhydride	$(CH_3CO)_2O$	ethanoic anhydride
R–CO–OR′	ester	$CH_3COOC_2H_5$	ethyl ethanoate
$R–CO–NH_2$	amide	CH_3CONH_2	ethanamide
R–CN	nitrile	CH_3CN	ethanenitrile

Carboxylic acids usually exist as hydrogen-bonded DIMERS (*see* HYDROGEN BOND), except in aqueous solution. They are weak acids, affected greatly by what is attached to the carboxyl group. They are able to neutralize alkalis and liberate carbon dioxide from CARBONATES. Carboxylic acids can be made by the OXIDATION of primary ALCOHOLS, RCH_2OH, or of ALDEHYDES, RCHO, using acidified DICHROMATE(VI):

$$RCH_2OH \rightarrow RCHO \rightarrow RCOOH$$

They can also be made by reacting GRIGNARD REAGENTS with carbon dioxide.

$$CO_2 + C_2H_5MgBr \rightarrow C_2H_5COOH + HOMgBr$$

Many carboxylic acids occur naturally in plants and animals. The lower carboxylic acids have strong smells; for example, ethanoic (acetic) acid gives the smell to vinegar, while rancid butter smells of butanoic acid. Ethanoic acid is important in the manufacture of vinegar and VISCOSE. Animal and vegetable fats and oils (which are mostly esters of GLYCEROL), along with fatty acids, are important in food products and diet.

carboxylic acid derivatives *See* CARBOXYLIC ACID.

carcinogen Any factor known to be linked with an increased incidence of cancer. Many chemicals are carcinogenic, including tobacco smoke, asbestos dust, benzpyrene in soot and tar and many other organic compounds (particularly amino aromatic or nitro-compounds). Workers in such industries need protection. Other carcinogens include IONIZING RADIATION, viral infections and dietary or genetic factors.

Carius method A technique for determining the amount of sulphur and HALOGENS in an organic compound. The compound is heated in a sealed tube (Carius tube) with silver nitrate in concentrated nitric acid. The compound is completely oxidized and precipitates of silver sulphides and halides are formed, which can be separated and weighed.

carnallite A mineral form of hydrated potassium magnesium chloride, $KCl.MgCl_2.6H_2O$. The mineral is mined as an important source of potassium.

Carnot cycle An ideal cycle of operations, leading to the greatest efficiency of conversion of heat energy into WORK attainable by any reversible HEAT ENGINE.

The Carnot cycle has four reversible stages, operating between two HEAT RESERVOIRS. An IDEAL GAS starts at a high temperature and pressure, and expands isothermally (at a constant temperature, *see* ISOTHERMAL), converting energy from a high temperature reservoir into mechanical work. The gas then expands further adiabatically (with no exchange of heat energy to its surroundings, *see* ADIABATIC) cooling to a lower temperature. In the third stage, the gas is compressed isothermally, releasing any remaining heat it possess. Finally the gas is compressed adiabatically to its original state. Heat energy is taken from the high temperature reservoir and a smaller amount given up to the lower temperature. The difference between these two energies is the amount of mechanical work done.

Since every stage in the Carnot cycle is reversible, there is no overall change in

ENTROPY and the Carnot engine must be the most efficient heat engine permitted by the SECOND LAW OF THERMODYNAMICS. Realistic heat engines are far less efficient, but are designed to approach the Carnot cycles as closely as possible.

Carnot engine A hypothetical HEAT ENGINE operating on the CARNOT CYCLE.

carnotite A yellowish ore of uranium potassium vanadate, $K_2(UO_2)_2(VO_4)_2$, mined, particularly in the USA and Australia, for its uranium content.

carotene ($C_{40}H_{56}$) A natural CAROTENOID pigment. It is responsible for the orange, yellow and red colour of carrots, tomatoes and oranges. Several forms exist, the most common being β-carotene. In animals it is a precursor of VITAMIN A.

carotenoid One of a number of coloured pigments found in many living organisms. Carotenoid pigments are lipids and can be yellow, orange, red or brown. They frequently occur in the leaves, fruits, roots and petals of plants, giving them their colour and providing the autumn colours. They can also act as accessory pigments in PHOTOSYNTHESIS. The group includes CAROTENE, LYCOPENE and the XANTHOPHYLLS.

carrageenan A naturally occurring POLYSACCHARIDE consisting of GALACTOSE units. It can be isolated from red algae and is used as a gelling agent and in foodstuffs.

casein The main protein constituent of milk. Caseinogen is its soluble form, which is precipitated by RENNIN to casein. This process is particularly important in young animals. Casein is also a major component of cheese.

caseinogen *See* CASEIN.

cassiterite An important ore of tin, SnO_2, brownish black in colour and widely mined.

cast A metal object formed by pouring molten metal into a suitably shaped mould. *See* CAST IRON.

cast iron Iron with a high proportion of carbon (typically 3 per cent). It is made by melting PIG IRON, often with steel scrap to reduce the proportion of carbon. This material is then poured into moulds and allowed to cool. Cast iron is very hard, but is BRITTLE, so difficult to machine and prone to cracking.

catabolism The breaking down of larger molecules in living organisms into smaller ones, with the release of energy and the formation of waste products. Respiration is an example of catabolism. *Compare* ANABOLISM. *See* METABOLISM.

catalase An enzyme that catalyses (*see* CATALYST) the breakdown of hydrogen peroxide to water and oxygen. It is found in the body within metabolically active cells where hydrogen peroxide is a toxic by-product of several biochemical reactions.

catalyst Any substance that changes the rate of a chemical reaction, without itself being permanently altered chemically by that reaction. For example, hydrogen peroxide, H_2O_2, decomposes to water and oxygen only very slowly at room temperature:

$$2H_2O_2 \rightarrow 2H_2O + O_2$$

but in the presence of manganese(IV) oxide, MnO_2, the reaction proceeds much more rapidly. The manganese(IV) oxide is itself unchanged by the reaction.

A catalyst is described as homogeneous if it exists in the same phase as the reagents, for example a liquid catalysing a reaction between liquids. A catalyst is heterogeneous if the reagents are in a different phase to itself, for example a solid catalysing reactions between gases. The term catalyst usually applies to substances that increase the rate of a reaction (positive catalysts). Those that slow a reaction down are termed negative catalysts.

The action of catalysts is complex, though in general they lower the ACTIVATION ENERGY of a chemical process. Many catalysts are TRANSITION METALS, and platinum is particularly widely used (for instance in the CONTACT PROCESS for manufacturing sulphuric acid). The multiple OXIDATION STATES displayed by transition metals seem important in some catalytic processes. These allow a reaction to proceed via an intermediate step that involves the formation of compounds including the catalyst, which is regenerated at a later stage in the reaction.

Many catalytic processes use heterogeneous catalysts, for example, solid platinum is used as a catalyst in many gas phase reactions. It is important that the catalyst is finely divided to allow maximum contact with the gas. In such processes, VAN DER WAALS' FORCES between the reagents and incompletely bound atoms in the surface of the solid catalyst appear to be important to the function of the catalyst.

Many biological reactions rely on complex catalysts, often specific to a particular reaction. These are called ENZYMES.

See also AUTOCATALYSIS.

catalytic converter A platinum-based CATALYST placed in the flow of the burnt gases leaving a petrol engine to ensure that they are fully oxidized. This reduces the levels of carbon monoxide and nitrogen oxides in the exhaust gases.

catalytic cracking *See* CRACKING.

catalytic reforming A process similar to catalytic CRACKING except that conditions are chosen to minimize the formation of straight-chain ALKANES and instead encourage the formation of branched-chain alkanes. These alkanes burn more easily in petrol engines with a lesser tendency to KNOCKING. Thus catalytic reforming is used to reduce the knocking and improve the OCTANE NUMBER of petrol.

cataphoresis *See* ELECTROPHORESIS.

catechol *See* 1,2-DIHYDROXYBENZENE.

catecholamines AMINES that contain a catechol ($C_6H_4(OH)_2$) ring. Examples include DOPAMINE, ADRENALINE and NORADRENALINE.

cathode A negatively charged ELECTRODE.

cathodic reduction The process that takes place at the CATHODE of an ELECTROLYSIS, regarded as a REDUCTION. Since the cathode is negatively charged, electrons are added to the ELECTROLYTE at this point and thus the electrolyte is reduced. For example, in the electrolysis of water, hydrogen ions are reduced to gaseous hydrogen:

$$2H^+ + 2e^- \rightarrow H_2$$

See also ANODIC OXIDATION.

cation A positively charged ion. So called because it will be attracted to the CATHODE in ELECTROLYSIS. *Compare* ANION.

cation pump *See* SODIUM PUMP.

caustic (*adj.*) Describing a highly reactive ALKALI, such as sodium hydroxide.

caustic potash *See* POTASSIUM HYDROXIDE.

caustic soda *See* SODIUM HYDROXIDE.

cell, *electrochemical cell* A device that connects chemical reagents to an electrical circuit. A cell consists of two ELECTRODES in contact with an ELECTROLYTE – a conducting liquid or jelly. An electrolytic cell is one in which a chemical change takes place when a current is passed through the electrolyte (*see* ELECTROLYSIS). In a voltaic cell, a chemical reaction between the electrolyte and the electrodes produces an ELECTROMOTIVE FORCE, which drives a current around a circuit.

Some voltaic cells can be used once only: examples are the ZINC-CARBON CELL, which is cheap but stores relatively little energy, and the more expensive but longer lasting MANGANESE-ALKALINE CELL. Others are rechargeable by passing a current through them in the reverse direction, the chemical reactions can be reversed and electrical energy converted back to chemical energy. Examples of rechargeable cells are the LEAD-ACID CELL, used to provide high currents in cars for example, and the NICKEL-CADMIUM (or Nicad) cell, often used in electronic appliances. The amount of charge that a voltaic cell can drive around a circuit is measured in ampere-hours.

See also HALF-CELL.

cellobiose, *4-(ß-D-glucopyranosido)-D-glucopyranose* ($C_{12}H_{22}O_{11}$) The DISACCHARIDE that forms the repeating unit of CELLULOSE.

cellular plastic A type of PLASTIC that is formed using gases to give an expanded or foamed structure. Cellular plastics include POLYSTYRENE, POLYURETHANE and PVC. They are lightweight and are used in packaging and for insulation.

cellulase An ENZYME that breaks down CELLULOSE into its constituent sugars.

celluloid A THERMOPLASTIC made from cellulose nitrate plasticized with CAMPHOR. It is highly flammable and therefore not used much now.

cellulose ($C_6H_{10}O_5$) A POLYSACCHARIDE made of long chains of GLUCOSE molecules, linked by GLYCOSIDIC BONDS. It is a major component of plant cell walls. Cellulose is unbranched with many chains running parallel to each other, which allows hydrogen bonds to form cross-linkages between the chains. This gives cellulose great stability and strength, which allows it to provide the structural support needed for plant cell walls. The strength of cellulose has been utilized by humans, for example in the use of cotton, RAYON, paper and plastics. Cellulose is an important constituent in the diet of many animals, although few possess the enzyme CELLULASE needed to digest it. For those organisms that can digest cellulose it forms a major component of their diet. For others, including humans, it provides a vital source of dietary fibre.

cellulose acetate, *cellulose ethanoate* A substance formed by treating CELLULOSE (cotton or wood pulp) with ethanoic anhydride, ethanoic acid and concentrated sulphuric acid. The cellulose is acetylated to produce a number of useful materials used in varnishes and textiles.

cellulose ethanoate *See* CELLULOSE ACETATE.

cellulose nitrate, *nitrocellulose* A substance prepared by treating CELLULOSE (cotton or wood pulp) with concentrated nitric acid. It is used in expolsives (as guncotton) and in CELLULOID.

Celsius A temperature scale in which the freezing point of water is defined as zero degrees Celsius (0°C), whilst the boiling point of water is 100°C, both temperatures being measured at ATMOSPHERIC PRESSURE. The degree Celsius is the same size as the KELVIN. *See also* ABSOLUTE TEMPERATURE.

cement Any substance used to hold two solid objects together. Portland cement, which is widely used in the building industry, is made by heating LIMESTONE with clay, which is then ground to a fine powder. The mixture contains anhydrous calcium aluminates and silicates, which form a rigid HYDRATED material when water is added and allowed to evaporate.

centi- Prefix used to denote one hundredth. For example, one centimetre is one hundredth of a metre (0.01 m).

centrifuge A machine for separating two different materials on the basis of their relative densities. A centrifuge may be used to separate the components in an EMULSION or a solid suspended in a liquid. The mixture is placed in a tube and rotated very rapidly in a horizontal circle. The denser component is forced outwards along the tube, displacing the less dense component, and collecting at the bottom of the tube. Centrifugation can be used for separating different cell types or for separating blood plasma from red blood cells. *See also* ULTRACENTRIFUGE.

ceramics A wide range of materials, mostly based on SILICATES and made by heating some form of clay, which contains fine silicate particles produced by the erosion of rocks.

cerium (Ce) The element with atomic number 58; relative atomic mass 140.1; melting point 798°C; boiling point 3,433°C; relative density 6.8. Cerium is the most abundant of the LANTHANIDE metals and is used in some specialist alloys.

cermet A COMPOSITE MATERIAL made by heating together a clay and a powdered metal. Cermets can be pressed into any required shape before heating and the finished material is hard and resistant to heat and corrosion.

cerussite A lead ore produced from the weathering of natural deposits of lead sulphide (galena),

$$PbS + CO_2 + H_2O \rightarrow PbCO_3 + H_2S$$

cesium *See* CAESIUM.

CFC *See* CHLOROFLUOROCARBON.

c.g.s. system A system of physical units derived from the metric system and based on the centimetre, gram and second. It has been replaced by SI UNITS.

chain reaction Any reaction in which the products of the reaction initiate further reactions of the same type, leading to an exponential growth in the rate of reaction.

In chemistry, chain reactions usually involve RADICALS, atoms or molecules having unpaired electrons. There are three stages in a chain reaction – initiation, propagation and termination. During initiation the reactive species are formed, often radicals. These are then involved in subsequent reactions that yield similar or different reactive species, themselves able to repeat the process, thus creating a chain of reactions. The chain is terminated when a reaction occurs in which the product is unable to react further.

An example of a chain reaction is the reaction between hydrogen and bromine to form hydrogen bromide (HBr). The reactive species here is the bromine radical (.Br), formed by the splitting of bromine (Br_2). This attacks hydrogen molecules

$$.Br + H_2 \rightarrow HBr + .H$$

and the .H generated reacts with bromine

$$.H + Br_2 \rightarrow HBr + .Br$$

The .Br can then react with another hydrogen so continuing the chain. The reaction is terminated if .Br reacts with another .Br to form Br_2, thereby removing the reactive species. Chain reactions can be inhibited (the rate of production of the product reduced) by one of the products formed. Many organic reactions involving radicals (*see* HOMOLYTIC FISSION) proceed by a chain reaction, for instance the reaction of methane with chlorine.

The speed at which chain reactions occur is variable, some slowly and some accelerating with increasing number of reactive species to the point of an EXPLOSION.

chalcopyrite A yellow/gold mineral, iron copper sulphide, $CuFeS_2$, a major source of copper.

chalk A soft SEDIMENTARY rock made from the remains of microscopic sea creatures and composed mainly of CALCIUM CARBONATE.

change of state Any process in which substance changes from one of the STATES OF MATTER to another. In any change of state, LATENT HEAT is taken in or given out. *See* BOILING POINT, MELTING POINT, SUBLIMATION.

char (*vb.*) To reduce an organic material to carbon, either by the action of heating or by the action of a strong dehydrating agent (*see* DEHYDRATE), such as concentrated sulphuric acid.

charcoal A porous, AMORPHOUS form of CARBON, formed by heating organic matter, usually wood, in a limited supply of air. Historically, charcoal was widely used as a fuel. The porous nature of charcoal makes it a strong absorber of gases.

charge A fundamental property of matter. Charges are of two types, positive and negative. Charged particles exert forces on one another: charges of the same type repel one another whilst opposite charges attract. The SI UNIT of charge is the COULOMB (C).

Electrons are negatively charged, whilst protons have an equal and opposite positive charge of 1.6×10^{-19} C. These are the basic units of electrically charged matter. If an object has equal numbers of electrons and protons it is electrically neutral, if it has an excess of electrons it has an overall negative charge, while if it has an excess of protons it has a positive charge. The flow of charged particles, in particular the flow of electrons, is what constitutes an electric current.

charge carrier A charged particle that moves through a substance to carry a current. In metals, the charge carriers are electrons. In semiconductors they are electrons or HOLES. In IONIC liquids and IONIZED gases the charge carriers are ions.

charge-transfer complex A COMPLEX ION in which there is only a weak force holding the LIGANDS in place, in particular a complex held together by ELECTROSTATIC forces arising from the redistribution of charge in AROMATIC rings.

Charles' law For a fixed mass of an ideal gas at constant pressure, the volume is proportional to the ABSOLUTE TEMPERATURE; that is, the volume divided by the absolute temperature is a constant. For a fixed mass of an ideal gas with an absolute temperature T in volume V, at constant pressure: V/T = constant. *See also* BOYLE'S LAW, GAS LAWS, IDEAL GAS EQUATION, PRESSURE LAW.

chelate *See* CHELATING AGENT.

chelating agent Any compound that forms a complex consisting of charged ion(s) or metal atom(s) chemically bonded to and surrounded by chains of organic residues. This linkage of the ion or atom joins the organic chains to form complete rings. The compound formed is called the chelate and the process is called chelation.

An example of a chelating agent is EDTA. EDTA complexes calcium ions by surrounding them with four oxygen atoms and two nitrogen atoms. In this configuration the metal ions are strongly attached to the organic molecules. Another example of chelation is the formation of complexes similar to copper tetrammine between copper ions and 1,2 diaminoethane, $[(H_2NCH_2CH_2NH_2)_2Cu]^{2+}$.

Chelating agents are used in chemical analysis and in determining the hardness of water. They are also used in the removal of toxic metal ions from the body, to combat lead poisoning (lead ions are chelated and since the complexes are soluble they can be safely excreted) and other diseases. Chelates can act as carriers of essential metal ions, useful in the agricultural industry.

chelation *See* CHELATING AGENT.

chemical bond The force of attraction that holds atoms together in a molecule or lattice. Chemical bonds are of sufficient force that they can only be broken by a chemical reaction and not by thermal vibrations at the temperatures under consideration. Bonds are broadly categorized as IONIC BONDS and COVALENT BONDS. Ionic, or electrovalent bonds, arise from the ELECTROSTATIC forces of attraction between oppositely charged ions. In covalent bonding, pairs of atoms share electrons to form a bond that is directed in space.

Many molecules have chemical bonds intermediate between ionic and covalent. These are described as POLAR: the bond is regarded as covalent but the electron pair in the bond spends more time with one atom than the other. This polarizes the molecule, so it has a negative charge on one end, which acts as an ionic bond.

See also BOND ENERGY, CO-ORDINATE BOND, ELECTRONEGATIVITY, HYDROGEN BOND, METALLIC BONDING, POLAR BOND, SHELL, VALENCY.

chemical combination The combination of elements to give compounds. There are three laws that govern chemical combination: the LAW OF CONSTANT PROPORTIONS, the LAW OF EQUIVALENT PROPORTIONS, and the LAW OF MULTIPLE PROPORTIONS. These laws were originally based on empirical evidence, but were later recognized to be a consequence of the different masses of atoms and the simple ratios in which atoms combine to form compounds. *See also* DALTON'S ATOMIC THEORY.

chemical energy The energy that can be released in forming or breaking chemical bonds.

chemical engineering The study of chemical processes applied to the commercial manufacture of materials, often on a large scale. It is also the design, construction and control of equipment for carrying out large-scale chemical processes.

chemical equation A shorthand for showing the reagents and products in a chemical reaction and the number of molecules of each substance involved. For example, the reaction between hydrochloric acid and magnesium to give hydrogen and magnesium sulphate could be denoted by the equation

$$2HCl + Mg \rightarrow MgCl_2 + H_2$$

The equation shows that two molecules of hydrogen chloride are needed for every atom of magnesium. It is also common to attach state symbols to the equation to indicate whether a substance is a solid (s), liquid (l), aqueous solution (aq) or a gas (g). With state symbols, the above equation would be written:

$$2HCl\,(aq) + Mg\,(s) \rightarrow MgCl_2\,(aq) + H_2\,(g)$$

chemical equilibrium The state of a system of containing two or more chemical substances, where the concentration of all substances is constant over time. The term is particularly used with reference to a chemical reaction that could proceed in either direction, such as the formation of ammonia:

$$N_2 + 3H_2 \Leftrightarrow 2NH_3$$

In this reaction, a chemical equilibrium exists when the concentrations of all three gases are constant. Chemical equilibrium is an example of a dynamic equilibrium: individual molecules continue to react, but the reactions in each direction take place at the same rate. *See also* CHEMICAL POTENTIAL, EQUILIBRIUM CONSTANT, LE CHATELIER'S PRINCIPLE, REVERSIBLE REACTION.

chemical equivalent The mass of a specified substance that can displace or react with 1 g of hydrogen or 8 g of oxygen, either directly or indirectly. For example, the chemical equivalent of chlorine is 35.5 g, since that much chlorine reacts with hydrogen in the reaction

$$Cl_2 + H_2 \rightarrow 2HCl$$

For a compound where more than one reaction is possible, or an element with more than one VALENCY, the equivalent weight will not be unambiguously defined, thus carbon has an equivalent weight of 3 g in

$$C + 2H_2 \rightarrow CH_4$$

but 12 g in

$$2C + H_2 \rightarrow C_2H_2$$

chemical potential In any system with two or more chemical components, the chemical potential of each component is the rate of change of the Gibbs free energy (*see* FREE ENERGY) of the system with the amount of that component, all other factors being kept constant. Chemical potential is a useful quantity in studying the CHEMICAL EQUILIBRIUM of a system: equilibrium is achieved when all components have equal chemical potential. *See also* WATER POTENTIAL.

chemical reaction Any process in which one or more COMPOUNDS or ELEMENTS interact, producing different compounds or converting a compound into its constituent elements.

See also ADDITION REACTION, CHAIN REACTION, CHEMICAL COMBINATION, CHEMICAL EQUILIBRIUM, COMBUSTION, CONDENSATION REACTION, DISPLACEMENT REACTION, ELIMINATION, HEAT OF REACTION, HESS'S LAW, KINETICS, OXIDATION, RATE OF REACTION, REACTIVITY, REDOX REACTION, REDUCTION, REVERSIBLE REACTION, SUBSTITUTION REACTION.

chemical shift A change in the ENERGY LEVELS of an atomic nucleus, or the inner electrons of an atom, in response to changes in the chemical bonding or IONIZATION state of the atom. The existence of such shifts means that processes such as X-ray absorption or NUCLEAR MAGNETIC RESONANCE can be used to give information about the bonding in which an atom is involved.

chemiosmotic theory *See* ELECTRON TRANSPORT SYSTEM.

chemisorption The ADSORPTION of a gas by a solid in which the molecules of the adsorbed gas are held on the surface of the adsorbing solid by the formation of chemical bonds.

chemistry The science of the ELEMENTS and the ways in which they interact with one another, particularly in the formation of CHEMICAL BONDS. Chemistry is usually divided into physical chemistry, the study of the physical properties of substances and how they are affected by chemical change; organic chemistry, the study of the carbon compounds (except oxides of carbon and metal carbides); and inorganic chemistry, the study of all other compounds.

Chile saltpetre A mineral form of SODIUM NITRATE extracted from CALICHE deposits, particularly in Chile. A major source of nitrate for fertilizer manufacture.

China clay *See* KAOLIN.

chiral (Greek *cheir* = hand) A molecule that cannot be superimposed on its mirror image because it is asymmetrical (like the relationship of the right hand to the left hand). Any compound with four different atoms or groups attached to a single carbon is chiral. The central carbon atom is termed the chiral centre. The term 'chirality' describes the property of a molecule to exhibit non-identity with its mirror image ('handedness'). The term can be used to refer to any asymmetric object or molecule. A chiral molecule would exhibit OPTICAL ISOMERISM.

chirality *See* CHIRAL.

chitin A structural POLYSACCHARIDE that consists of long chains of N-acetyl-D-glucosamine units (a derivative of GLUCOSE) linked together. Structurally and chemically, chitin is similar to CELLULOSE (with acetyl-amino groups, $NH.OCCH_3$, instead of the HYDROXYL GROUPS, –OH, of cellulose). Like cellulose, chitin has great strength and forms the hard, protective outer layer of insects and crustaceans. Chitin is insoluble in water and provides protection from many solvents, acids and alkalis.

chloral *See* TRICHLOROETHANAL.

chlorate Any salt containing an ANION that contains chlorine and oxygen. Most common is the chlorate(V) ion, ClO_3^-, but chlorate(I) (or hypochlorite) ClO^-, chlorate(III) (or chlorite) ClO_2^- and chlorate(VII) (or perchlorate) ClO_4^- ions also exist.

chloric acid Any acid comprising hydrogen, oxygen and chlorine. They are: chloric(I) acid (hypochlorous acid), HOCl; chloric(III) acid (chlorous acid), $HClO_2$; chloric(V) acid $HClO_3$; and chloric(VII) acid (perchlorous acid) $HClO_4$. Chloric(I) and chloric(III) acids are stable only in solution, chloric(V) and chloric(VII) acids can be isolated as ANHYDROUS liquids, but are unstable and explode on heating. All are OXIDIZING AGENTS with oxidizing strengths that increase with the oxygen content of the acid. Chloric(I) and(III) acids are weak acids, chloric(V) and(VII) acids are strong acids.

chloride Any BINARY COMPOUND containing chlorine. Non-metallic chlorides are covalently bonded (*see* COVALENT BOND). CARBON TETRACHLORIDE (tetrachloromethane), CCl_4, is an important solvent. The metallic chlorides are IONIC.

chlorination The incorporation of chlorine into an organic compound, by ADDITION REACTIONS or SUBSTITUTION REACTIONS.

chlorine (Cl) The element with atomic number 17; relative atomic mass 35.5; melting point –101°C; boiling point –35°C. Chlorine is a greenish yellow gas, the most widely used of the HALOGENS. It is found in large quantities in seawater, where it occurs as dissolved sodium chloride, from which it is extracted by ELECTROLYSIS in a MERCURY-CATHODE CELL. It is a powerful OXIDIZING AGENT and is widely used for its bleaching properties. Industrially, it is used in the manufacture of many chlorinated organic compounds, such as polyvinyl chloride (PVC) which is an important PLASTIC.

chlorite Any salt containing the chlorate(III) (or chlorite) ion, ClO_3^-.

chlorobenzene (C_6H_5Cl) An AROMATIC compound; boiling point 132°C. It is used in the manufacture of DDT. Chlorobenzene consists of a BENZENE RING in which a hydrogen atom has been replaced by a chlorine atom, by ELECTROPHILIC SUBSTITUTION. It is prepared by the direct chorination of benzene in the presence of an iron catalyst.

2-chlorobuta-1,3-diene, *chloroprene* ($CH_2=CCl.CH=CH_2$) A colourless liquid; boiling point 59°C. It is a chlorinated DIENE. It is polymerized to make the synthetic rubber NEOPRENE.

chlorocruorin A green-coloured protein found in certain polychaete worms. It can bind weakly to oxygen in the blood or other tissues, to increase the uptake, transport or unloading of oxygen. It contains iron and is similar to HAEMOGLOBIN.

chloroethane, *ethyl chloride* (C_2H_5Cl) A colourless, gaseous HALOGENOALKANE; melting point –136°C; boiling point 12.5°C. It is made by the reaction of ETHENE with hydrogen chloride. Its main use is in the production of LEAD TETRAETHYL for petrol.

chloroethanoic acids, *chloroacetic acids* Acids in which chlorine atoms have replaced hydrogen atoms in the methyl group, CH_3O– of ETHANOIC ACID. If one hydrogen group is replaced then monochloroethanoic acid results ($CH_2ClCOOH$); two replaced gives dichloroethanoic acid ($CHCl_2COOH$) and three replaced gives trichloroethanoic acid (CCl_3COOH). The chloroethanoic acids are stronger than their corresponding ethanoic acids, trichloroethane being the strongest. They have various uses including being used as weedkillers.

chloroethene *See* VINYL CHLORIDE.

chlorofluorocarbon (CFC) A synthetic chemical used in AEROSOLS as a propellant, in refrigerators and air conditioners as a coolant, and in foam packaging. Although CFCs are chemically inert, odourless, non-flammable and non-toxic, they can remain in the Earth's atmosphere for more than 100 years. This is destroying the OZONE LAYER. When CFC's are released into the atmosphere, ultraviolet radiation from the sun causes them to break down into chloride atoms that combine with OZONE. This decreases the ozone concentration, which allows harmful radiation from the Sun to reach the Earth. The more destructive CFCs have now been banned and replacements are being developed, and safer methods of disposal are being investigated for existing CFCs. *See also* GREENHOUSE EFFECT.

chloroform, *trichloromethane* ($CHCl_3$) A colourless liquid with a pleasant, sweet odour; boiling point 60–61°C. It is made by the CHLORINATION of methane. It was used as an ANAESTHETIC but has been replaced by less toxic chemicals. Chloroform is a good solvent for fats, waxes and rubber, and is used chiefly in the manufacture of CHLOROFLUOROCARBON refrigerants.

chloromethane, *methyl chloride* (CH_3Cl) A colourless, gaseous HALOGENOALKANE; melting point –97°C; boiling point 24°C. It is produced by the reaction of METHANOL with hydrogen chloride or by the direct chlorination of methane. It is used in the production of SILICONES, as a refrigerant and in the manufacture of anti-knocking (*see* KNOCKING) additives in fuel.

chlorophyll A green pigment present in most plants. It is responsible for the capture of light during PHOTOSYNTHESIS and for the coloration of plants.

Several types of chlorophyll exist. Chlorophyll-α ($C_{55}H_{72}MgN_4O_5$) is common to all plants and is obtained as a blue/black powder; melting point 150–153°C. Chlorophyll-β ($C_{55}H_{70}MgN_4O_6$) is a dark green powder; melting point 120–130°C. Chlorophyll is one of the PORPHYRIN group of pigments and is chemically similar to HAEMOGLOBIN but contains magnesium instead of iron.

cholecalciferol *See* VITAMIN D.

cholesterol ($C_{27}H_{46}O$) A STEROL that occurs throughout the body of animals, but is not usually present in higher plants or bacteria. Cholesterol is a component of all cell membranes and a precursor of the STEROID HORMONES. In the diet, cholesterol is obtained from dairy products and meat. A high level of cholesterol in the blood is thought to contribute to atherosclerosis, a condition whereby lipids build up inside the walls of arteries and reduce blood flow. It is broken down by the liver and excess is secreted in the bile.

choline An amino alcohol, sometimes considered to be part of the VITAMIN B COMPLEX. It is found in certain PHOSPHOLIPIDS, particularly LECITHIN, which is a component of all animal and plant cells.

chondroitin A GLYCOSAMINOGLYCAN similar to HYALURONIC ACID except that it contains galactosamine instead of glucosamine. It is found in cartilage, bone and tendons.

chromate(VI) Any salt containing the chromate(VI) ION, CrO_4^{2-}. The ion has a yellow colour. In solution, chromate(VI) salts are stable only under alkaline conditions. In the presence of acids, the chromate(VI) ion is converted to the DICHROMATE(VI) ion:

$$2CrO_4^{2-} + H^+ \rightarrow Cr_2O_7^{2-} + H_2O$$

chromatogram The pattern of separated chemicals along a separating medium in CHROMATOGRAPHY.

chromatography An analytical technique used to separate the components of a mixture, which flow at different rates along some separating medium. The fixed material over which the mixture passes is called the stationary phase and is usually a solid or GEL. The moving fluid, often a liquid, but sometimes a gas, is called the moving phase. The mixture is dissolved in a solvent, called an ELUENT in this context, which then diffuses along the separating medium. In adsorption chromatography, a column of aluminium oxide is often used as the solid phase. The different strengths of the interatomic force between the molecules being separated and the molecules of the stationary phase and the eluent are responsible for the separation of the mixture. The different components may then be identified, and in some cases may be determined quantitatively. *See also* AFFINITY CHROMATOGRAPHY, CHROMATOGRAM, COLUMN CHROMATOGRAPHY, GAS CHROMATOGRAPHY, GEL FILTRATION, ION EXCHANGE CHROMATOGRAPHY, PAPER CHROMATOGRAPHY, THIN-LAYER CHROMATOGRAPHY.

chrome alum *See* POTASSIUM CHROMIUM SULPHATE.

chromic (*adj.*) Describing compounds containing the CHROMIUM ion in its higher oxidation states; in other words, as the ions Cr^{3+} or Cr^{6+}. An example is chromic oxide, CrO_3 *Compare* CHROMOUS.

chromic acid The hypothetical acid H_2CrO_4, containing the chromate(VI) ion, CrO_4^{2-}. Salts with this ion do exist, but the acid itself has not been isolated.

chromite The chief ore of chromium, ion chromium oxide, $FeCr_2O_4$.

chromium (Cr) The element with atomic number 24; relative atomic mass 52.0; melting point 1,900°C; boiling point 2,640°C; relative density 7.2. Chromium is a TRANSITION METAL and shows multiple VALENCY. It is fairly resistant to corrosion and is often used to electroplate steel parts, giving them a hard shiny finish (*see* ELECTROPLATING). It is also incorporated into many steels, particularly stainless steel. *See also* THERMIT PROCESS.

chromium oxide Any of the BINARY COMPOUNDS of chromium with oxygen: chromium(II) oxide, CrO, a black powder; chromium(III) oxide, Cr_2O_3, a green crystalline solid; chromium(IV) oxide (chromium dioxide), CrO_2, a black powder; and chromium(VI) oxide (chromium trioxide), CrO_3, a red crystalline solid.

Chromium(III) oxide is the most stable of the oxides. Chromium(VI) oxide is an extremely strong OXIDIZING AGENT and decomposes partially on heating:

$$4CrO_3 \rightarrow 2Cr_2O_3 + 3O_2$$

Chromium(VI) oxide is the only chromium oxide to be soluble in water, forming an acidic solution believed to contain a mixture of chromic acid, H_2CrO_4, and dichromic acid, $H_2Cr_2O_7$.

chromophore The group responsible for the colour of a compound, such as a DYE. Such groups include –C=C– and –N=N–.

chromous (*adj.*) Describing compounds containing the Cr^{2+} ion, for example chromous chloride, $CrCl_2$. *Compare* CHROMIC.

chromyl chloride (CrO_2Cl_2) A deep red liquid; melting point –97°C; boiling point 117°C; relative density 1.9. Chromyl chloride can be prepared by the action of concentrated sulphuric acid on potassium dichromate and sodium chloride:

$$K_2Cr_2O_7 + 2H_2SO_4 + 4NaCl \rightarrow 2CrO_2Cl_2 + 2Na_2SO_4 + K_2SO_4 + 3H_2O$$

It is a powerful OXIDIZING AGENT.

cinnabar A mineral form of mercury sulphide, HgS. It is an important ore of mercury.

cinnamic acid, *3-phenylpropenoic acid* (C_6H_5-CH:CHCOOH) A white crystalline AROMATIC CARBOXYLIC ACID; melting point 135°C; boiling point 300°C. Derivatives are used as flavourings and in perfumery.

cis/trans isomerism *See* GEOMETRIC ISOMERISM.

citric acid An organic acid found in many plants. It exists in particularly high concentrations in citrus fruits, such as oranges and lemons. It has a sharp, sour taste. Citric acid is an intermediate in the KREBS CYCLE.

citric acid cycle *See* KREBS CYCLE.

Claisen condensation A CONDENSATION REACTION in which two ESTERS react, in the presence of sodium ethoxide, to form a keto-ester with the elimination of an alcohol. An example of a general reaction is

$$2CH_3COOR \longrightarrow CH_3COCH_2COOR + ROH$$

The reaction is important in the synthesis of many organic chemicals. It is similar to an ALDOL CONDENSATION.

Clapeyron Clausius equation An equation relating the rate of change of the SATURATED VAPOUR PRESSURE with temperature to the LATENT HEAT of vaporization of a material:

$$dP/dT = -LT/\Delta V$$

where P is the saturated vapour pressure, T the absolute temperature, L the molar latent heat of vaporization and ΔV the change in molar volume associated with the PHASE change.

clathrate A solid mixture in which individual molecules of one compound are trapped within the crystal lattice of another.

Claude process A process for liquefying air or other gases, in which gas is compressed, then cooled by making it expand ADIABATICALLY. The gas then flows through a heat exchanger, where it cools incoming compressed gas before itself being re-compressed and repeating the cycle of ISOTHERMAL compression and adiabatic expansion After passing around the cycle several times, the air is cooled to its boiling point.

Clausius Clapeyron equation *See* CLAPEYRON CLAUSIUS EQUATION.

Clausius statement of the second law of thermodynamics The SECOND LAW OF THERMODYNAMICS forbids any system in which heat energy only flows from a region of low temperature to one of higher temperature. For example, a refrigerator cannot work without a supply of extra energy, usually electrical, which is also deposited in the warm part of the system.

clay A fine-grained deposit, consisting chiefly of silicates of aluminium and/or magnesium.

cleavage plane In a non-metallic crystal, a surface along which the crystal will break cleanly and relatively easily. If an attempt is made to break the crystal in other directions, it may shatter into smaller fragments rather than breaking cleanly. A knowledge of the cleavage planes of diamond is important for breaking naturally occurring diamonds, which usually have no particular shape, into gemstones for jewellry. *See also* SLIP PLANE.

clinker Non-VOLATILE waste products, particularly in a furnace or boiler. In particular the solid residue from burning coal or coke.

close packed (*adj.*) Describing a pattern in a crystal that maximizes the numbers of atoms that will fit into a given volume. The CO-ORDINATION NUMBER of a close packed structure is 12. *See also* CUBIC CLOSE PACKED, HEXAGONAL CLOSE PACKED.

cluster compound A compound in which a group of TRANSITION METAL atoms, either all of the same element or of several different metals, are held together by the METALLIC BONDING more commonly found in pure metals and alloys.

coagulation The process of particles coming together, particularly of particles in a LYOPHOBIC sol, which tend to clump into larger particles.

coal A soft black SEDIMENTARY rock formed by the compaction of dead plant material growing over 200 million years ago. Coal burns slowly, producing heat and is therefore a useful domestic fuel and is also used in the chemical industry. It contains carbon in varying proportions, giving rise to a number of different types of coal; for example anthracite contains more than 90 per cent carbon, bituminous coal contains 80 per cent and peat about 50 per cent.

Many important products are derived from coal, including COAL GAS, COAL TAR, COKE and many by-products that form the basis of organic chemistry. The burning of coal contributes to atmospheric pollution and ACID RAIN.

coalesce (*vb.*) To stick together.

coal gas An inflammable gas produced when coal is heated in the absence of air. It contains many compounds, but chiefly hydrogen, methane and carbon monoxide. It is sometimes used as a fuel and also for gas lighting. Coal gas has been mostly replaced by NATURAL GAS as a domestic energy source.

coal tar A black, oily substance produced by the DESTRUCTIVE DISTILLATION of bituminous coal (*see* COAL). Further DISTILLATION yields various fractions of oil with the residue forming PITCH. Products such as ANTHRACENE and tar are derived from coal tar. A large number of other products for use in the medical and dyeing industries can be obtained by yet further distillation.

cobalt (Co) The element with atomic number 27; relative atomic mass 58.9; melting point 1,495°C; boiling point 2,870°C; relative density 8.9. Cobalt is a light-grey TRANSITION METAL. Its ions are pink when HYDRATED, but blue in the ANHYDROUS state – this feature is sometimes used as a test for small quantities of water.

Cobalt is a FERROMAGNETIC metal and its alloys are widely used in the manufacture of magnets. The radioactive ISOTOPE cobalt–60 is widely manufactured and used as a source of GAMMA RADIATION, particularly for radiotherapy (the BETA PARTICLES also produced are stopped by an absorbing layer).

cobalt oxide Any of the binary compounds of cobalt and oxygen. They are: cobalt(II) oxide, CoO, cobalt(III) oxide, Co_2O_3, and cobalt(II) cobalt(III) oxide, Co_3O_4.

Cobalt(II) oxide is a pink solid; melting point 1,795°C; decomposes on further heating; relative density 6.5. It can be formed by reacting cobalt(II) nitrate with sodium hydroxide to precipitate cobalt(II) hydroxide, which loses water on gentle heating.

$$Co(NO_3)_2 + 2NaOH \rightarrow Co(OH)_2 + 2NaNO_3$$

$$Co(OH)_2 \rightarrow CoO + H_2O$$

Cobalt(II) oxide is readily reduced to cobalt metal by hydrogen,

$$CoO + H_2 \rightarrow Co + H_2O$$

On heating in air it is oxidized to the mixed oxide Co_3O_4:

$$6CoO + O_2 \rightarrow 2Co_3O_4$$

Cobalt(III) oxide is a dark grey solid; decomposes on heating; relative density 5.2. It can be obtained by the gentle heating of cobalt(II) nitrate,

$$4Co(NO_3)_2 \rightarrow 2Co_2O_3 + 8NO_2 + O_2$$

On further heating oxygen is given off and the mixed oxide Co_3O_4 is formed,

$$6Co_2O_3 \rightarrow 4Co_3O_4 + O_2$$

Cobalt(II) cobalt(III) oxide is a black solid; decomposes on heating; relative density 6.0. It is readily formed by the OXIDATION of cobalt(II) oxide or the REDUCTION of cobalt(III) oxide. It is readily reduced to cobalt metal by hydrogen,

$$Co_3O_4 + 4H_2 \rightarrow 3Co + 4H_2O$$

coenzyme An organic molecule, often a vitamin derivative, that acts as a COFACTOR in an enzyme reaction. It acts without binding, or only temporarily binding, to the enzyme, unlike a PROSTHETIC GROUP. Examples include COENZYME A, FAD and NAD. Coenzymes are frequently essential for the removal of end-products of enzyme reactions that would otherwise cause inhibition of the enzyme.

coenzyme A A derivative of PANTOTHENIC ACID that is a COENZYME acting as a carrier of ACYL GROUPS in the KREBS CYCLE and FATTY ACID OXIDATION.

coenzyme Q *See* ELECTRON TRANSPORT SYSTEM.

cofactor A non-protein substance that is needed for some enzymes to function efficiently. Cofactors can be COENZYMES, PROSTHETIC GROUPS or activators. Activators are substances other than coenzymes and prosthetic groups that are needed to activate an enzyme, for example, calcium ions are needed to activate certain enzymes involved in the clotting of blood. An enzyme–cofactor complex is called a haloenzyme and the inactive enzyme on its own is called an apoenzyme.

coke The residue remaining after coal has been strongly heated in an airtight oven. Coke consists of 90 per cent carbon mixed with inorganic material and is a widely used domestic and industrial fuel.

collagen A major structural protein of vertebrates that occurs in connective tissue of skin, tendons, ligaments, cartilage and bone. Collagen contains the AMINO ACIDS glycine, proline and hydroxyproline, which coil together to form fibres. These provide great strength but little elasticity. Collagen is essentially insoluble but can be boiled with water, in which case the strands separate to form GELATIN.

colligative properties Those properties of a solution that are dependent on the concentration of particles in the solution, but not on the nature of those particles. Examples are the LOWERING OF VAPOUR PRESSURE, the DEPRESSION OF FREEZING POINT, the ELEVATION OF BOILING POINT and OSMOSIS.

colloid A mixture containing small particles of one material suspended in another, often of a different phase. The particles in a colloid have sizes between one and 100 nanometres, so are larger than the individual molecules that occur in a solution and smaller than the particles that occur in precipitates and can be removed by a filter. Examples include AEROSOLS and FOAMS (gas and liquid mixtures), EMULSIONS (liquid mixtures such as milk and paint), and SOLVENTS and GELS (solids dispersed in a liquid). A colloid containing solid particles suspended in a liquid is more accurately called a SOL.

Particles in a colloid can be separated by passing them through a porous material. The material that forms the separate individual particles is sometimes called the dispersed phase, to distinguish it from the continuous phase, which forms a single connected body of material.

See also LYOPHILIC, LYOPHOBIC.

colloid mill A MILL used to create a COLLOID by grinding a suspension of particles between fast moving, closely spaced surfaces with the liquid phase of the colloid used as a lubricant.

colorimeter A device for comparing the colour, usually of a solution, against a calibrated filter. Colorimeters are used to measure the concentration of a coloured component in a solution.

colour index (CI) A definitive listing of DYES and PIGMENTS, including their commercial names, method of application, colourfastness etc.

column chromatography CHROMATOGRAPHY in which the ELUENT runs down a glass column containing the stationary phase, such as tightly packed powdered alumina.

combustion Any chemical process that is self-supporting and releases heat. In particular, the rapid combination of any compound with oxygen from the air with the production of heat and light. *See also* EXPLOSION, FLAME, HEAT OF COMBUSTION, SPONTANEOUS COMBUSTION.

common ion effect A salt will be less soluble if the solution already contains ions of a type produced when the salt is dissolved. *See also* SOLUBILITY PRODUCT.

common salt *See* SODIUM CHLORIDE.

complex ion An ion, usually a metal ion, that has a number of molecules or ions bound to it by CO-ORDINATE BONDS. HYDRATED ions of TRANSITION METALS are an example. Some of these ions will form similar complexes with ammonia molecules in place of water, for example the copper tetrammine ion, $[Cu(NH_3)]^{2+}$. Another example is the hexacyanoferrate(III) ion, $Fe(CN)_6^{3-}$, which consists of an Fe^{3+} ion surrounded by six CN^- ions. *See also* LIGAND.

complexometric analysis Any form of VOLUMETRIC ANALYSIS in which a COMPLEX ION is involved. The fact that many complexes are coloured may be used to indicate the END POINT of a TITRATION, for example.

component The number of distinct chemical species (compounds or unreacted elements) present in a MIXTURE less the number of reactions taking place in the mixture between those components. Thus a mixture of oxygen, carbon dioxide and ammonia in which no reactions take place has three components, whilst a mixture of hydrogen, nitrogen and ammonia in equilibrium has only two components.

composite material Any engineering material made from two or more different materials, designed to exploit the advantages of each without suffering their weaknesses. *See also* CARBON FIBRE, GLASS-FIBRE REINFORCED PLASTIC.

compound A substance made up of two or more elements that cannot be separated by physical means. In a compound, the quantity of the elements present is fixed and is a simple ratio, though more than one compound may exist containing the same elements, for example, water (H_2O) and hydrogen peroxide (H_2O_2). *See also* CHEMICAL COMBINATION.

concentrated (*adj.*) Describing a solution having a concentration at or close to the maximum that can be achieved at a particular temperature; that is, a solution that is SATURATED or nearly so.

concentration The amount of SOLUTE dissolved in a specified amount of solution. Concentration is normally expressed in terms of the number of MOLES of solute per decimetre cubed of solution. A solution with a concentration of 1 mol dm^{-3} is described as molar. *See also* HENRY'S LAW, MOLALITY.

concentration gradient The difference in concentration of a substance between two regions. *See also* DIFFUSION.

condensation 1. The process by which a gas or vapour turns into a liquid.

2. Droplets, usually of water, formed by this process.

condensation polymerization *See* POLYMERIZATION.

condensation reaction In organic chemistry, a reaction in which two compounds are joined together to make a larger molecule with the loss of a small molecule, such as water. Such reactions can be considered to be addition-elimination reactions. Many POLYMERS are made by condensation POLYMERIZATION. *See also* ADDITION REACTION, ELIMINATION.

condense (*vb.*) To turn from a gas or vapour into a liquid.

condensed matter Solids and liquids; those states of matter where the spacing between the molecules is of the same order as their size.

condenser A device for the condensing of a vapour, often in the form of a coiled glass tube surrounded by an outer glass jacket through which cold water is passed. *See also* LIEBIG CONDENSER.

conduction The flow of electricity or heat through a material.

conduction band In the BAND THEORY of solids, an energy band that is not completely occupied by electrons, so an electron in that band can gain a small amount of energy and move through the material, carrying heat or electricity. *See also* VALENCE BAND.

conductometric titration A TITRATION in which the electrical conductivity of the solution is measured, with a sudden change in conductivity indicating the end point of the titration. This technique is particularly useful for coloured solutions, where conventional INDICATORS cannot be used.

conductor, electrical Any material through which current can flow. All metals are conductors, as their structure contains VALENCE ELECTRONS that are free to move through a lattice of positive IONS. Metals conduct less well as their temperature increases, since thermal vibrations make the lattice less regular, making collisions between the electrons and the lattice more likely. For similar reasons, pure metals, which have highly regular lattice structures, usually conduct electricity (and heat) better than alloys. Graphite is one of the few non-metallic solids that conducts electricity.

Molten IONIC materials and solutions of ionic salts all conduct electricity. In all these cases there are both positive and negative ions free to move through the liquid and carry the charge. Once these ions reach an ELECTRODE, the ions gain or lose electrons and chemical reactions may take place (*see* ELECTROLYSIS). In IONIC SOLIDS, these ions are locked into a rigid lattice and so are not free to carry charge. Molten metals also conduct electricity.

Gases do not normally conduct electricity, but may do so if they are IONIZED – that is, if they contain a number of ions as well as uncharged atoms or molecules.

See also BAND THEORY, INSULATOR, SEMICONDUCTOR.

configuration 1. The spatial arrangement of atoms, particularly LIGANDS around a central molecule, or atoms in a covalent bond. *See also* ISOMER, STEREOCHEMISTRY.

2. *electron configuration* The arrangement of ELECTRONS in ORBITALS in an ion or atom.

conformation Any of the possible CONFIGURATIONS that a molecule can take on as a result of rotation about a single COVALENT BOND.

conjugated diene *See* DIENE.

conjugated protein A PROTEIN with a non-protein group incorporated into its structure, which plays a vital role in the functioning of the protein. For example, HAEMOGLOBIN contains a haem group. *See also* PROSTHETIC GROUP.

conjugate solutions The pair of liquids formed when two liquids that are only partially MISCIBLE are mixed together. If the liquids are A and B, with A being denser, the conjugate solutions will form with a SATURATED solution of A in B floating above a saturated solution of B in A.

conjugation A term describing double or triple bonds in an organic compound that are separated by a single bond. An example of conjugated double bonds is in the compound buta-1,3-diene ($CH_2{=}CH{-}CH{=}CH_2$).

constantan A copper nickel ALLOY typically containing 55 per cent copper and noted for its small change in electrical resistance with temperature. It is used in the manufacture of electrical devices.

constitutive (*adj.*) Describing an enzyme that is synthesized all the time, regardless of the availability of SUBSTRATE. *Compare* INDUCIBLE.

contact process An industrial process for the manufacture of sulphuric acid. A mixture of sulphur dioxide, SO_2, and air are passed over a hot vanadium(V) oxide catalyst. The sulphur dioxide is oxidized to sulphur trioxide, SO_3:

$$2SO_2 + O_2 \rightarrow 2SO_3$$

The sulphur trioxide is then dissolved in sulphuric acid producing OLEUM:

$$H_2SO_4 + SO_3 \rightarrow H_2S_2O_7$$

The oleum is then diluted with water to give concentrated sulphuric acid.

co-ordinate bond, *dative bond* A COVALENT BOND in which both the bonding electrons are donated to the bond by one of the atoms. An example is the bond between boron trichloride and ammonia:

$$H_3N{:} + BCl_3 \rightarrow H_3NBCl_3$$

The nitrogen atom donates a LONE PAIR of electrons (shown as :) to form a bond with the boron atom.

co-ordination compound Any compound containing CO-ORDINATE BONDS, particularly a salt containing an inorganic COMPLEX ION, such as potassium hexacyanoferrate(III), $K_3[Fe(CN)_6]$ in which the cyano groups (CN^-) form co-ordinate bonds with the iron atom.

co-ordination number In a crystalline structure, the number of nearest neighbours surrounding any atom or ion. In the sodium chloride structure, the co-ordination number is 6: each sodium ion has six nearby chlorine ions and vice versa. In a BODY CENTRED CUBIC structure, the co-ordination number is 8, whilst CLOSE PACKED structures have a co-ordination number of 12.

copolymerization The POLYMERIZATION of two or more different MONOMERS, resulting in a compound with properties distinct from the original monomers.

copper (Cu) The element with atomic number 29; relative atomic mass 63.5; melting point 1,083°C; boiling point 2,582°C; relative density 8.9. Copper is a TRANSITION METAL with a characteristic reddish brown colour. It is extracted from its ores by ELECTROLYSIS. Many copper compounds have a characteristic blue or green colour, though copper oxide is black.

Copper is a good conductor of heat and electricity and is widely used as the conductor in electrical cables. It is also used to manufacture domestic water pipes. Brass, a hard alloy of copper and zinc, is used for the manufacture of small metal objects, but the high cost of copper limits its use for larger structures.

See also BRONZE.

copper chloride Either of the salts copper(I) chloride (cuprous chloride), CuCl, or copper(II) chloride (cupric chloride), $CuCl_2$.

Copper(I) chloride is the less stable of the two compounds. It is a white solid; melting point 430°C; boiling point 1,490°C; relative density 3.4. Copper(I) chloride can be produced in a reaction of copper(II) chloride with excess copper in concentrated hydrochloric acid, to form the complex ion $CuCl_2^-$:

$$CuCl_2 + Cu + 2HCl \rightarrow 2H^+ + 2CuCl_2^-$$

On diluting the acid, this solution produces a PRECIPITATE of the insoluble chloride:

$$CuCl_2^- \rightarrow CuCl + Cl^-$$

Copper(II) chloride is a brownish yellow powder; melting point 620°C; decomposes on further heating; relative density 3.4. It also occurs in a blue-green HYDRATED form, $CuCl_2.2H_2O$. Copper(II) chloride can be formed by burning copper in chlorine:

$$Cu + Cl_2 \rightarrow CuCl_2$$

Copper(II) chloride dissolves in water to give a solution that is blue when dilute but changes through green to brown at higher concentrations, due to the replacement of the blue hydrated copper ion $[Cu(H_2O)_6]^{2+}$, by brown chlorinated complexes, such as $CuCl_4^{2-}$.

copper nitrate ($Cu(NO_3)_2$) Usually found in the deep blue DELIQUESCENT crystalline form, $Cu(NO_3)_2.3H_2O$; melting point 115°C; decomposes on further heating; relative density 2.3. The anhydrous form sublimes on heating. Copper nitrate can be obtained by the action of nitric acid on copper oxide,

$$CuO + 2HNO_3 \rightarrow Cu(NO_3)_2^- + H_2O$$

It decomposes on heating to give copper oxide, nitrogen dioxide and oxygen,

$$2Cu(NO_3)_2 \rightarrow 2CuO + 4NO_2 + O_2$$

copper oxide Either of the two oxides of copper, copper(I) oxide, Cu_2O, or copper(II) oxide, CuO. Copper(I) oxide is a red powder; melting point 1,235°C; decomposes on further heating; relative density 3.4. It is formed by the partial thermal decomposition of copper(II) oxide:

$$4CuO \rightarrow 2Cu_2O + O_2$$

Copper(I) oxide disproportionates (*see* DISPROPORTIONATION) in dilute acids to form a solution of copper(II) ions plus a precipitate of metallic copper, for example:

$$Cu_2O + H_2SO_4 \rightarrow Cu\ SO_4 + Cu + H_2O$$

Copper(II) oxide is a black powder; decomposes on heating; relative density 6.3. It can be formed by heating copper carbonate, which occurs naturally:

$$CuCO_3 \rightarrow CuO + CO_2$$

Copper(II) oxide is insoluble in water, but reacts with acids to give blue solutions of copper(II) salts, for example:

$$CuO + 2HNO_3 \rightarrow Cu(NO_3)_2 + H_2O$$

It can be reduced to metallic copper by heating in a stream of hydrogen:

$$CuO + H_2 \rightarrow Cu + H_2O$$

copper(II) sulphate ($CuSO_4$) A white powder in its ANHYDROUS form, but more familiar as HYDRATED blue crystals, $CuSO_4.5H_2O$. It decomposes on heating; relative density 3.6 (anhydrous), 2.3 (hydrated). Copper(II) sulphate can be prepared as an aqueous solution by mixing copper carbonate with dilute sulphuric acid:

$$CuCO_3 + H_2SO_4 \rightarrow CuSO_4 + CO_2$$

Copper(II) sulphate is widely used as a fungicide and wood preservative.

correlation diagram A diagram showing how the ENERGY LEVELS of a series of atoms change with increasing ATOMIC NUMBER. It can be used to understand the FINE STRUCTURE of atomic spectra and the changes in the sequence in which atomic energy levels are filled. *See also* AUFBAU PRINCIPLE.

corrode (*vb.*) Of a metal, to form compounds by reaction with surrounding materials, particularly with oxygen, moisture and acidic gases in the atmosphere. This reduces the strength and electrical conductivity of metallic objects and limits their life when exposed to the atmosphere and atmospheric pollutants, such as sulphur dioxide.

Corrosion can be reduced either by coating the metal with an unreactive layer (such as paint, oil or plastic – *see* POWDER COATING) or by SACRIFICIAL CORROSION. Aluminium is remarkable in that the aluminium oxide, Al_2O_3, formed by the initial corrosion of an exposed aluminium surface is hard enough to prevent further reaction.

corrosion 1. Powdery material produced when a metal corrodes.

2. The process by which a metal corrodes.

corticoid *See* CORTICOSTEROID.

corticosteroid, *corticoid* A collective term for a group of STEROID HORMONES secreted by the adrenal gland. They are either GLUCOCORTICOIDS, concerned with GLUCOSE metabolism (for example CORTISOL) or MINERALOCORTICOIDS, concerned with mineral metabolism (for example ALDOSTERONE). They are produced in response to stress situations.

cortisol, *hydrocortisone* ($C_{21}H_{30}O_5$) A major GLUCOCORTICOID hormone of humans and other mammals secreted by the adrenal gland in response to internal stress (for example, low blood temperature or volume). It raises blood pressure and promotes the conversion of protein and fat into GLUCOSE. It can also be made synthetically from naturally occuring STEROIDS.

cortisone ($C_{21}H_{28}O_5$) A GLUCOCORTICOID hormone secreted by the adrenal gland that promotes the synthesis and storage of GLUCOSE. Cortisone is produced as part of the body's normal response to stress. The action of cortisone is similar to that of CORTISOL.

corundum A mineral form of ALUMINIUM OXIDE, Al_2O_3. In its pure form it is clear, but small amounts of chromium give rise to the red form known as ruby, and iron and titanium produce the blue form sapphire, both of which are used as gemstones. Corundum is the hardest mineral known apart from diamond, and is widely used as an ABRASIVE.

coulomb The SI UNIT of electric charge. One coulomb is the amount of charge carried in one second by a current of one AMPERE.

coumarone *See* BENZFURAN.

covalency The number of COVALENT BONDS an atom can form. Generally this is equal to the number of unpaired electrons in the outer ORBITALS. Thus carbon, with four electrons in the 2s and 2p orbitals, forms four covalent bonds, whilst nitrogen, which has an extra electron, has a covalency of 3.

covalent bond A chemical bond in which electrons are shared between two atoms, giving each one a share in the other's electrons.

Molecules with covalent bonds tend to be relatively small (unless they are giant structures) and forces between them tend to be weak. Therefore the compounds tend to have low melting and boiling points, and are likely to be gases, volatile liquids or solids with low melting points. Covalent compounds do not conduct electricity, and are generally insoluble in water (although some are hydrolysed by water) but soluble in organic solvents.

In a covalent bond, electron ORBITALS in two neighbouring atoms overlap to form a MOLECULAR ORBITAL of lower energy. Since the orbitals involved in covalent bonding are P-ORBITALS or sp HYBRID ORBITALS, these bonds are aligned in space in a way that is related to the directions of these orbitals. This means that covalent bonds between two atoms are at specified angles to one another. For example, each covalent bond in a carbon atom that forms four such bonds has a bond angle of 109.5°. If the STEREOCHEMISTRY of a complex

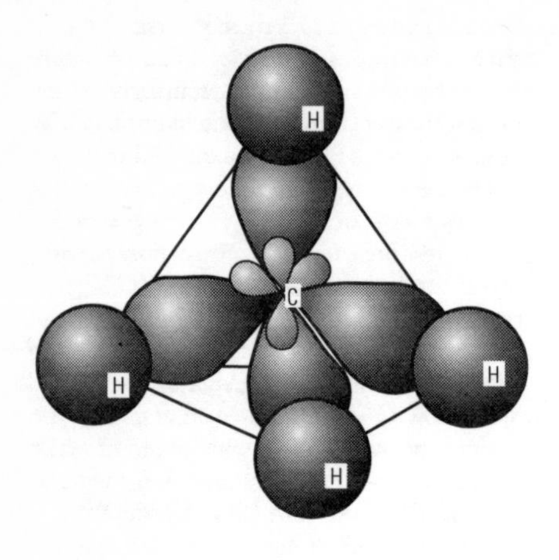

Covalent bond.

molecule demands a different bond angle, the bond will be substantially weaker.

Where two atoms share one pair of electrons a single covalent bond is formed, represented by a single line drawn between the atoms, such as in the structure of the ALKANES. Where the two atoms share two pairs of electrons a DOUBLE BOND forms, such as exists in the ALKENES, represented by two lines, and a TRIPLE BOND forms if three pairs of electrons are shared, such as in the ALKYNES.

Compare IONIC BOND. *See also* COVALENT CRYSTAL, FAJAN'S RULES, POLAR BOND, RESONANCE HYBRID, SHELL.

covalent crystal A crystalline MACROMOLECULE with all molecules being attached to their neighbours by a regular pattern of COVALENT BONDS. Diamond and silicon dioxide are important examples. Covalent crystals are hard and have high melting points. *See also* IONIC SOLID.

covalent radius An effective size associated with an atom involved in a COVALENT BOND. The covalent radius is half the separation between the nuclei of two similar atoms bonded together covalently. The bond length in a bond between two different atoms is then found by adding their covalent radii. This calculation works well for non-POLAR BONDS, but must be used with caution. The bond length in the POLAR MOLECULE HCl, for example, is not equal to the sum of the covalent radii of hydrogen and chlorine as measured by studying molecules of H_2 and Cl_2.

cracking A process used in the petrochemical industry for breaking larger ALKANES obtained from the FRACTIONAL DISTILLATION of crude oil into smaller alkanes and ALKENES. This is achieved by heating the alkanes to high temperatures of around 800°C, which breaks the carbon-carbon bonds. Often a CATALYST is used that allows lower temperatures (around 500°C) to be employed. This is termed catalytic (cat) cracking. Cracking is the method by which alkenes and also petrol are manufactured.

cream of tartar *See* POTASSIUM HYDROGENTARTRATE.

creosote A mixture of PHENOLS obtained by the destructive distillation of COAL TAR or wood. Wood creosote is traditionally used as an antiseptic, whilst coal-tar creosote is widely used as a wood preservative.

critical point *See* CRITICAL TEMPERATURE.

critical solution temperature The temperature above which (for an upper critical solution temperature) or below which (for a lower critical solution temperature) two liquids are completely MISCIBLE and can be mixed together in any proportions.

critical temperature, *critical point* The temperature at which there is no distinction between the liquid and gas states of a substance. Above the critical temperature a gas cannot be liquefied by pressure alone. *See also* GAS.

crown ethers Organic compounds with molecules consisting of large rings of carbon and oxygen atoms. They contain the repeating unit $(O–CH_2CH_2)_n$. The first crown ether to be synthesized was 18-crown-6, which consists of a ring of six $–CH_2–CH_2–O–$ units ($C_{12}H_{24}O_6$). The number 18 refers to the total number of atoms in the ring and 6 is the total number of oxygen atoms in the ring.

Crown ethers are able to form strong complexes with metal ions by co-ordination (*see* CO-ORDINATION COMPOUND) through the oxygen atoms. They can be used to increase solubility of chemicals or act as catalysts. They can also be used to extract specific ions from a mixture. *See also* CRYPTANDS.

crucible An open vessel made of a REFRACTORY material used to heat materials to high temperatures.

crude oil Unrefined PETROLEUM.

crust The solid surface layer of the Earth.

cryogenics The study of materials and processes at temperatures close to ABSOLUTE ZERO. *See* SUPERCONDUCTIVITY, SUPERFLUIDITY.

cryolite A mineral form of sodium aluminoflouride, Na_3AlF_6. It is used in the production of aluminium from its ore BAUXITE.

cryoscopic constant *See* DEPRESSION OF FREEZING POINT.

cryoscopy The use of DEPRESSION OF FREEZING POINT to determine RELATIVE MOLECULAR MASS. If a material is dissolved in a SOLVENT, the reduction in the freezing point of the solution compared to the solvent is a COLLIGATIVE PROPERTY, depending on the molar concentration of dissolved material. Since the mass of material dissolved is known, it is possible to obtain a value for the relative molecular mass.

cryostat A vessel for storing a material at low temperatures, usually immersed in a liquefied gas such as liquid helium or liquid nitrogen. Cryostats are generally many-walled vessels with a vacuum between the walls to prevent heat entering by conduction or convection, and also usually incorporate reflective coatings to reflect thermal radiation.

cryptands Large, three-dimensional, polycyclic organic compounds containing ether chains linked by nitrogen atoms. They are able to form co-ordination complexes (*see* CO-ORDINATION COMPOUND) with ions that can fit into the centre of their three-dimensional strcture. Thus cryptands can bind metals strongly and can be designed to bind specific ions. They can be used to stabilize unusual ionic species. They have similar properties and uses to CROWN ETHERS.

crystal A piece of solid material throughout which the atoms are arranged in a single regular arrangement called a lattice. This arrangement is apparent in many naturally occurring crystals, which have symmetrical shapes reflecting the long-range ordering of their atoms.

Crystals are grouped into seven CRYSTAL SYSTEMS, each one based on a different geometric shape. A given mineral will always crystallize in the same system, although the crystals may not always be the same shape.

Compare DISORDERED SOLID. *See also* BODY CENTRED CUBIC, CLEAVAGE PLANE, CLOSE PACKED, COVALENT CRYSTAL, CRYSTALLIZATION, CRYSTALLOGRAPHY, CUBIC CLOSE PACKED, HEXAGONAL CLOSE PACKED, IONIC SOLID, ISOMORPHIC, SLIP PLANE, WATER OF CRYSTALLIZATION.

crystal field theory A theory of the behaviour of COMPLEX IONS, particularly the coloured nature of TRANSITION METAL complexes, in terms of the effect of the electromagnetic field produced by the surrounding LIGANDS and their influence on the electrons in the central atom, particularly those in D-ORBITALS. The name arises because the ligands are treated as being regularly placed around the central atom rather like the arrangement of atoms in a CRYSTAL. The energy of the d-orbitals now varies depending on their orientation relative to the ligands.

crystalline (*adj.*) Describing a material having the structure of a crystal, though the material itself may be POLYCRYSTALLINE, i.e. made of many small crystals with irregular shapes.

crystallization The process of forming crystals, particularly when used to purify a material or extract it from a solution. Since the SOLUBILITY of most materials increases with temperature, a typical crystallization process may begin with a warm solution that is cooled until it becomes SATURATED, at which point crystals will usually start to form. Some solutions become supersaturated, in which case it may be necessary to add a SEED CRYSTAL to start the process. *See also* FRACTIONAL CRYSTALLIZATION.

crystallography The study of crystals. Early crystallography was aimed at discovering the arrangement of atoms or molecules within a crystal, but more recently similar techniques have been used to map the density of electrons within the molecules forming organic crystals. X-RAY DIFFRACTION and NEUTRON DIFFRACTION are important techniques in this study. *See also* X-RAY CRYSTALLOGRAPHY.

crystal system Any one of the seven basic CRYSTALLINE structures classified according to the geometry of the UNIT CELL of the crystal and how many of the lengths and angles which define this geometry are equal. If the lengths are a, b and c, and the angles are a, b and g, then the possible systems are:

cubic	$a = b = c$ and $\alpha = \beta = \gamma = 90°$
tetragonal	$a = b \neq c$ and $\alpha = \beta = \gamma = 90°$
orthorhombic	$a \neq b \neq c$ and $\alpha = \beta = \gamma = 90°$
hexagonal	$a = b \neq c$ and $\alpha = \beta = 90°$, $\gamma = 120°$
trigonal	$a = b = c$ and $\alpha = \beta = \gamma \neq 90°$
monoclinic	$a \neq b \neq c$ and $\alpha = \gamma = 90°$, $\beta \neq 90°$
triclinic	$a \neq b \neq c$ and $\alpha \neq \beta \neq \gamma \neq 90°$

cubic close packed, *face-centred cubic* (*adj.*) Describing a crystal structure in which each layer of atoms is CLOSE PACKED, with each atom surrounded by six others in that layer. The next layer is placed so that it lies above gaps in the previous layer with the third layer again lying in gaps in the second layer, but also above the gaps in the first layer. The atoms in the fourth layer lie directly above those in the first layer, so labelling the layers of atoms as A, B and C, the packing can be described as ABCABC...

Seen from an angle, this structure can be seen as being cubic, with the UNIT CELL having an eighth of an atom at each corner and half an atom in the middle of each face. This accounts for the alternative name for this structure: face centred cubic. Many metals occur with this structure, including calcium and copper, though the HEXAGONAL CLOSE PACKED structure is also common.

cumene process An industrial method used for the manufacture of PHENOL and PROPANONE. It is more economical to make both products together than separately (from BENZENE and PROPENE respectively). Cumene itself, which is (1-methylethyl)benzene, $PhCH(CH_3)_2$, is obtained by the reaction of benzene with propene in the presence of a catalyst, such as hydrogen fluoride. The cumene is then oxidized and roughly equal amounts of both phenol and propanone result.

cuprammonium The copper tetrammine COMPLEX ion $[Cu(NH_3)_4]^{2+}$.

cupric (*adj.*) Describing a compound containing the higher OXIDATION STATE (+2) of copper. Such compounds usually contain the Cu^{2+} ion, for example cupric chloride $CuCl_2$. They are more stable than CUPROUS compounds.

cupric chloride *See* COPPER CHLORIDE.

cuprite A mineral form of COPPER(I) OXIDE, reddish brown in colour. It is mined as an ore of copper.

cuprous (*adj.*) Describing a compound containing the lower (+1) OXIDATION STATE of copper, for example cuprous chloride, CuCl. Such compounds are sometimes described as containing the Cu^+ ion, but are often covalent in character.

cuprous chloride *See* COPPER CHLORIDE.

curie (Ci) A unit of ACTIVITY, now superseded by the BECQUEREL. One curie is equivalent to 3.7×10^{10} Bq.

curium (Cm) The element with atomic number 96; melting point 1,340°C (approx.). Curium does not occur naturally, but has been synthesized in fairly large amounts. The longest lived ISOTOPE (curium–247) has a HALF-LIFE of 17 million years. Curium is used as a heat source to generate electricity in some artificial satellites.

current The flow of electric charge through a conductor. The size of the current through a certain cross-section of conductor is equal to the rate of flow of charge. The unit of current is the AMPERE. The amount of charge is measured in a unit called the COULOMB (C) and the size of the charge on one electron or proton is 1.6×10^{-19} C. The current is the number of coulombs passing a given point in one second.

cyanamide 1. Any salt containing the cyanamide ion CN_2^{2-}.

2. The parent acid of cyanamide salts, H_2NCN. Cyanamide is a colourless crystalline solid; melting point 42°C; boiling point 140°C; relative density 1.3. It is formed by the reaction of carbon dioxide on sodium amide:

$$2NaNH_2 + CO_2 \rightarrow H_2NCN + 2NaOH$$

It is a weak acid, and in acidic solutions it hydrolyses (*see* HYDROLYSIS) to form UREA:

$$H_2NCN + H_2O \rightarrow H_2NCONH_2$$

cyanic acid (HOCN) An unstable colourless liquid that polymerizes rapidly on heating to form a colourless solid called cyamelide. It is hydrolysed (*see* HYDROLYSIS) by water,

$$HOCN + H_2O \rightarrow CO_2 + NH_3$$

cyanide Any salt or COMPLEX ION containing the cyanide ion, CN^-, such as potassium cyanide, KCN, and potassium hexacyanoferrate(III), $[Fe(CN)_6]^{3-}$. Cyanide salts can be formed by the action of bases on hydrogen cyanide, for example:

$$KOH + HCN \rightarrow KCN + H_2O$$

Cyanides are highly toxic because the ion has the ability to form stable complexes with the iron in haemoglobin, preventing the uptake of oxygen.

cyanine dyes A group of DYES containing the –CH= group linking two nitrogen-containing rings. They are used as sensitizers in colour photography.

cyanocobalamin, *vitamin B_{12}* ($C_{63}H_{90}CoN_{14}O_{14}P$) A VITAMIN of the VITAMIN B COMPLEX. Cyanocobalamin is important in synthesis of RNA and is needed for red blood cell formation. It also works with FOLIC ACID to synthesize the AMINO ACID methionine. It is usually made by micro-organisms in the gut, and occurs in meat and dairy products. Deficiency is usually caused by inability to absorb the vitamin from the stomach, resulting in pernicious anaemia.

cyanogen ($(CN)_2$) A gas with a strong characteristic odour; melting point –28°C; boiling point –21°C. Highly toxic, it can be produced by the oxidation of hydrogen cyanide in the presence of a silver catalyst:

$$4HCN + O_2 \rightarrow (CN)_2 + H_2O$$

It is an important FEEDSTOCK in the manufacture of some fertilizers.

cyanuric acid ($(HNCO)_3$) A white crystalline material, unstable on heating. Its structure contains alternate carbon and nitrogen atoms in a six-membered ring.

cyclic (*adj.*) Describing any organic compound in which any of the carbon atoms are linked in a ring structure. Cyclic compounds can be ALICYCLIC, AROMATIC or HETEROCYCLIC.

cyclic AMP (cAMP) A cyclic NUCLEOTIDE produced from ATP by the action of the ENZYME adenylate cyclase. It has an important role in many biochemical reactions, and is produced in response to other substances, such as hormones. It determines the rate of many biochemical pathways.

cycloalkane Any SATURATED HYDROCARBON that contains a ring of carbon atoms. The general formula is C_nH_{2n}. An example of a cycloalkane is CYCLOHEXANE.

cyclohexadiene-1,4-dione, *quinone, benzoquinone* ($C_6H_4O_2$) A yellow solid; melting point 116°C. It is used in dye manufacture and can be made by the oxidation of aniline with chromic acid.

cyclohexane (C_6H_{12}) A colourless, inflammable liquid; boiling point 81°C; melting point 6.5°C.

H_2
C
H_2C CH_2
H_2C CH_2
C
H_2

It is produced by the REDUCTION of benzene with hydrogen in the presence of a nickel catalyst. It is an intermediate in the preparation of nylon and is used as a solvent and paint remover.

cyclopentadiene (C_5H_6) A colourless liquid with a sweet smell; melting point –97°C; boiling point 40°C. It is a cyclic ALKENE prepared during the CRACKING of petroleum. It easily polymerizes at room temperature to give the dimer dicyclopentadiene. Cyclopentadiene can undergo CONDENSATION REACTIONS with KETONES to form coloured compounds called fulvenes. It undergoes typical DIELS-ALDER REACTIONS.

cyclopentadienyl ion $(C_5H_5)^-$ A stable ion formed by the removal of a hydrogen atom from CYCLOPENTADIENE. *See* FERROCENE.

cysteine ($C_3H_7NO_2S$) A sulphur-containing AMINO ACID found in most proteins.

cystine ($C_6H_{12}N_2O_4S_2$) The main sulphur-containing AMINO ACID present in proteins.

cytidine A PYRIMIDINE NUCLEOSIDE consisting of the organic base CYTOSINE and the sugar RIBOSE.

cytochrome A protein forming part of the ELECTRON TRANSPORT SYSTEM. Electrons are transferred to the next cytochrome in a series of electron carriers, resulting in the reduction of oxygen, O_2, to oxygen ions, O^{2-}, which combine with hydrogen ions to form water, H_2O, during aerobic respiration. The passage of electrons along the carrier chain results in the release of energy that is used to make ATP.

cytochrome oxidase An ENZYME involved in the ELECTRON TRANSPORT SYSTEM. Cytochrome oxidase is a complex consisting of the final CYTOCHROMES of the electron transport system and a copper PROSTHETIC GROUP. It is responsible for combining two hydrogen atoms with an oxygen to form water. Cyanide is a respiratory inhibitor since it can attach to the copper prosthetic group of cytochrome oxidase blocking its action.

cytosine, *2-oxy-4-amino pyramidine* ($C_4H_5N_3O$) An organic base called a PYRIMIDINE occurring in NUCLEOTIDES. *See also* DNA, RNA.

cytotoxic (*adj.*) Capable of killing cells. The term cytotoxic is used particularly to describe drugs that destroy cells.

D

2,4-D, *2,4-dichlorophenoxyacetic acid* ($C_8H_6Cl_2O_3$) A white solid; melting point 138°C. It is a selective weedkiller that kills broad-leafed plants (most crops are narrow-leafed and most weeds broad-leafed).

Dalton's atomic theory The earliest contribution to an understanding of chemistry in terms of ATOMS. In 1805, John Dalton (1766–1844) postulated that: (i) elements are all composed of indivisible small particles (atoms); (ii) the atoms of an element are all identical to one another, but differ from the atoms of other elements; (iii) these atoms can neither be created nor destroyed; and (iv) compounds are formed by the combination of atoms of different elements in fixed ratios. *See also* LAW OF CONSTANT PROPORTIONS.

Dalton's law of partial pressures In any mixture of gases, the total PRESSURE exerted by the mixture is equal to the sum of the PARTIAL PRESSURE of each gas on its own (that is, the pressure that each gas would exert if it were present alone).

Daniell cell An early form of primary VOLTAIC CELL. It has a copper ANODE and a zinc/mercury AMALGAM as the CATHODE. The anode is surrounded by an ELECTROLYTE of dilute sulphuric acid, in which is immersed a POROUS pot containing the anode surrounded by a solution of copper sulphate. The cell produces an EMF of 1.1 V.

dative bond *See* CO-ORDINATE BOND.

d-block element A TRANSITION METAL; that is, an element with the outer electrons in D-ORBITALS. *See also* PERIODIC TABLE.

DDT, *dichlorodiphenyltrichlorethane* (systematic name ***1,1,1-trichloro-2,2-di(4-chlorophenyl)ethane***) A synthetic pesticide discovered in 1939 that has been used world-wide to kill organisms such as lice, fleas and mosquitoes. It has caused many problems because it is persistent and accumulates along food chains.

The use of DDT is banned in most countries, although it is still in use in developing countries where insect-borne diseases are a problem. Despite being banned, the persistence of DDT means that it is still found in many organisms. Many insects have developed DDT resistance.

deactivation A reduction in REACTIVITY, as in the poisoning of a catalyst, for example.

deamination Removal of the AMINO GROUP, $-NH_2$, from an AMINO ACID to form ammonia, urea or uric acid, depending on the animal, for excretion in the urine. In vertebrates this process occurs in the liver, to remove unwanted amino acids.

de Broglie wavelength The wavelength that a particle appears to have when it is exhibiting wave-like properties. The de Broglie wavelength λ is given by

$$\lambda = h/p$$

where h is PLANCK'S CONSTANT and p the momentum of the particle.

debye A measurement of DIPOLE MOMENT used to state the degree to which a molecule is POLAR. One debye is equal to 3.34×10^{-30} Cm. Highly POLAR MOLECULES, such as caesium chloride, have dipole moments of about 10 debye, whilst a more typical polar molecule, such as hydrogen chloride, has a dipole moment of about 1 debye.

decahydrate A HYDRATE containing 10 parts water to one part of the compound. For example, sodium carbonate forms the decahydrate $NaCO_3.10H_2O$.

decant (*vb.*) To pour a liquid carefully from one vessel to another, so only the required part of the liquid is transferred, leaving behind any solid material, or the denser of two IMMISCIBLE liquids.

decarboxylation The removal of carbon dioxide from a molecule, especially in biological reactions.

decay constant In RADIOACTIVITY, the probability per second of an atomic nucleus decaying. In a sample of radioactive nuclei, which originally contains N_0 nuclei with decay constant λ, after a time t, the number of nuclei which remain in their undecayed state will be N, where

$$N = N_0 e^{-\lambda t}$$

deci- Prefix denoting one tenth. For example, a decimetre is one tenth of a metre (0.1 m).

decomposition A chemical reaction in which a compound is broken down into its elements or into simpler compounds, usually under the action of heat.

decomposition voltage The voltage that must be applied to make an electric current flow in ELECTROLYSIS.

defect Any discontinuity in the regular pattern of atoms within a CRYSTAL. Line defects are better known as DISLOCATIONS. Point defects are those produced by a vacancy in the crystal lattice, by a different type of atom at a particular location or by the presence of an INTERSTITIAL atom. Frenkel defect are those caused by exposure to IONIZING RADIATION, in which an atom removed from its normal location in the lattice creates a vacancy and occupies an interstitial position.

deflagration A sudden and spontaneous bursting into flames, such as observed with phosphorous in oxygen.

degassing The process of removing dissolved gases from a liquid or gases absorbed into a solid, usually by heating. In particular, degassing is the removal of gas from the walls of a vacuum vessel.

degenerate (*adj.*) Describing ENERGY LEVELS or ORBITALS that have the same energy as one another. If some outside influence, such as an electric or magnetic field from a neighbouring atom, alters the energies of the orbitals so they are no longer degenerate, the degeneracy is said to be 'lifted'.

degree of freedom 1. Any one of a set of independent variables needed to define the state of a physical system (such as pressure, temperature, etc.)

2. A way in which a molecule can possess energy independent of any other degree of freedom. Thus a MONATOMIC gas – one composed of molecules each containing only one atom – has three degrees of freedom, motion in each of three independent directions. A molecule containing two or more atoms in a line has an extra two degrees of freedom, due to rotational motion about two axes at right angles to one another and to the line of the molecule. Molecules with three or more atoms not arranged in a straight line have a further degree of freedom. The MOLAR HEAT CAPACITY of an IDEAL GAS at constant pressure is $R/2$ times the number of degrees of freedom active in the material, where R is the MOLAR GAS CONSTANT.

In solids, there are potentially six degrees of freedom: three due to KINETIC ENERGY from motion in three dimensions and three due to POTENTIAL ENERGY from departure from equilibrium in three dimensions. QUANTUM effects mean that these six degrees of freedom are not fully active in most solids at ordinary temperatures – the molecules can have only certain quantized energies. This makes the molar heat capacity less than the expected value of $3R$. *See also* PHASE RULE, DULONG AND PETIT'S LAW.

dehydrate (*vb.*) To remove water from a substance. This may be from a solid containing WATER OF CRYSTALLIZATION, or from any substance where the removal of water requires a chemical reaction, for example the dehydration of ethanol by concentrated sulphuric acid:

$$C_2H_5OH \rightarrow C_2H_4 + H_2O$$

See also HYDRATION.

dehydrogenase Any ENZYME that catalyses the removal of a hydrogen atom in a DEHYDROGENATION reaction. Dehydrogenases are important in biological systems.

dehydrogenation A chemical reaction in which hydrogen is removed from a compound. In organic compounds, this results in an increase in the saturation of the compound; that is, single bonds are converted into double bonds. A metal catalyst is usually used, such as platinum, or DEHYDROGENASE enzymes in biological systems.

deliquescent (*adj.*) Describing a solid that absorbs water vapour from the atmosphere and dissolves in this water to form a concentrated solution. Sodium hydroxide is an example of a deliquescent material.

delivery tube A glass tube, bent at one end, used with a GAS JAR to collect gas produced in a chemical reaction.

delocalized orbital One of a number of MOLECULAR ORBITALS that overlap to effectively produce a single large orbital that can hold as many electrons as the original orbitals. This overlap increases the stability of the molecule.

In metals, the VALENCE ELECTRON orbitals overlap to form a CONDUCTION BAND. In

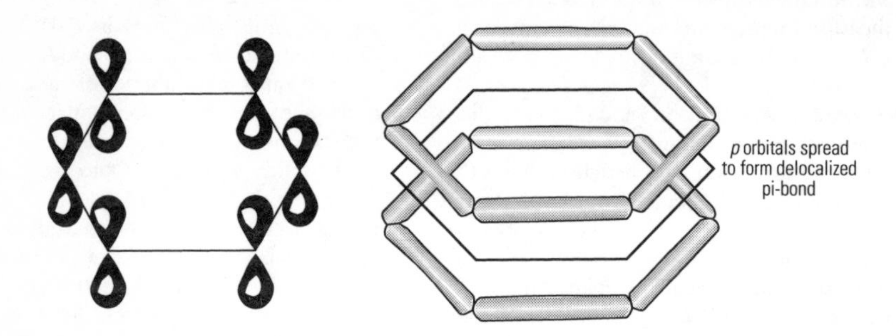

Delocalized orbital.

BENZENE, and other AROMATIC compounds, the P-ORBITALS at right angles to the plane of the aromatic ring overlap from one atom to the next, forming a ring around which electrons can move freely.

delta bond A very weak COVALENT BOND formed by the overlap of D-ORBITALS to form a MOLECULAR ORBITAL similar to a PI-BOND, but with four regions of high electron density instead of two. Delta bonds are generally too weak to have any significant chemical effect.

denature (*vb.*) **1.** Of ETHANOL, to add a toxic or unpleasant substance to make it unfit for human consumption. methylated spirits, for example, is ethanol denatured with methanol for industrial use.

2. Of PROTEINS and some other complex organic molecules, to produce a structural change that makes the molecule incapable of performing its usual biological function. Heat and some inorganic chemicals, particularly strong acids and bases, can denature a wide range of proteins.

dendrite A CRYSTAL whose growth has repeatedly branched into two separate sections, producing a tree-like final structure. Many metals show dendrite structures in their crystals.

dendrochronology The dating of fallen and fossilized trees by comparison of their growth rings. Dendrochronology is used to provide a check on radiocarbon dates, enabling them to be recalibrated to take account of the changes in carbon–14 concentration in the atmosphere. *See also* RADIOCARBON DATING.

denitrification The process by which NITRATES in the soil are converted back to atmospheric nitrogen. Anaerobic bacteria, such as *Pseudomonas denitrificans* and *Thiobacillus denitrificans,* bring about this conversion, particularly in waterlogged soil. Because most organisms cannot utilize atmospheric nitrogen (*see* NITROGEN CYCLE) denitrification reduces soil fertility. To avoid this, soil is ploughed and dug to improve drainage and aeration. *See also* NITRIFICATION.

density The mass of a substance contained in a given volume. This depends both on the mass of the molecules or atoms from which the material is made and on their separation.

Most solids or liquids have densities of a few thousand kilograms per metre cubed, whilst gases under ATMOSPHERIC PRESSURE have densities of just a few kilograms per metre cubed. This difference is due to the fact that in solids and liquids, which are not easily compressed, molecules are pretty closely packed, whilst in gases, which are much more easily compressed, the spacing between the molecules is far greater, typically ten times as great. *See also* RELATIVE DENSITY, SPECIFIC VOLUME.

deoxyribonuclease *See* DNASE.

deoxyribonucleic acid *See* DNA.

deoxyribose ($C_5H_{10}O_4$) A PENTOSE sugar that is a component of DNA; melting point 91°C.

depression of freezing point The amount by which the freezing point of a solvent is reduced by the presence of dissolved molecules or ions. It is a COLLIGATIVE PROPERTY, independent of the nature of the dissolved particles, but proportional to their molar concentration, with a constant of proportionality known as the cryoscopic constant.

Measurements of freezing point depression for measured masses of a material dissolved in a solvent can be used to estimate RELATIVE MOLECULAR MASSES.

derivative A chemical compound derived from some other compound by a straightforward reaction, which usually retains the structure and some of the chemical properties of the original compound.

desalination The removal of salt, particularly from sea water to make it fit to drink or for irrigation. Whilst many different techniques have been used, all require a cheap source of energy to make them economically worthwhile. Evaporation and re-condensation of pure water, driven by solar energy is often used in hot, dry countries.

desiccant A substance, usually a solid, that can be used to desiccate (remove all water from) a substance. Silica gel and ANHYDROUS calcium chloride are common desiccants.

desiccate (*vb.*) To completely remove all water from a substance.

desiccator A vessel with a sealed lid used to remove water from a material by exposing it to a DESICCANT.

destructive distillation A type of DISTILLATION process in which the substance being heated partly or fully decomposes, leaving a solid or viscous liquid. Volatile liquids can be collected as usual after condensation. Destructive distillation is used, for example, in the heating of coal to produce COAL TAR.

desulphuration The removal of sulphur from a compound.

desulphurization The removal of sulphur compounds from petroleum fractions.

detergent A cleansing agent that is capable of wetting a variety of surfaces and removing dirt, usually found in oily or greasy deposits. Detergents are said to be surface active. They consist of molecules with long HYDROCARBON chains ('tails'), which are soluble in oil, attached to a salt group (the 'head'), such as SULPHATE, which is water-soluble. The detergent molecules surround the oil or grease drops and the hydrocarbon chain is able to penetrate the droplet whilst the salt group remains in the water. The salt groups become negatively charged and so then do the oil droplets, which thus repel one another and remain in suspension until they are removed by rinsing.

SOAP is a detergent but has the disadvantage of forming insoluble salts with the magnesium and calcium in HARD WATER, leaving a scum. A range of synthetic detergents derived from petroleum have soluble magnesium and calcium salts to avoid this problem. Another problem with detergents is their disposal. Early detergents could not be degraded by the bacteria in sewage works and therefore caused foaming in rivers. New detergents have now been developed that are more easily broken down by bacteria, although there is still a problem if they escape the normal sewage process.

In addition, many detergents contain added PHOSPHATES, bleaches or fluorescent substances as whiteners. These can cause pollution and EUTROPHICATION. More recently, detergents are being used which contain no phosphates or bleaches and are thus more environmentally friendly. *See also* SURFACTANT.

deuterium The naturally occurring heavy ISOTOPE of hydrogen, hydrogen–2 (one proton, one neutron).

deuteron A nucleus of the ISOTOPE hydrogen-2, containing one proton and one neutron. *See also* DEUTERIUM.

devitrification The CRYSTALLIZATION of glass and other usually AMORPHOUS materials, which happens slowly over time. In glass, the formation of crystals makes the glass less transparent and more brittle.

Dewar structure A variation on the structure of BENZENE (C_6H_6) in which the two opposite carbon atoms in the hexagonal ring are joined (across the middle of the ring) by a single bond with two double bonds on each side of the hexagon. This is an ISOMER of benzene (Dewar benzene) and on heating it will revert to benzene. Dewar benzene is a colourless liquid.

dextrin An intermediate POLYSACCHARIDE product, formed during the HYDROLYSIS of STARCH. Dextrin is used as an adhesive.

dextrorotatory (*adj.*) Describing the ENANTIOMER of a compound exhibiting OPTICAL ISOMERISM that rotates PLANE-POLARIZED light to the right (clockwise). This used to be denoted by the prefix *d*- but (+) is now used. This is not to be confused with the prefix D- used to indicate the configuration of CARBOHYDRATES and AMINO ACIDS.

dextrose *See* GLUCOSE.

diagonal relationship A similarity in chemical properties between two elements in the

PERIODIC TABLE where one element is diagonally below and to the right of the other. The most striking examples are the chemical similarities between lithium and magnesium, beryllium and aluminium, and boron and silicon. In each case, the relationships arise because the first element in each group of the periodic table forms ions that are so small that they are more typical in size to those of the next group and similarly the first element in a group tends to have an increased ELECTRONEGATIVITY, again more typical of the next group.

dialysate *See* DIALYSIS.

dialysis The method by which small and large molecules in a mixed solution are separated. The mixed solution is placed inside a semipermeable bag (one that allows molecules of a certain size to pass through its pores) and surrounded by water. Small molecules diffuse out of the bag into the water, which is repeatedly changed, leaving the larger molecules inside the bag. In kidney failure, this principle is used in renal dialysis to remove toxic substances from the bloodstream.

1,6-diaminohexane, *hexamethylene diamine* ($H_2N(CH_2)_6NH_2$) A colourless solid AMINE; melting point 41°C; boiling point 204°C. It is used in the manufacture of NYLON-6,6

diamond A crystalline ALLOTROPE of CARBON. Diamond is the hardest known mineral and is widely used in cutting and drilling tools, and as a gem. Diamond has a macromolecular structure (*see* MACROMOLECULE) with each carbon atom making four COVALENT BONDS with its neighbours. The rigidity of this structure is responsible for the hardness of diamond and its high thermal conductivity.

diastereoisomers STEREOISOMERS that are not ENANTIOMERS. They are not identical nor are they mirror images. For example the D-form of tartaric acid and mesotartaric acid are diastereoisomers, differing only in the configuration at one carbon atom.

mesotartaric acid	D-tartaric acid
CO_2H	CO_2H
H–C–OH	OH–C–OH
H–C–OH	H–C–OH
CO_2H	CO_2H

Diastereoisomers have different chemical and physical properties, unlike enantiomers.

diatomic (*adj.*) Describing a molecule comprising just two atoms, such as chlorine, Cl_2, or hydrogen chloride, HCl.

diazo compounds *See* DIAZONIUM COMPOUNDS.

diazonium compounds, *diazo compounds* A class of compounds containing the RN=NX group, where R is an ALKYL or ARYL GROUP, and X is some other ion or group. The most important diazo-compounds are DIAZONIUM SALTS, which form the basis of the manufacture of AZO DYES.

diazonium salts DIAZONIUM COMPOUNDS derived from the base RN=NOH, where R is an ARYL GROUP. They are made from the reaction of aromatic AMINES with nitrous acid (HNO_2) in a process called DIAZOTIZATION.

The $-N^+{\equiv}N$ group formed in the process of diazotization can be easily replaced. For example, if a diazonium salt in water is heated, the $-N^+{\equiv}N$ group is replaced by an OH (hydroxyl) group, resulting in the formation of a PHENOL.

$$C_6H_5N^+{\equiv}N\ HOSO_3^- + H_2O \rightarrow C_6H_5OH + N_2 + (HO)_2SO_2$$

The $-N^+{\equiv}N$ group can also be replaced by chlorine or bromine by treatment with copper(I) chloride or bromide in what is termed the Sandmeyer reaction. Similarly, treating with potassium iodide substitutes iodine.

Diazonium salts undergo coupling reactions with phenols or amines in which the $-N^+{\equiv}N$ group of the salt is not replaced but is instead added to by a second BENZENE RING. The products thus formed are brightly coloured and form the basis of the manufacture of AZO DYES.

$$C_6H_5N^+{\equiv}N + C_6H_5OH \rightarrow C_6H_5N{=}NC_6H_4OH$$

diazotization The process by which DIAZONIUM SALTS are made. A reaction mixture of dilute sulphuric and aqueous sodium nitrate is used to dissolve an AMINE. This results in the production of nitrous acid, HNO_2, which reacts with the amine. AROMATIC amines react at low temperatures to produce diazonium salts. The diazonium salts produced by ALIPHATIC amines are unstable and decompose rapidly.

$$C_6H_5NH_2 + HONO + (HO)_2SO_2 \rightarrow C_6H_5N{\equiv}N^+HOSO_3^- + 2H_2O$$

dibasic acid Any acid having a BASICITY of 2, such as sulphuric acid (H_2SO_4) and carbonic acid (H_2CO_3). The acid can form a NORMAL SALT (such as a SULPHATE, SO_4^{2-}) or an ACID SALT (such as a hydrogensulphate, HSO_4^-).

dibenzo-4-pyrone *See* XANTHONE.

1,2-dibromoethane, ***ethylene dibromide*** ($BrCH_2CH_2Br$) A colourless liquid with a sweet smell; melting point 10°C; boiling point 132°C. It is a HALOGENOALKANE made by passing ethene through bromine or a mixture of bromine and water at 20°C. It is used in petrol to combine with the lead formed by the decomposition of LEAD TETRAETHYL.

dicarbide, ***acetylide*** A BINARY COMPOUND formed by ALKALI METALS and ALKALINE EARTHS in combination with the dicarbide ion, C_2^{2-}. They react with water to produce ethyne (acetylene, C_2H_2). For example, calcium dicarbide, CaC_2:

$$CaC_2 + 2H_2O \rightarrow C_2H_2 + Ca(OH)_2$$

dicarboxylic acid A CARBOXYLIC ACID that contains two CARBOXYL GROUPS (-COOH). This is indicated by the addition of –dioic to the systematic name.

dichlorodiphenyltrichloroethane *See* DDT.

2,4-dichlorophenoxyacetic acid *See* 2,4-D.

dichloryl diethyl sulphide *See* MUSTARD GAS.

dichromate(VI) Any salt containing the dichromate(VI) ion, $Cr_2O_7^{2-}$. The ion has an orange colour. In solution, the dichromate(VI) ion is stable only under acid conditions. In the presence of alkalis, they are converted to the CHROMATE(VI) ion:

$$Cr_2O_7^{2-} + 2OH^- \rightarrow 2CrO_4^{2-} + H_2O$$

Diels–Alder reaction The addition of an ALKENE or ALKYNE across a conjugated DIENE (that is, a compound containing two double bonds separated by a single bond) to give a ring compound. The alkene or alkyne are known as dienophiles and to be suitable must have a double bond with a CARBONYL GROUP on each side. The reaction was discovered by two German chemists Otto Diels (1876–1954) and Kurt Alder (1902–58).

diene An ALKENE containing two double bonds. Compounds such as 2-chlorobuta-1,3-diene are conjugated dienes since the two double bonds are separated by a single bond.

Dieterici equation A modified form of VAN DER WAALS' EQUATION. It takes into account the decrease in pressure near the boundary of a gas as a result of the imbalance in attractive INTERMOLECULAR FORCES, meaning that a molecule tends to slow down before it collides with the wall of its container. The Dieterici equation is

$$p(v - b) = RT\exp(-RT/va)$$

where a and b are constants representing the intermolecular attraction and EXCLUDED VOLUME respectively, p is the pressure, v the MOLAR VOLUME, R the MOLAR GAS CONSTANT, and T the absolute temperature.

diethylamine *See* ETHYLAMINE.

1,4-diethylene dioxide *See* DIOXAN.

diffuse (*vb.*) To spread out by DIFFUSION.

diffusion The spontaneous and random movement of molecules or particles in a fluid (gas or liquid) from a region where they are at a high concentration to one where they are at a low concentration, until a uniform concentration or dynamic EQUILIBRIUM is achieved. Once at a uniform concentration, the molecules will continue to move in random motion, but there is no net diffusion. The concentration gradient is the difference in concentration of a substance between two regions. The rate at which one material diffuses into another is determined by the average speed of its molecules and by the MEAN FREE PATH. *See also* GRAHAM'S LAW OF DIFFUSION.

diffusion pump A vacuum pump for reducing the pressure of gas in a vessel from which most of the gas has already been removed. Mercury or oil vapour enters the vessel through a nozzle and is then condensed on the cooled walls of a nearby part of the vessel. Gas molecules tend to become attached to the vapour molecules and the pressure of gas in the vessel is further reduced. Pressures as low as 10^{-7}Pa can be achieved.

dihedral The angle formed between two intersecting planes, particularly the planes of atoms in a CRYSTAL.

dihydrate A HYDRATE containing two parts water to one part of the compound. For example, calcium sulphate forms the dihydrate $CaSO_4.2H_2O$.

dihydric (*adj.*) Describing a compound with two HYDROXYL GROUPS. For example, GLYCOL is a dihydric ALCOHOL.

1,2-dihydroxybenzene, *catechol* ($C_6H_4(OH)_2$) A colourless, crystalline PHENOL; melting point 105°C; boiling point 240°C. It is a strong reducing agent and is used as a photographic developer.

2,3-dihydroxybutendioic acid *See* TARTARIC ACID.

dilead(II) lead(IV) oxide, *red lead* (Pb_3O_4) A red powder (black when hot); decomposes on heating; relative density 9.1. Dilead(II) lead(IV) oxide can be made by heating lead(II) oxide to about 400°C in air:

$$6PbO + O_2 \rightarrow Pb_3O_4$$

The oxide decomposes back to lead(II) oxide on heating. The structure is complex and largely covalent (*see* COVALENT BOND), often containing less oxygen than the formula would suggest. The material was once widely used as a pigment in paints, but this has been discontinued because of concerns about the toxicity of lead.

dilute 1. (*adj.*) Having a low concentration.

2. (*vb.*) To add more solvent to a solution, reducing the concentration. *See also* HEAT OF DILUTION, OSTWALD'S DILUTION LAW.

dilution The factor by which the volume of a dilute solution is greater than the original concentrated solution, or liquid solute. Thus a dilution of 5 indicates a solution in which 5 litres of dilute solution has been produced from 1 litre of the original material.

dimer 1. A structure in which two identical molecules are held together, either by COVALENT BONDS or more loosely by HYDROGEN BONDS. Dinitrogen tetroxide, N_2O_4, is a dimer of nitrogen dioxide, NO_2, for example.

2. Two MONOMERS joined together by POLYMERIZATION.

4-dimethylamino-4′-azobenzene sodium sulphonate *See* METHYL ORANGE.

dimethylbenzene, *xylene* (C_8H_{10}) A colourless liquid with a characteristic smell. It is a mixture of three ISOMERS and is widely used as a solvent.

dimethylformamide (DMF) ($HCON(CH_3)_2$ A colourless liquid; melting point –61°C; boiling point 153°C. It is used widely as a solvent for organic compounds.

dimethylsulphoxide (DMSO) ($(CH_3)_2SO$) A colourless, odourless solid, melting point 18°C, boiling point 189°C. It is used as a solvent for organic compounds and for low temperature preservation in biology and medicine.

dinitrogen oxide, *nitrous oxide* (N_2O) An odourless gas; melting point –91°C; boiling point –89°C. It can be prepared by heating ammonium nitrate:

$$NH_4NO_3 \rightarrow N_2O + 2H_2O$$

Dinitrogen oxide will support the combustion of many compounds, for example:

$$Mg + N_2O \rightarrow MgO + N_2$$

It decomposes on heating:

$$2N_2O \rightarrow 2N_2 + O_2$$

Inhalation of dinitrogen oxide causes a sense of well-being, followed by unconsciousness, hence its alternative name of laughing gas, and its use as an ANAESTHETIC.

dinitrogen tetroxide (N_2O_4) A pale brown liquid; melting point 11°C; boiling point 21°C; relative density 1.5. Dinitrogen tetroxide exists in equilibrium with nitrogen dioxide:

$$N_2O_4 \Leftrightarrow 2NO_2$$

In the liquid phase the mixture consists almost entirely of the DIMER, whereas in the gas the MONOMER predominates. Dinitrogen tetroxide can be produced by condensing nitrogen dioxide produced in the reaction between concentrated nitric acid and copper:

$$Cu + 4HNO_3 \rightarrow Cu(NO_3)_2 + 2NO_2 + 2H_2O$$

dioxan, *1,4-diethylene dioxide* ($C_4H_8O_2$) A colourless liquid; melting point 11°C; boiling point 101°C. It consists of a six-membered ring with four CH_2 groups and two oxygen atoms opposite one another. It is used as a solvent and is toxic.

2,6-dioxytetrahydropyrimidine See URACIL.

diphosphane (P_2H_4) A yellow liquid, spontaneously inflammable on exposure to air. It can be obtained by the HYDROLYSIS of calcium phosphide in an inert atmosphere,

$$Ca_3P_2 + 3H_2O \rightarrow P_2H_4 + 2CaO + Ca(OH)_2$$

dipole An ELECTROSTATIC system, such as a POLAR MOLECULE, containing two equal but opposite charges.

At a distance from the dipole that is large compared to the separation of the charges or poles, the field produced decreases with increasing distance by a factor equal to the inverse third power of the distance from the dipole. Thus if the distance is doubled, the field falls to one eighth of its original strength.

dipole moment The strength of a DIPOLE. The size of the charges multiplied by the distance between them. *See also* DEBYE.

disaccharide A double sugar formed by the combination of two MONOSACCHARIDES, with the loss of a water molecule (a CONDENSATION REACTION). A disaccharide can be split into its single sugars by the addition of water (a HYDROLYSIS reaction). Disaccharides are sweet, soluble and crystalline. Examples include SUCROSE (cane sugar or table sugar, $C_{12}H_{22}O_{11}$; GLUCOSE plus FRUCTOSE), maltose (glucose plus glucose) and LACTOSE (glucose plus GALACTOSE). *See also* CARBOHYDRATE, POLYSACCHARIDE.

dislocation An imperfection in the lattice structure of a crystalline solid, particularly a metal. A dislocation arises as a result of a misalignment of atomic planes as the crystals of the material are forming, with a plane of atoms coming to a halt in the middle of the atom or forming a 'spiral staircase' structure. *See also* ALLOY.

disordered solid A solid in which the atoms are not arranged in a regular crystalline lattice. Some materials, such as glass, are naturally disordered; others form disordered solids only if they are cooled so rapidly that crystals do not have time to form.

dispersed phase *See* COLLOID.

displacement reaction A chemical reaction in which one element, particularly a metal, is replaced by another, more reactive, element. For example, iron filings will displace copper in copper sulphate solution, with the formation of a PRECIPITATE of metallic copper:

$$CuSO_4 + Fe \rightarrow FeSO_4 + Cu$$

This reaction shows that iron is more reactive than copper. *See also* ELECTROCHEMICAL SERIES.

disproportionation A process in which the same chemical compound is both oxidized and reduced (*see* OXIDATION, REDUCTION). For example, copper(I) compounds in aqueous solution will form copper(II) compounds (an oxidation) and a precipitate of metallic copper (a reduction), for instance:

$$Cu_2SO_4 \rightarrow Cu + CuSO_4$$

dissociate (*vb.*) To break up, for example into ions or into the elements from which a substance is made.

dissociation The breaking up of a compound into simple compounds or the breaking up of an IONIC compound into separate ions when it is dissolved or melted. For example, copper(II) sulphate, $CuSO_4$, dissociates into copper ions, Cu^{2+}, and sulphate ions, SO_4^{2-}. *See also* DISSOCIATION CONSTANT, HEAT OF DILUTION, OSTWALD'S DILUTION LAW.

dissociation constant The EQUILIBRIUM CONSTANT for the dissociation of a molecule into ions, $AB \Leftrightarrow A^+ + B^-$. The dissociation constant is $[A^+][B^-]/[AB]$, where the square brackets, [ ], denote CONCENTRATION. *See also* PK.

dissolve (*vb.*) The breaking up of a solid into individual ions or molecules when placed in a solvent, thus producing a liquid solution. The term also refers to the distribution of the molecules of a gas through the liquid, again producing a liquid solution. *See also* HEAT OF SOLUTION, SOLUBILITY, SOLUBLE.

distil (*vb.*) To perform a DISTILLATION process.

distillate *See* DISTILLATION.

distillation A process used to purify liquids or to separate mixtures of liquids by virtue of their differing boiling points. The liquid is heated to the required temperature and the vapour is collected into a separate vessel (the condenser), where it is cooled and condensed back to a liquid. The liquid collected (the distillate) is pure, leaving solid impurities in the distillation vessel. For separating mixtures of liquids with similar boiling points, such as crude oil, a process called FRACTIONAL DISTILLATION is used. Distillation is also used in the manufacture of alcoholic beverages. *See also* DESTRUCTIVE DISTILLATION, MOLECULAR DISTILLATION.

disulphuric(VI) acid ($H_2S_2O_7$) A colourless crystalline solid; melting point 35°C; decomposes on further heating; relative density 1.9. Disulphuric(VI) acid is strongly HYGROSCOPIC. Commonly found in association with sulphuric acid, it is formed by dissolving sulphur trioxide in sulphuric acid:

$$H_2SO_4 + SO_3 \rightarrow H_2S_2O_7$$

The solution of sulphuric and disulphuric acids is called OLEUM and is an important material in the CONTACT PROCESS. Oleum is also used in the SULPHONATION of organic molecules.

dithionate Any salt containing the dithionate ion, $S_2O_6^{2-}$. Dithionates can be formed from the OXIDATION of sulphites by manganese(IV) oxide, for example:

$$4Na_2SO_3 + MnO_2 + 2H_2O \rightarrow 2Na_2S_2O_6 + 4NaOH + Mn$$

divalent (*adj.*) Having a VALENCY of 2.

DNA, *deoxyribonucleic acid* A complex, very large molecule that contains all the information for building and controlling a living organism. It is a double-stranded NUCLEIC ACID made of NUCLEOTIDES with the bases ADENINE, GUANINE, CYTOSINE and THYMINE (never URACIL, which is found only in RNA) and a PENTOSE sugar that is always DEOXYRIBOSE. Except in bacteria, DNA is found in the nuclei of cells, arranged in chromosomes. The molecular structure of DNA was first proposed by James Watson (1928–) and Francis Crick (1916–) in 1953, for which they were awarded the Nobel Prize for Medicine or Physiology.

The molecule consists of two polynucleotide strands (each of millions of nucleotides) linked to each other by base pairing (*see* BASE PAIR) and HYDROGEN BONDS. The bases adenine and thymine always link and cytosine and guanine always link. The linking of one PURINE and one PYRIMIDINE in this way allows the same spacing between the strands throughout the length of the molecule. So the DNA is like a ladder, with the base pairs forming the rungs and the deoxyribose and PHOSPHATE groups forming the uprights. In addition, the two chains forming the upright run in opposite directions and are called anti-parallel. The ladder is then twisted into a double helix.

The hereditary information is stored as a specific sequence of bases. Individual AMINO ACIDS are coded for by a set of three bases, the precise sequence of which determines the

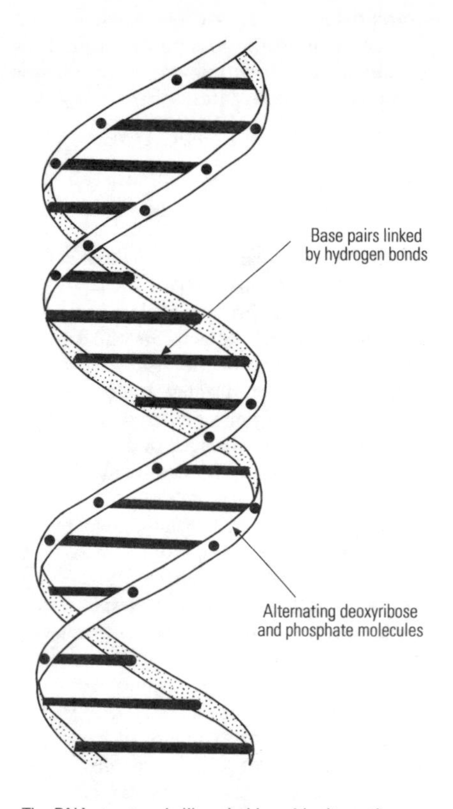

The DNA structure is like a ladder with alternating deoxyribose and phosphate molecules forming the uprights and base pairs forming the rungs. The ladder is twisted to form a double helix. The uprights run in opposite directions to each other. Base pairings are always adenine-thymine and cytosine-guanine.

amino acids that are made and therefore the PROTEINS that are produced by the cell. By controlling PROTEIN SYNTHESIS in cells, DNA is able to determine the hereditary characteristics and control the whole organism. In order for the genetic information to be passed on from cell to cell and generation to generation, DNA must be able to replicate. This is under the control of the enzyme DNA polymerase, which breaks down the hydrogen bonds between the two strands, allowing them to separate. Each strand joins with free necleotides present in the nucleus to make new complementary strands. Thus two identical DNA molecules are formed, also identical to the original parent. *See also* GENETIC ENGINEERING, X-RAY CRYSTALLOGRAPHY.

DNAse, *deoxyribonuclease* One of many ENZYMES that hydrolyse DNA by breaking down the sugar-phosphate bonds. *See also* RESTRICTION ENDONUCLEASES.

DNA ligase *See* LIGASE.

dodecahydrate A HYDRATE that contains 12 parts water to one part of the compound. For example, trisodium phosphate forms the dodecahydrate $Na_3PO_4.12H_2O$.

dolomite A common METAMORPHIC rock containing calcium and magnesium carbonates.

Donnan equilibrium The EQUILIBRIUM established between the concentrations of the IONS of an ELECTROLYTE, such as sodium chloride solution, on each side of a SEMIPERMEABLE MEMBRANE when a molecule, such as a protein, which cannot pass through the membrane, is introduced on one side of the membrane. In such cases, a POTENTIAL DIFFERENCE, called the membrane potential is produced across the membrane.

donor atom An atom that contributes a pair of electrons to form a CO-ORDINATE BOND, for example the nitrogen atom in the ammonium RADICAL, NH_4^+.

dopamine ($C_8H_{11}NO_2$) In biochemistry, a NEUROTRANSMITTER that is an intermediate in the synthesis of ADRENALINE. There are special areas in the brain that particularly use dopamine to transmit nerve impulses. Patients with the tremors of Parkinson's disease show degeneration of such areas.

Doppler effect The apparent observed change in observed frequency (or wavelength) of a wave due to the relative motion between the source and the observer. An example is a police car siren that increases in pitch (frequency) as it moves towards a stationary observer and decreases in pitch as it moves away.

d-orbital The third lowest energy ORBITAL for a given PRINCIPAL QUANTUM NUMBER. They exist only for principal quantum number of 3 or greater. There are five d-orbitals for each principal quantum number. Four of these consist of four lobes in an X-shape, each lying in a single plane, whilst the fifth d-orbital consists of two lobes surrounded by a TORUS.

dose The amount of IONIZING RADIATION absorbed by a living organism, usually human, over a specified period of time. It is usually specified in terms of the amount of energy deposited by the radiation, measured in GRAY.

double bond A COVALENT BOND formed when two atoms share two pairs of electrons. Two sets of ORBITALS overlap to form a bond stronger than the usual, single, covalent bond. This involves the formation of a PI-BOND and a SIGMA-BOND.

double decomposition A reaction between two soluble salts in aqueous solution, in which an exchange of ions forms an insoluble salt, which is produced as a PRECIPITATE, for example:

$$Na_2CO_3\ (aq) + CuSO_4\ (aq) \rightarrow CuCO_3\ (s) + Na_2SO_4\ (aq)$$

double salt A crystalline structure containing two NORMAL SALTS, such as potassium aluminium sulphate, $KAl(SO_4)_2.12H_2O$.

doublet A pair of lines in the ABSORPTION SPECTRUM or EMISSION SPECTRUM of many elements, such as sodium, indicating the presence of two ENERGY LEVELS which are approximately DEGENERATE.

dross The coating of solid impurities that forms on the surface of a metal when it is melted. The metal can be refined by skimming off the dross before being allowed to re-solidify, usually in moulds.

dry cell *See* ZINC-CARBON CELL.

dry ice The solid form of CARBON DIOXIDE, particularly when used to cool other materials. Dry ice sublimes at –78°C. It can be produced by allowing compressed carbon dioxide to undergo a FREE EXPANSION.

drying oil Any naturally occurring OIL, such as linseed oil, that hardens on exposure to air. Such materials form the base of many varnishes and gloss paints. The hardening occurs as a result of the OXIDATION and POLYMERIZATION of unsaturated FATTY ACIDS in the oil.

ductile (*adj.*) Able to be drawn out into a long strand, such as a wire. Ductility is an important property of many metals.

Dulong and Petit's law An experimental rule that states that the MOLAR HEAT CAPACITY of a solid tends to $3R$ at high temperatures, where R is the MOLAR GAS CONSTANT. The theoretical basis for this lies in EQUIPARTITION OF ENERGY between six DEGREES OF FREEDOM, elastic and kinetic energy in each of three dimensions.

Dumas' method A technique for finding the RELATIVE MOLECULAR MASS of a volatile substance. A sample of the substance is placed in the bottom of a glass vessel and then vaporized by heating to a known temperature in a water or oil bath, driving out all the air. The vessel is then sealed and weighed. By comparing this with the mass of the vessel when full of air, the density of the substance in its gaseous form at a known temperature can be found, and from this its molecular mass.

Duralumin (*Trade mark*) A class of aluminium alloys containing about 4 per cent copper and smaller amounts of magnesium, manganese and silicon. Duralumin and similar alloys combine the low density of pure aluminium with improved strength and are widely used in the manufacture of aircraft.

dyes, *dyestuffs* Coloured compounds that can be applied to fibres, paper, hair, etc. to give colour that is resistant to washing. They absorb PHOTONS in the visible region of the ELECTROMAGNETIC SPECTRUM. There are numerous dyes, some natural, such as indigo and cochineal, and many synthetic.

Dyes vary in their chemical composition and the method in which they are applied. Most are organic compounds but can be applied with inorganic substances. Direct dyes (such as Congo red), chemically bond to the fabric forming a coloured compound. Indirect dyes require the fabric to be first treated with another compound. Vat dyes are insoluble coloured compounds formed from colourless soluble compounds on exposure to air in the dyeing vat. These then become enmeshed in the fibres of the fabric. In acid dyes (such as naphthol green) the acidic group of the dye attaches to the basic AMINO GROUP of protein fibres, such as in wool or silk. AZO DYES form the most important group of industrial dyes.

dyestuffs *See* DYES.

dynamic equilibrium *See* EQUILIBRIUM.

dynamite A type of explosive originally manufactured by absorbing NITROGLYCERINE onto a porous base of KIESELGUHR. Nowadays the term tends to be used for any explosive used for rock blasting where the active ingredient is mixed with an inert base to form a stick-shaped charge that will not explode until required.

dysprosium (Dy) The element with atomic number 66; relative atomic mass 162.5; melting point 1,412°C; boiling point 2,567°C; relative density 8.6. Dysprosium is a member of the LANTHANIDE series. The element is highly magnetic, but it is too scarce and too difficult to purify for this property to be used commercially.

E

ebonite *See* VULCANITE.

ebullioscopic constant *See* ELEVATION OF BOILING POINT.

ebullioscopy The technique of using the ELEVATION OF BOILING POINT to measure the RELATIVE MOLECULAR MASS of a substance. The increase in boiling point is a COLLIGATIVE PROPERTY, so by dissolving a known mass of material in a suitable solvent and measuring the increase in boiling point of the solution compared to the pure solvent, the relative molecular mass can be found.

EDTA *See* ETHYLENEDIAMINETETRAACETIC ACID.

effervescence The production of a gas in a liquid by a chemical reaction, with the formation of many small bubbles and a characteristic fizzing sound.

efflorescence The loss of WATER OF CRYSTALLIZATION to the atmosphere from a crystalline solid, leading to the formation of a powdery solid, for example, sodium carbonate:

$$Na_2CO_3.10H_2O \rightarrow Na_2CO_3.H_2O + 9H_2O$$

effusion The flow of a gas through a small opening, comparable in size to the spacing between the molecules in the gas, so that the gas cannot be considered to flow through the opening as a fluid, but rather passes through one molecule at a time. The rate of effusion is proportional to the ROOT MEAN SQUARE speed of the molecules, and thus at a given temperature is inversely proportional to the square root of the molecular mass.

einsteinium (Es) The element with atomic number 99. It does not occur naturally but has been synthesized by the bombardment of lighter nuclei with charged particles. All known ISOTOPES have HALF-LIVES of less than one year. Einsteinium has not found uses to justify its manufacture in commercial quantities.

elastic (*adj.*) Describing a substance or a process by which the substance is deformed by an applied force, but returns to its original shape once the force is removed. *See also* PLASTIC.

elastin A protein of animal connective tissue that is very elastic. It occurs in ligaments, arteries and lungs. Elastin is structurally related to COLLAGEN but contains 30 per cent GLYCINE, 30 per cent LEUCINE and 15 per cent proline.

elastomer A group of natural and synthetic materials, which includes RUBBERS, that have the ability to recover their shape after extensive deformation.

electrical double layer The structure of the boundary between charged particles in a SOL and the surrounding liquid. A simple model suggests that the particle would be surrounded by a layer of oppositely charged ions from the liquid phase. A fuller treatment recognizes that this surface layer only partially neutralizes the charge on the particle and a diffuse polarized (*see* POLARIZATION) region also exists around such particles.

electric arc furnace A device used to melt iron as part of a steel making process. The furnace is a cylindrical vessel with a lid containing graphite ELECTRODES. An ELECTRIC ARC between the electrodes and the surface melts the material in the furnace, and the molten material is then removed by tipping the furnace onto one side.

The process does not introduce any oxygen and so can be closely controlled chemically. Steel scrap is often added to the furnace, the chemical composition of which is determined from a sample of the molten steel and then adjusted as required.

See also BESSEMER CONVERTER, OPEN HEARTH FURNACE, OXYGEN FURNACE.

electricity The general term describing all effects caused by electric charge, whether at rest (electrostatic) or in motion (electric current).

electrochemical cell *See* CELL.

electrochemical equivalent The mass of a specified material released or dissolved when one COULOMB of charge flows in an ELECTROLYSIS experiment. The electrochemical equivalent is the mass of the atom in grams divided by the charge of the ion in coulombs ($g\,C^{-1}$).

electrochemical series, ***electromotive series*** The list of metallic elements, plus hydrogen, in order of increasing (more positive) STANDARD ELECTRODE POTENTIAL. Elements above hydrogen are described as electropositive, whilst those below hydrogen are said to be electronegative. For the more common metals, the electrochemical series is: potassium, calcium, sodium, magnesium, aluminium, zinc, iron, lead, hydrogen, copper, silver, platinum.

The standard electrode potentials effectively compare the energy needed to form one coulomb of ions in aqueous solution from the solid metal. Since many other processes involve a similar chemical change, the electrochemical series has applications well beyond the considerations of electrochemical CELLS, and this series is effectively the REACTIVITY SERIES for metals. It also gives a far better indication of the relative energies in chemical processes than the IONIZATION ENERGIES of the metals, since ionization energies relate to gaseous atoms being turned into gaseous ions, whilst the electrochemical series relates to the more common process of solid metals being converted to aqueous ions.

Any metal that lies higher than another one in the electrochemical series (that is, has a more negative standard electrode potential) will displace the lower metal in a DISPLACEMENT REACTION. Those metals that lie below hydrogen can be deposited on a CATHODE in an ELECTROLYSIS of an aqueous solution. The metals above hydrogen will release hydrogen in an electrolysis of an aqueous solution, and electrolysis can only be used to extract such metals if a molten salt is used rather than an aqueous solution.

electrochemistry The part of chemistry that deals with the interaction between electricity and chemical materials, particularly electrochemical CELLS. *See also* ELECTROLYSIS.

electrode A conductor through which an electric current passes in or out of an ELECTROLYTE (as in an electrolytic CELL), a gas (as in a discharge tube) or a vacuum (as in a thermionic valve).

electrode potential The POTENTIAL DIFFERENCE between an IONIC solution and a metal ELECTRODE. The difference between the electrode potential of the two electrodes in a CELL is what gives rise to voltage of that cell. *See also* HALF-CELL, REDOX POTENTIAL, STANDARD ELECTRODE POTENTIAL.

electrodialysis A method of obtaining a pure SOLVENT from an IONIC solution. Electrodialysis is used for the DESALINATION of sea water. The solution is ELECTROLYSED in a cell containing a sequence of SEMIPERMEABLE MEMBRANES that are alternately permeable to only positive and negative ions. The result is an increase in concentration of the ions in one set of gaps between these membranes, whilst the other gaps contain the solvent with a reduced concentration of ions.

electrolyse (*vb.*) To bring about chemical decomposition by ELECTROLYSIS.

electrolysis The chemical change effected by the passage of an electric current through an IONIC liquid (*see* ELECTROLYTE). In electrolysis, ions are attracted to the oppositely charged ELECTRODE (*see* ANODE, CATHODE, ANION, CATION). When the ions reach the electrode they may gain electrons at the cathode or lose them at the anode and form neutral atoms or molecules. Alternatively, they may react with the electrode, which is then IONIZED and dissolves in the liquid. This latter reaction generally takes place at the anode, as the electrodes tend to be metals that readily give up electrons to form positive ions. Which of the possible reactions takes place depends on the relative reactivities of the substances involved. The more reactive substance will generally form ions or remain in its ionic form. When water acidified with sulphuric acid (to ionize the water so it will conduct) is electrolysed using platinum electrodes, hydrogen gas is formed at the cathode and oxygen gas and water are liberated at the anode, since hydrogen is more reactive than platinum and the HYDROXIDE ion (OH^-) gives up electrons more readily than the SULPHATE ion (SO_4^{2-}).

Electrolysis is used to extract some metals from their ores – for example the extraction of aluminium from bauxite. Electrolysis is also used in ELECTROPLATING, where the object to be coated is the negative electrode in a solution of a salt of the coating metal.

See also ANODIC OXIDATION, CATHODIC REDUCTION, ELECTROCHEMICAL EQUIVALENT, FARADAY'S LAWS OF ELECTROLYSIS, VOLTAMMETER.

electrolyte Any conducting liquid, through which electric charge flows by movement of IONS. Electrolytes are molten IONIC compounds or solutions of ionic salts or of compounds that ionize in solution. *See also* MOLAR CONDUCTIVITY.

electrolytic cell *See* CELL.

electromagnetic radiation Energy resulting from the acceleration of electric charge that propagates through space in the form of ELECTROMAGNETIC WAVES. Alternatively, electromagnetic radiation can be thought of as a stream of PHOTONS travelling at the speed of light (c) each with energy hc/λ, where h is PLANCK'S CONSTANT and λ is the wavelength of the associated wave. *See also* ELECTROMAGNETIC SPECTRUM, RADIATION.

electromagnetic spectrum The spread of different frequencies and wavelengths of ELECTROMAGNETIC WAVES with particular reference to the similarities and differences between different parts of this range. In order of increasing frequency, it comprises RADIO WAVES, INFRARED radiation, visible LIGHT, ULTRAVIOLET radiation, X-RAYS and GAMMA RADIATION.

electromagnetic waves Waves of energy composed of oscillating electric and magnetic fields, at right angles to one another, propagating through space. Since the associated fields are capable of existing in empty space, electromagnetic waves can travel through a vacuum. The detailed properties of the wave depend upon its frequency and wavelength: wavelengths from several kilometres down to 10^{-15} m have been observed. This range is called the ELECTROMAGNETIC SPECTRUM. Electromagnetic waves travel through a vacuum at a constant speed of 3×10^8 m s^{-1}. In other materials they travel more slowly than this, depending on the medium.

electromotive force (e.m.f.) The amount of electrical energy given to each COULOMB of charge that is driven around a circuit by the source of e.m.f. The SI UNIT of electromotive force is the VOLT. An e.m.f. may be provided by an electrolytic CELL (which converts chemical energy into electrical energy) or an electrical generator, which relies on the relative motion of a wire though a magnetic field.

electromotive series *See* ELECTROCHEMICAL SERIES.

electron The negatively charged elementary particle that occurs in all atoms. Atoms consist of a central NUCLEUS surrounded by orbiting electrons. The number and distribution of the electrons in an atom (the electron structure) is responsible for the chemical properties of the atom. An electron that has become detached from an atom is known as a free electron.

The mass of an electron is 9.1×10^{-31} kg and it has a charge of 1.6×10^{-19} C. This charge is the fundamental unit of negative charge, and the electron is the basic particle of electricity. In metals, an electric current consists of a movement of free electrons.

See also CHEMICAL BOND, NEUTRON, ORBITAL, PROTON.

electron affinity 1. Qualitatively, the tendency of an atom or molecule to gain electrons and form a negative ion.

2. Quantitatively, the energy required for the formation of negative ions from one mole of an element. Energy is sometimes released in the formation of such an ion, so for the more reactive nonmetals, the electron affinities are negative. If the electron affinity is positive, the reaction can only proceed if the energy required is released at some other stage of the reaction, such as the formation of an IONIC SOLID.

electron-deficient compound A covalently bound compound in which the number of electrons involved in chemical bonding is less than the normal two electrons per bond. The boranes are electron-deficient compounds; in diborane, B_2H_6, for example, two of the hydrogen atoms effectively share their electrons with both the boron atoms rather than with a single atom as is normally the case.

electron diffraction The spreading of a beam of electrons as they pass through a narrow aperture or crystal lattice. This effect can only be explained in terms of the wave nature of particles. Electron diffraction is used to measure bond lengths and angles, and in the study of solid surfaces. *See also* NEUTRON DIFFRACTION, X-RAY DIFFRACTION.

electronegativity A measure of the tendency of an atom in a COVALENT BOND to attract the electrons in the bond towards itself. Thus if the two elements in the bond have equal electronegativities the bond will not be POLAR, otherwise it will be polar with the more electronegative atom located at the negatively charged end.

Electronegativities can be calculated in a number of ways, but the most commonly used is the PAULING ELECTRONEGATIVITY based on the energies of covalent bonds.

See also ELECTROPOSITIVITY.

electron microscope An instrument, developed in 1933, that magnifies objects using a beam of electrons. The electrons are accelerated by a high voltage through the object (held in a vacuum) and focused by powerful electromagnets (iron bars with coils of wire around through which an electric current flows, inducing magnetism), instead of optical lenses, onto a fluorescent screen for viewing. A camera may be built in to record what is seen on the screen. A beam of electrons of a DE BROGLIE WAVELENGTH of typically 0.005 nm is used, compared to the wavelength of about 500 nm of the light rays used in a light microscope. This vastly improves the resolving power of the electron microscope so that objects only 1 nm apart can be distinguished, and objects can be magnified more than 500,000 times.

The sample to be viewed is fixed and embedded in ARALDITE (an EPOXY RESIN), which enables ultra-thin sections to be cut. Living specimens cannot be viewed. There are several types of electron microscope.

See also ELECTRON-PROBE MICROANALYSER, SCANNING ELECTRON MICROSCOPE, TRANSMISSION ELECTRON MICROSCOPE.

electron-probe microanalyser A modified ELECTRON MICROSCOPE in which the object emits X-rays when it is hit by electrons, the different intensities of which indicate the presence of different chemicals. The specimens can be examined without being destroyed.

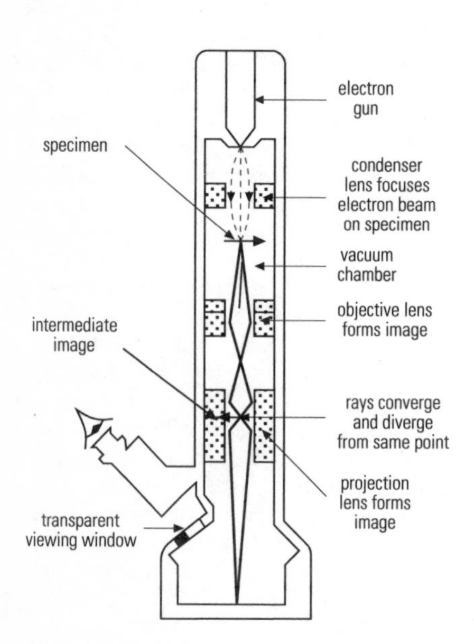

The electron microscope.

electron-spin resonance A technique used to locate unpaired electrons, such as are found in paramagnetic materials (see PARAMAGNETISM) and FREE RADICALS. In a magnetic field, a molecule containing an unpaired electron will have an energy that depends on the orientation of the SPIN of that electron relative to the magnetic field. Energy will be absorbed from a microwave beam if the frequency of that beam corresponds to the energy difference between the two energy states. Measurement of this energy difference gives information about the location of the unpaired electron within the molecule.

electron transport system, *respiratory chain* A series of oxidation-reduction reactions that results in the transfer of electrons or hydrogen atoms through a series of carriers, with the release of energy. The electron transport system is the third and final stage in aerobic respiration. Hydrogen atoms are carried from the KREBS CYCLE by the electron carriers NAD or FAD, and transferred to a series of other carriers at lower energy levels. At each transfer the energy released is used to produce ATP. The other carriers are COENZYME Q and a series of CYTOCHROMES. The hydrogen atoms are split into their protons and electrons. The electrons are passed along the carrier chain and recombine to hydrogen at the end, which then links with oxygen to form water. This whole process of forming ATP through oxidation of hydrogen atoms is termed OXIDATIVE PHOSPHORYLATION and occurs in living cells. *See also* GLYCOLYSIS.

electron-volt (eV) A unit of energy used in atomic and nuclear physics. One eV is the energy gained by an electron when it moves through a POTENTIAL DIFFERENCE of one volt. It is equivalent to 1.6×10^{-19} JOULES.

electrophile (Greek = electron loving) Any ion or molecule lacking in electrons and so tends to gain electrons in a chemical reaction, acting as a LEWIS ACID, or a REDUCING AGENT. In organic chemistry, electrophiles often attach themselves to negatively charged parts of a molecule. *See also* ELECTROPHILIC ADDITION, ELECTROPHILIC SUBSTITUTION.

electrophilic addition An ADDITION REACTION that is triggered by attack from an ELECTROPHILE. This type of reaction is typical of the

ALKENES, for example the addition of hydrobromic acid (HBr) or bromine to ethene, giving bromoethane and 1,2-dibromoethane respectively:

$$CH_2{=}CH_2 + HBr \rightarrow CH_3CH_2Br$$

$$CH_2{=}CH_2 + Br_2 \rightarrow BrCH_2CH_2Br$$

See also MARKOWNIKOFF'S RULE.

electrophilic substitution A SUBSTITUTION REACTION in which an atom or group in a molecule is replaced by an ELECTROPHILE (which accepts electrons to form a new COVALENT BOND). The general reaction can be considered to be as follows:

$$RX + E^+ \rightarrow RE + X^+$$

where R is a HYDROCARBON, metal or METALLOID group and the electrophile (E) and X can be a variety of organic or inorganic CATIONS. Electrophilic substitution involving AROMATIC compounds is termed electrophilic aromatic substitution. Examples of this include NITRATION, SULPHONATION and FRIEDEL-CRAFTS REACTION of aromatic compounds.

electrophoresis, *cataphoresis* A technique used for separating molecules on the basis of their charge. The mixture to be separated is placed on a gel or paper in a BUFFER solution at a given pH, and an electric current is applied through the buffer. Substances of different sizes can be separated because they diffuse at different rates, and can be compared with known standards. Electrophoresis is widely used in biology as an analytical technique, to separate different length strands of DNA for example. The process is also sometimes used as a way of attracting paint particles to metallic objects. *See also* POLYACRYLAMIDE GEL ELECTROPHORESIS.

electroplating An ELECTROLYSIS technique in which metal is deposited on the CATHODE. For example, when copper sulphate is electrolysed with a copper ANODE, the anode dissolves and copper is deposited on the cathode. Metal objects are often electroplated with silver.

electropositivity The tendency of an atom to lose electrons and form positive ions, or, in a COVALENT BOND, to lose hold of the electrons in the bond, leading to that atom finding itself at the positive end of a POLAR BOND. It is the opposite of ELECTRONEGATIVITY.

electrostatic (*adj.*) Related to, producing or caused by electric charges at rest.

electrovalency The number of electrons that may be lost or gained by the atoms of a given element in forming an ion. For S-BLOCK and P-BLOCK elements, the electrovalency is determined by the number of electrons that must be lost or gained to achieve a NOBLE GAS configuration. Thus the electrovalency of sodium is 1 (it loses an electron), as is that of chlorine (though in this case an electron is gained). The TRANSITION METALS often form more than one type of ion, for example iron exists with electrovalency of 2, Fe^{2+}, and 3, Fe^{3+}.

element A substance that cannot be broken down into more fundamental constituents by normal chemical means, being made of ATOMS all of the same type (that is, having the same ATOMIC NUMBER). About 110 elements are known, though there are claims for a few more. Of these, 91 are found in nature.

elementary particle, *fundamental particle* Any one of the particles from which all MATTER, such as the ELECTRON, PROTON and NEUTRON.

elevation of boiling point The increase in the boiling point of a solvent caused by the presence of dissolved ions or molecules. This is a COLLIGATIVE PROPERTY, independent of the nature of the particles present, and proportional to their molar concentration, with a constant of proportionality called the ebullioscopic constant. Measurement of the elevation of boiling point caused by dissolving a known mass of substance in a solvent can be used to estimate its RELATIVE MOLECULAR MASS.

elimination A reaction in which an atom or group in a molecule is eliminated. HALOGENOALKANES can undergo elimination reactions in the presence of a NUCLEOPHILE, for example the formation of an ALKENE from 2-bromo-2-methylpropane in the presence of a hydroxide ion by the elimination of H^+ and Br^- from the latter:

```
   H  CH3
   |   |
H—C —C— Br + HO⁻ → H2O + Br⁻
   |   |
   H  CH3              H\       /CH3
                    +    C = C
                       H/       \CH3
```

Tertiary, and to a lesser extent secondary,

halogenoalkanes undergo elimination reactions. *See also* NUCLEOPHILIC SUBSTITUTION REACTIONS.

eluent A SOLVENT used in CHROMATOGRAPHY.

elution The process of using a SOLVENT to remove an adsorbed substance (*see* ADSORPTION) from an adsorbing medium, particularly in CHROMATOGRAPHY. In chromatography, the rate at which material is dissolved and redeposited is the key to separation of different substances.

elutriation A technique for sorting particles of different sizes from a SUSPENSION. An adjustable upward flowing fluid stream is used. Particles larger than a certain size will sink under their own weight whilst smaller particles are carried upward by the fluid flow.

emanation An old name for the ISOTOPES of RADON that are produced as a gas from the decay of the radioactive elements RADIUM, THORIUM and ACTINIUM.

emerald The green gemstone form of the mineral beryl.

emery A naturally occurring rock containing aluminium oxide (CORUNDUM), Al_2O_3, together with impurities, mostly iron oxide, Fe_3O_4. In its powdered form, it is widely used as an abrasive, often glued to a paper backing to form emery paper.

e.m.f. *See* ELECTROMOTIVE FORCE.

emission spectrum A SPECTRUM produced as excited atoms or molecules return to lower ENERGY LEVELS, giving off electromagnetic radiation of specific wavelengths, with all other wavelengths being absent.

empirical (*adj.*) Describing a result that is known purely from experiment, without any theoretical understanding or explanation.

empirical formula The simplest type of chemical formula. It gives the relative proportions of the elements present in a compound, but gives no indication of the RELATIVE MOLECULAR MASS or the structure. For example, the empirical structure of ethane is CH_3, giving no indication of the presence of two carbon atoms and six hydrogen atoms in each molecule. The full MOLECULAR FORMULA of ethane is C_2H_6. The empirical formula is obtained by QUANTITATIVE ANALYSIS of the proportion of each element present. *See also* STRUCTURAL FORMULA.

emulsion A mixture of two liquids, with one liquid forming small droplets suspended in the other. *See also* COLLOID, IMMISCIBLE.

enantiomer (Greek *enantios* = face-to-face) One of a pair of optical ISOMERS. *See* OPTICAL ISOMERISM.

enantiotropy A form of ALLOTROPY in which one of the allotropes is stable at high temperature and the other is stable at lower temperatures. Transitions between one state and the other can be brought about by heating or cooling the material, though such transitions may happen only slowly. The allotropy of SULPHUR is an example of this, with the rhombic form being stable at low temperatures and the monoclinic form at high temperatures. *See also* MONOTROPY.

encephalin One of a variety of PEPTIDES produced by the brain that acts as a natural pain-killer in a similar way to OPIATE drugs. It is, however, not addictive as it is quickly degraded by the body. *See also* ENDORPHIN.

endorphin A natural PEPTIDE that modifies the action of nerve cells. Endorphins are considered to be natural morphine-like pain-killers, reducing the perception of pain by reducing the transmission of nerve impulses. They also affect the release of sex hormones, mood and hunger. Their effects are short-lived since they are rapidly degraded by the body. OPIATES act in a similar way to endorphins but are not so rapidly degraded. *See also* ENCEPHALIN.

endothermic (*adj.*) Describing a chemical reaction in which heat energy is taken in; that is, one in which the HEAT OF REACTION is positive. The formation of hydrogen iodide is an endothermic reaction:

$$^1/_2H_2 + {}^1/_2I_2 \rightarrow HI, \quad \Delta H = 26 \text{ kJ mol}^{-1}$$

end point In a TITRATION, the point in a reaction at which an INDICATOR shows that enough of a second REAGENT has been added to have used up all of a fixed quantity of the first. For example, the indicator may show that enough acid has been added to completely neutralize a given quantity of alkali. In practice, a slight excess of the second reagent will be needed to bring about the change in the indicator. The reaction is actually complete a little sooner, at a point called the EQUIVALENCE POINT.

energy A measure of the ability of an object or a system to do WORK. Energy can be broadly divided into two classes: POTENTIAL ENERGY, which arises from an objects state or position (in a gravitational field, for example); and KINETIC ENERGY, which is a function of motion.

The amount of work done, or the energy that gives the capability to do work, is measured by a unit called the JOULE.

The importance of the idea of energy comes from the LAW OF CONSERVATION OF ENERGY, which states that the total energy of any closed system is conserved. *See also* CHEMICAL ENERGY, HEAT, INTERNAL ENERGY.

energy band A range of energies that electrons can have in a solid. *See* BAND THEORY.

energy barrier The amount of energy that may need to be provided before a physical or chemical process can take place, even though the process as a whole may release energy. *See also* ACTIVATION PROCESS.

energy level Any one of the permitted energy states in which an atom or molecule may exist. Under the rules of QUANTUM MECHANICS, a system can only have certain fixed energies, and each different atom or molecule has a series of possible energy levels. When the system moves from one energy level to another, the energy is absorbed or emitted, often as a PHOTON of light or other electromagnetic radiation. The distinct energies of the allowed levels mean that only photons of certain discrete energies are involved. This gives rise to absorption and emission spectra characteristic of the atoms or molecules involved. *See also* ATOMIC ABSORPTION SPECTROSCOPY, ATOMIC EMISSION SPECTROSCOPY, BAND THEORY, SPECTROSCOPY.

enols Compounds containing the group –CH=C–OH–. They are the tautomeric (*see* TAUTOMERISM) forms of certain KETONES. The keto form contains the group $-CH_2-CO-$ and the enol form occurs when there is migration of a hydrogen atom between a carbon atom and the oxygen on an adjacent carbon.

enthalpy A measure of energy, commonly used in the study of chemical reactions. It is the INTERNAL ENERGY of the molecules of a material plus the pressure at which the material is held multiplied by the volume occupied. It is a useful quantity because it is constant in any chemical reaction that takes place at a constant pressure. For a substance where the molecules have energy U and occupy a volume V at a pressure p, the enthalpy is H, where

$$H = U + pV$$

The internal energy of the molecules alone is not constant if there is any change in volume during the reaction. If a gas is produced, for example, work will have to be done at the expense of the internal energy of the molecules in pushing back the surrounding atmosphere to make room for this gas. Enthalpy is also constant during the FREE EXPANSION of a gas.

enthalpy of combustion *See* HEAT OF COMBUSTION.

enthalpy of formation *See* HEAT OF FORMATION.

enthalpy of reaction *See* HEAT OF REACTION.

entropy A measure of the degree of disorder in a thermodynamic system. Typically, this concept is applied to a system containing many molecules. The most probable state in which those molecules are found is the state that can be achieved in the largest number of ways. Entropy can be defined by the equation

$$S = k \ln W$$

where S is the entropy, k is the BOLTZMANN CONSTANT and W is the number of ways in which the molecules in the system can be arranged to produce the specified state.

For any system containing a large number of particles, the most probable state becomes overwhelmingly likely to be the state in which the system is found. This leads to the statement that any irreversible change (not able to happen in reverse) is one which produces an increase in the total entropy of the system. Since such systems tend to move from ordered (low entropy) to disordered (high entropy) states, the entropy of a system can be used as a measure of the extent to which the energy in a system is available for conversion to WORK.

See also FREE ENERGY, SECOND LAW OF THERMODYNAMICS.

E number One of a group of ADDITIVES (not including flavourings) approved by the European community. E numbers do not have to be listed with the ingredients of a product. Some E numbers are more often referred to by their name, for example E102 is TARTRAZINE. Some can cause side-effects in some people, for example tartrazine, used to provide an orange colour, is known to cause hyperactivity and worsen asthma.

enzyme A biological molecule that alters the rate of a reaction (usually speeds it up) without undergoing a chemical change itself (and can therefore be used repeatedly). An enzyme is a natural CATALYST that does not alter the final CHEMICAL EQUILIBRIUM of a reaction, only the speed at which it is achieved. Enzyme reactions are always reversible.

Most enzymes are GLOBULAR PROTEINS with a three-dimensional structure that provides a specific ACTIVE SITE where the SUBSTRATE molecule it acts upon fits, like a LOCK-AND-KEY MECHANISM. Modern interpretation of this theory suggests that the three-dimensional shape of the active site changes as the substrate binds. This is called 'induced fit'. An enzyme-substrate complex forms until the substrate is altered or split, and then its shape changes and it no longer fits into the active site, so the enzyme falls away. All enzymes act on specific substrates but some will bind a variety of similar substrates while others are very specific and bind only one.

Enzymes need different and precise conditions in which to function. Deviations from the optimum pH and temperature will cause the shape of the enzyme to change, which will eventually make it non-functional. The concentration of both enzyme and substrate has an effect on enzyme activity. A very low concentration of enzyme is needed for a reaction, and if there is an excess of substrate the RATE OF REACTION is proportional to the enzyme concentration.

The affinity of the enzyme for the substrate is variable. Some enzymes are CONSTITUTIVE while others are INDUCIBLE. Some enzymes need a COFACTOR for them to function. There are six recognized categories of enzymes based on their functions: OXIDOREDUCTASES, TRANSFERASES, HYDROLASES, LYASES, ISOMERASES and LIGASES. The nomenclature often gives information regarding an enzyme's activity, for example, peptidases break down PEPTIDES.

Enzymes have many uses outside the body, in medicine and industry and as research tools in molecular biology, and they can be extracted from bacteria and even modified by GENETIC ENGINEERING for a particular purpose.

See also COENZYME, PROSTHETIC GROUP.

enzyme inhibition The reduction of the RATE OF REACTION of an ENZYME by an INHIBITOR, which can be reversible or non-reversible. Heavy metal ions, for example mercury and silver, cause non-reversible inhibition and therefore permanent damage to enzymes by altering their shape when they break the SULPHIDE bonds. Reversible inhibition is temporary and the enzyme function returns once the inhibitor is removed.

Competitive inhibitors function by having a structure similar to the SUBSTRATE, and they bind and remain in the ACTIVE SITE. Non-competitive inhibitors attach elsewhere on the enzyme molecule and change the enzyme's shape so that the substrate can no longer bind. Cyanide is a non-competitive inhibitor that attaches to the copper PROSTHETIC GROUP of the enzyme cytochrome oxidase, inhibiting respiration. In many metabolic reactions the end-product of a pathway may inhibit the enzyme at the start of the pathway; this is an example of NEGATIVE FEEDBACK.

epimerism A type of OPTICAL ISOMERISM exhibited by molecules that have two CHIRAL centres but differ in the arrangment about one of these. Carbohydrate epimers often exist, for example D-gluconic acid and D-mannonic acid are epimers differing only in the orientation of the attached to one carbon atom.

epinephrine *See* ADRENALINE.

epoxy resin A synthetic RESIN that is a THERMOSETTING PLASTIC. Epoxy resins are used as tough adhesives and in paints. Araldite is a trade name for a commonly used household adhesive that is also used to mount specimens to be viewed in an ELECTRON MICROSCOPE.

equation of state Any equation describing the behaviour of a material under different conditions. In particular, the change in volume under differing temperatures and pressure.

equilibrium The state of an object or system in which all effects – forces, interactions, reactions, etc. – are balanced so that there is no net change. *See also* CHEMICAL EQUILIBRIUM.

equilibrium constant In a REVERSIBLE REACTION,

$$xA + yB \Leftrightarrow mC + nD$$

the reaction will proceed in one direction until a CHEMICAL EQUILIBRIUM is reached. The reaction then proceeds at the same rate in the forward and backward directions. At this point, the CONCENTRATIONS of the REAGENTS, indicated by square brackets [ ], will obey the equation

$$[A]^x[B]^y/[C]^m[D]^n = K$$

where K is the equilibrium constant.

equipartition of energy The principle that states that thermal energy will be distributed equally between the available DEGREES OF FREEDOM.

equivalence point In a TITRATION, the point at which sufficient REAGENT has been added to have reacted completely with the fixed quantity of the second reagent. For example, in adding an acid to an alkali, the equivalence point is reached when the alkali has been completely neutralized. *See also* END-POINT.

equivalent weight An obsolete term for the mass of a substance that can react, directly or indirectly, with one gram of hydrogen. For an element, it is the RELATIVE ATOMIC MASS of the compound divided by its VALENCY, for example, 8 g of oxygen can combine with 1 g of hydrogen to form 9 g of water (0.5 mole). *See also* LAW OF EQUIVALENT PROPORTIONS.

erbium (Er) The element with atomic number 68; relative atomic mass 167.3; melting point 1,529°C; boiling point 2,863°C; relative density 9.0. It is a member of the LANTHANIDE series. Difficulties in purifying it from the other chemically similar lanthanides has meant that it has found little commercial use.

essential amino acid An AMINO ACID that is needed by humans but cannot be made by them and must therefore be included in their diet. The nine essential amino acids are LYSINE, PHENYLALANINE, LEUCINE, THREONINE, METHIONINE, ISOLEUCINE, TRYPTOPHAN, HISTIDINE and VALINE.

essential element Any chemical substance required by living organisms for normal growth and development. They include MACRONUTRIENTS and MICRONUTRIENTS.

essential fatty acid One of several FATTY ACIDS required in the diet of humans and certain other animals for normal growth. They include LINOLEIC and LINOLENIC and ARACHIDONIC ACIDS. The latter is a precursor of PROSTAGLANDINS.

essential oil A natural oil with a pleasant odour obtained from plants and used in perfumes and flavourings.

ester An organic compound formed when an acid and ALCOHOL react together with the elimination of water. For example, ethanoic acid, CH_3COOH, reacts with ethanol, C_2H_5OH, to give the ester ethyl ethanoate:

$$CH_3COOH + C_2H_5OH \rightarrow CH_3COOC_2H_5 + H_2O$$

The general formula of an ester is RCOOR′, where R and R′ are HYDROCARBONS. Esters occur naturally in fruit and many have a characteristic fruity odour. Esters are used as solvents, food flavourings and to provide the scent in perfumes. *See also* ESTERIFICATION, SAPONIFICATION.

esterification A term used to describe the formation of an ESTER from an ALCOHOL and an acid. If equal amounts of acid and alcohol are used an equilibrium is set up such that equal amounts of the four species (acid, alcohol, ester and water) exist. The equilibrium can be altered to give good yields of ester and also the ester can be hydrolysed (*see* HYDROLYSIS) back to a mixture of acid and alcohol (such as in SAPONIFICATION).

ethanal, *acetaldehyde* (CH_3CHO) A colourless, inflammable liquid ALDEHYDE; boiling point 20.8°C. It is made from the oxidation of ethanol or ethene. Ethanal is miscible with water, ALCOHOL and ETHER. It can be oxidized to give ethanoic acid and reduced to give ethanol. It is used in the manufacture of many organic chemicals.

ethanamide, *acetamide* (CH_3CONH_2) A colourless solid, crystallizing in long white crystals with a strong smell of mice; melting point 82°C; boiling point 222°C. It is manufactured by the dehydration of ammonium ethanoate or by the action of ammonia on ethanoyl chloride, ethanoic anhydride, or ethyl ethanoate.

ethane (CH_3CH_3) A gas with no colour or odour; boiling point –89°C. It is the second member of the ALKANE series of HYDROCARBONS. Ethane can be made by the reduction of ethene or ethyne by hydrogen under pressure, in the presence of a nickel catalyst. It is found in natural gas and forms an explosive mixture with air.

ethane-1,2-diol, *ethylene glycol, glycol* ($(CH_2OH)_2$) A thick, colourless, odourless, sweetish, liquid ALCOHOL; boiling point 197°C. It is used as an ANTIFREEZE and coolant for engines because it mixes with water and lowers the freezing point below 0°C. It is also used in the preparation of ETHERS and ESTERS, including POLYESTER fibres.

ethanedioic acid, *oxalic acid* ($C_2H_2O_4$) A poisonous substance found in rhubarb and as oxalate salts in wood sorrel. Ethanedioic acid causes paralysis of the nervous system. It is used as a bleaching agent in the textile and leather industries and domestically for removing stains such as rust and blood. It is also used for metal cleaning.

ethanoate, *acetate* Any salt or ESTER of ETHANOIC ACID. Ethanoates are used in the textile industry to make acetate RAYON, a synthetic fabric made from CELLULOSE treated with ethanoic acid, and in the photographic industry as acetate film, a clear sheet made from cellulose ethanoate.

ethanoic acid, *acetic acid* (CH_3COOH) A colourless liquid CARBOXYLIC ACID with a pungent odour (characteristic of vinegar); boiling point 119°C; melting point 17°C. It is made by oxidation of ethanal or ethanol. Vinegar contains 5 per cent or more ethanoic acid and is produced by FERMENTATION. Ethanoic acid forms large ice-like crystals when solid, and when in this form is sometimes called 'glacial acetic acid'.

ethanol, *ethyl alcohol* (C_2H_5OH) The ALCOHOL found in alcoholic drinks such as beer, wines and spirits. Ethanol is a colourless liquid with a pleasant odour; boiling point 78°C. It is miscible with water and burns in air with a pale blue flame.

Ethanol is produced in nature by the FERMENTATION of CARBOHYDRATES by yeast cells or bacteria. In the past, pure ethanol was manufactured by the fermentation of sugar or STARCH by yeast enzymes, but it now comes mainly from ethene derived from petroleum. In Brazil, where petroleum is expensive, it is economically advantageous to produce ethanol by fermentation of sugar cane, for use as a motor fuel.

Ethanol can be oxidized to ethanal or ethanoic acid. It reacts with acids to give ESTERS and with chlorine to give trichloroethanal (chloral). ETHER is produced by the action of sulphuric acid on ethanol. In industry, ethanol is used as a solvent and in the manufacture of other chemicals (particularly ethanal). It is also used in foodstuffs.

ethanoylation, *acetylation* A process for incorporating the ethanoyl (acetyl) group, CH_3CO–, into organic compounds that contain the groups –OH, $–NH_2$ or –SH.

ethanoyl chloride, *acetyl chloride* (CH_3COCl) A colourless, liquid ACYL CHLORIDE with a strong smell; melting point –112°C; boiling point 51°C. It is used to introduce ETHANOYL GROUPS into organic compounds containing hydroxyl or amino groups; a process termed ETHANOYLATION.

ethanoyl group, *acetyl group* The group CH_3CO–.

ethene, *ethylene* (C_2H_4) A gaseous HYDROCARBON; melting point –169°C; boiling point –105°C. It is the first of the series of ALKENES, obtained from petroleum by CRACKING. As with the other alkenes, ethene reacts by ELECTROPHILIC ADDITION reactions, adding groups across the carbon-carbon double bond. It can be reduced by hydrogen to give ethane.

Ethene is widely used in industry, in the manufacture of plastics, detergents, paints and pharmaceuticals. POLY(ETHENE) and POLYVINYL CHLORIDE are examples of plastic products derived from ethene.

Ethene is also a by-product of plant METABOLISM. It stimulates the ripening of fruit and is therefore useful commercially as a spray to ripen fruit, such as tomatoes and grapes. Ethene also stimulates abscission (drop) of leaves, fruit and flowers and is used commercially to promote fruit loosening.

ether One of a series of organic chemicals in which an oxygen atom is inserted between the carbon atoms of two HYDROCARBON groups. An ether thus has the general formula ROR′ (where R is a hydrocarbon group) or $C_nH_{2n+2}O$, which is the same as the formula for ALCOHOLS except for the position of the oxygen atom. Ethers are less reactive than alcohols.

The nomenclature for the ethers is based on the ALKANE forming the longest carbon chain in the molecule but with the prefix *alkoxy-* indicating the RO– group attached, for example $CH_3OC_3H_7$ is 1-methoxypropane, $C_2H_5OC_2H_5$ is ethoxyethane. The latter was called diethyl ether, commonly termed ether, and is used as an anaesthetic, antiseptic and a solvent.

See also WILLIAMSON ETHER SYNTHESIS.

ethoxyethane *See* ETHER.

ethyl acetate *See* ETHYL ETHANOATE.

ethyl alcohol The common name for ETHANOL.

ethylamine, *monoethylamine* ($CH_3CH_2NH_2$) A compound derived from a primary AMINE in which one of the hydrogen atoms of ammonia has been replaced by an ETHYL GROUP. Two hydrogen atoms of ammonia can be replaced by ethyl groups and the resulting compound is termed diethylamine $(CH_3CH_2)_2NH$. If three are replaced the compound is called triethylamine $(CH_3CH_2)_3N$. These ethylamines are all colourless with a strong smell of ammonia and show the typical properties of ALIPHATIC amines.

ethylbenzene ($C_6H_5C_2H_5$) A colourless liquid used in the manufacture of POLYSTYRENE; melting point –95°C; boiling point 136°C. It can be made by the addition of ethene to benzene or by the FRIEDEL-CRAFTS REACTION between C_2H_5Cl and benzene.

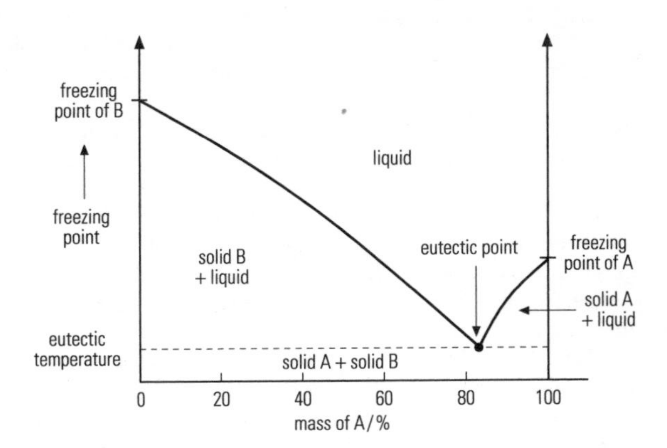

Eutectic mixture.

ethyl bromide *See* BROMOETHANE.

ethyl chloride *See* CHLOROETHANE.

ethylenediaminetetraacetic acid (EDTA) A CHELATING AGENT used in chemical analysis.

ethylene dibromide *See* 1,2-DIBROMOETHANE.

ethylene glycol, *glycol* The common name for ETHANE-1,2-DIOL.

ethyl ethanoate, *ethyl acetate, acetic ester, acetic ether* ($CH_3COOC_2H_5$) A colourless liquid ESTER with a fruity odour; boiling point 77°C. It is formed by the reaction of ethanol and ethanoic acid. It is an important solvent and is also used in cosmetics and as an artificial essence.

ethyl group In organic chemistry, the group C_2H_5, sometimes written as Et–.

ethyne, *acetylene* (C_2H_2) A colourless, inflammable gas; boiling point –84°C; melting point –82°C. The first member of the ALKYNE series of HYDROCARBONS, it is produced on mixing water with calcium carbide.

Ethyne was first used in early gas lamps. It gives off an intense heat on burning and is therefore used in some welding torches. It is used today as the starting point for the manufacture of a number of organic chemicals, such as PROPENONITRILE, VINYL CHLORIDE and ETHANOIC ACID.

eudiometer A vessel that measures changes in the volumes of gases during chemical reactions. From this, changes in the number of molecules present may be deduced. A simple eudiometer comprises a graduated glass tube closed at one end and sealed at the other by being immersed in mercury. A pair of ELECTRODES are provided, between which a spark can be formed to initiate a reaction.

europium (Eu) The element with atomic number 63; relative atomic mass 152.0; melting point 852°C; boiling point 1,529°C; relative density 5.2. Europium is a member of the LANTHANIDE series and is used to make PHOSPHORS for cathode ray tubes in colour television systems.

eutectic mixture A SOLID SOLUTION of two or more substances, with the lowest freezing point for any possible mixture of the components. The freezing point of a eutectic mixture is called the eutectic point. On a graph of temperature against composition for a mixture of two substances, which may be either solid or liquid, there are generally four regions. Below the eutectic point, the mixture is entirely solid. This temperature may be lower than the melting point of either substance alone. At higher temperatures, one substance may occur as a solid – generally the one present in higher proportions – together with a liquid that may be regarded as a saturated solution of the substance present as a solid. At higher temperatures, both substances will be liquid.

The FREEZING MIXTURE formed by adding sodium chloride to water is an example of a eutectic mixture. A lower liquid temperature can be achieved than would be achieved from either material alone. Not all mixtures form eutectics.

eutectic point The freezing point of of a EUTECTIC MIXTURE.

eutrophication Excessive enrichment of lakes and rivers by NITRATES and PHOSPHATES. Eutrophication can occur naturally or as a result of human activities. The natural accumulation of salts into a body of water is slow and is usually counter-balanced by loss of salts through natural drainage.

The more serious cause of eutrophication is artificial. Artificial eutrophication can be caused by the addition of nitrates and phosphates from fertilizers (washed from the soil by rain), sewage and detergents. This enrichment causes the growth of algal blooms, which prevent light reaching deeper regions of the water and so aquatic plants die because they are unable to photosynthesize. The dead plants and algae are decomposed by saprophytic bacteria (those that feed on dead organisms or excrement), which use all the available oxygen in the water, leading to the death of other species such as fish.

See also POLLUTION.

evaporation The process by which a liquid turns into a gas at a temperature below the boiling point as some of the faster moving molecules escape from the surface of the liquid. Unlike boiling, evaporation takes place only on the liquid surface. The rate of evaporation can be increased by heating, by blowing air over the liquid surface so molecules cannot re-enter the liquid, or by increasing the surface area. A liquid that evaporates easily is called volatile.

exciplex A state of two atoms that do not form a covalent bond in their ground state, but which do in an excited state. Xenon chloride, XeCl, is an example. Exciplex states are important in some lasers.

excited state A QUANTUM state, in an atom for example, with an energy higher than the lowest allowed energy. *See also* ENERGY LEVEL, GROUND STATE.

excluded volume In a gas, the volume taken up by the molecules themselves. *See* IDEAL GAS.

exothermic (*adj.*) Describing a chemical reaction in which heat is given out; that is, one in which the HEAT OF REACTION is negative. COMBUSTION processes are exothermic, for example:

$$H_2 + {}^1/_2O_2 \rightarrow H_2O, \Delta H = -289 \text{ kJ mol}^{-1}$$

experiment A set of measurements or observations, often performed on equipment designed specifically for this purpose, and designed to suggest or to test a theory. If the theory satisfies a sufficient range of experimental tests, it may then be used to predict what will happen in similar situations. If an experiment provides results that genuinely contradict the theory, the theory must be revised. In the design of any experiment, it is important to ensure that only those factors being studied can change, and that all other factors remain constant throughout the experiment.

explosimeter A device designed to measure the concentrations of explosive or inflammable gas. Current flows through a thin wire, which becomes hot. In the presence of inflammable gases, even in low concentrations, the OXIDATION of the gas heats the wire still further and increases its electrical resistance. By measuring the changes in resistance, the concentration of the gases may be determined.

explosion A violent and rapid COMBUSTION process in which the reaction accelerates to a rate that is limited only by the speed of movement of the molecules. Such rapid combustion produces a rise in temperature leading to an increase in pressure as the reaction products do not have time to escape the immediate vicinity of the explosion. This high pressure is responsible for the destructive power of explosive materials.

explosives Substances that react rapidly to produce a great increase in temperature and releasing one or more gases. It is the high pressure which this gas reaches in a confined space, or without having time to expand, which is responsible for the destructive power of explosives. Important explosives are gunpowder, nitroglycerine (often in the form of dynamite) and TNT.

extender An inert material added to a substance, such as washing powder, paint or glue, to increase its volume, and sometimes to alter its physical properties in a way that makes it easier to use.

extraction 1. The separation of a metal from its ore.

2. The separation of one substance from a MIXTURE. This may be carried out on the basis that one of the materials in the mixture will dissolve in a solvent that will not dissolve the others – this is called SOLVENT EXTRACTION.

F

face-centred cubic *See* CUBIC CLOSE PACKED.

FAD (flavine adenine dinucleotide) A derivative of the vitamin RIBOFLAVIN that acts as an electron carrier in the ELECTRON TRANSPORT SYSTEM. It is also a PROSTHETIC GROUP in some enzymes.

Fahrenheit A temperature scale obsolete in science but still in everyday use in the UK and US. The fixed points are taken as the freezing point of water, 32°F, and the boiling point of water, 212°F, both at a pressure of one ATMOSPHERE.

Fajan and Soddy's Group Displacement Law An EMPIRICAL law summarizing the changes in ATOMIC NUMBER that take place during ALPHA DECAY and BETA DECAY. In an alpha decay, the new element produced is two places to the left of the decaying element in the PERIODIC TABLE. In a beta decay, the element produced is one place to the right of the original element.

Fajan's rules A set of three rules indicating the circumstances in which an IONIC compound may show significant covalent characteristics. Bonding is likely to become covalent if: (i) the charge on the ions is high, tending to allow electrons to be transferred from the negative ion to the positive ion; (ii) the positive ion is small or the negative ion large, again favouring electron transfer; and (iii) the positive ion has an electron configuration that is not that of a noble gas, and so is not particularly stable.

Farad (F) The SI UNIT of electrical capacitance. One farad is the capacitance of a capacitor that stores one COULOMB of charge for each VOLT applied.

faraday A unit of CHARGE particularly used in the study of ELECTROLYSIS. One faraday is the charge on one MOLE of electrons, i.e. about 96,487 COULOMBS.

Faraday's constant (*F*) The amount of CHARGE in one FARADAY: 96,487 COULOMBS.

Faraday's laws of electrolysis Two laws governing the mass of substance dissolved or liberated from a solution in an ELECTROLYSIS. (i) The mass of the substance is proportional to the charge flowing. (ii) This mass is equal to the charge in FARADAYS multiplied by the number of electron charges (positive or negative) carried by the ion concerned multiplied by the relative atomic mass of the ion.

fat A LIPID mixture consisting of GLYCEROL and FATTY ACIDS, mostly TRIGLYCERIDES, which are solid at room temperature – unlike OILS, which are liquid fats. In many animals, fats are stored to provide an energy reserve and also to give some protection and insulation to the animal and its internal organs. Fats are an essential constituent of an animal diet, but too much fat has been linked to heart disease in humans.

fatty acid An organic compound (a CARBOXYLIC ACID) comprising a straight HYDROCARBON chain of up to 24 carbon atoms with a CARBOXYL GROUP (–COOH) at one end. Carbon atoms can be joined by double or single bonds. Where there is a DOUBLE BOND, only one hydrogen instead of two is carried by the carbon. If a fatty acid chain has a double bond, it is said to be unsaturated (polyunsaturates), for example, OLEIC and LINOLEIC ACIDS. Saturated fatty acids, for example, PALMITIC and STEARIC ACIDS, have a single bond between the carbon atoms. The carbon atoms therefore carry all the hydrogen atoms possible. The more double bonds the fatty acid chains contain, the lower the melting point of the fat, for example, oil has many double bonds and lard has none. Polyunsaturates (such as in margarine) are thought to be less likely to contribute to cardiovascular disease than saturated fats (butter has both). Fatty acids usually combine with GLYCEROL to form LIPIDS. *See also* PROSTAGLANDIN, SOAP.

fatty acid oxidation, *β-oxidation* The pathway by which fats are metabolized to release energy. Most energy comes from the breakdown of CARBOHYDRATES, but when these are exhausted, for example during starvation or dieting, fats and proteins are used. Fats are broken down to their constituents – GLYCEROL and FATTY ACIDS. Glycerol is phosphorylated and enters GLYCOLYSIS as glyceraldehyde-3-

phosphate. Fatty acids are progressively broken down into 2-carbon fragments, which combine with COENZYME A to form ACETYL COENZYME A, which then enters the KREBS CYCLE. This fatty acid oxidation occurs in living cells and is accompanied by the production of reduced NAD, which enters the ELECTRON TRANSPORT SYSTEM where it yields ATP.

f-block element A LANTHANIDE or ACTINIDE; that is, an element with outer electrons in an F-ORBITAL.

feedstock Any compound used as a starting point in an industrial chemical process.

Fehling's test A test used on organic substances to determine which are REDUCING AGENTS. It is usually used to detect REDUCING SUGARS and ALDEHYDES. The test involves heating the sample with a fresh solution of copper(II) sulphate, sodium hydroxide and sodium potassium tartrate. The presence of a reducing sugar is indicated by the production of a red precipitate. *See also* BENEDICT'S TEST.

feldspars A class of minerals containing aluminosilicate ions, $(Si,Al)O_4$, in a tetrahedral structure together with ions of potassium, sodium, calcium or barium. Feldspars are the most abundant minerals in the Earth's crust, and are the minerals from which GRANITE is composed.

fermentation The process by which sugars are broken down by bacteria or yeasts, in the absence of oxygen. This process is considered to be analogous to anaerobic respiration. In nature the role of fermentation is to remove the hydrogen ions formed at the end of GLYCOLYSIS to allow the process to continue, since in the absence of oxygen the hydrogen ions cannot be used further to yield energy. The products of fermentation can be alcohol and carbon dioxide (alcoholic fermentation) or LACTIC ACID (lactate fermentation).

The process of fermentation has been utilized by humans in the baking and brewing industries, in cheese and yoghurt manufacture and also in the production of some antibiotics. In brewing, the alcohol ethanol is the important product, and in baking the carbon dioxide is more important, the bubbles causing bread to rise.

fermium (Fm) The element with atomic number 100. It does not occur naturally, but has been synthesized in particle accelerator experiments. The longest lived ISOTOPE (fermium–257) has a HALF-LIFE of only 10 days – too short for the element to be commercially useful.

ferric (*adj.*) Describing a compound containing the Fe^{3+} ion, for example, ferric chloride, $FeCl_3$.

ferricyanide The old name for HEXACYANOFERRATE(III).

ferrite Any of a number of mixed oxides of the form $MO.Fe_2O_3$, where M is typically a FERROMAGNETIC element such as iron, cobalt or nickel. Ferrites have ceramic structures and are electrical insulators, but they are also ferromagnetic.

ferrocene, *di-π-cyclopentadienyl iron* $(C_{10}H_{10})Fe$ An orange/red solid; melting point 173°C; boiling point 249°C. It is a sandwich compound with the iron ion between two carbon rings. Ferrocene was the first of a class of compounds called METALLOCENES to be discovered. It is synthesized from $Na^+C_5H_5^-$ (cyclopentadienyl sodium, made from sodium and CYCLOPENTADIENE) and iron(III) chloride. Ferrocene can be oxidized to $(C_5H_5)_2Fe^+$, which is blue.

ferrocyanide Any compound containing the hexacyanoferrate(II) COMPLEX ION $[Fe(CN)_6]^{4-}$.

ferroelectric (*adj.*) Describing a substance that retains its electrical POLARIZATION when the polarizing electric field is removed. Many ferroelectric materials are also PIEZOELECTRIC.

ferromagnetic (*adj.*) Describing a strongly magnetic material, such as iron. In such materials the imbalance in angular momentum of the electrons in each atom causes the atom to behave like a tiny magnet. Neighbouring atoms line up to form domains, regions where all the atoms are pointing in the same direction.

fibreglass *See* GLASS-FIBRE REINFORCED PLASTIC.

fibrous protein An insoluble PROTEIN consisting of long, coiled strands or flat sheets. This structure confers strength and elasticity and so fibrous proteins provide structural roles. Examples include COLLAGEN, KERATIN, ACTIN and MYOSIN.

filler A solid, inert material incorporated into synthetic rubber or plastics to change their general properties or to dilute them for economy.

film badge A device that records the total level of IONIZING RADIATION to which it has been exposed. It contains a piece of photographic film covered by a number of 'filters' of different materials and thicknesses. Ionizing radiation has a similar effect on a photographic film to light. When processed, the badge gives information about the amount and type of radiation to which it has been exposed. Film badges are often worn by those who work with ionizing radiation to monitor the level of radiation they are receiving.

filter A fine POROUS material through which a liquid can pass, but solid particles cannot. Filters are widely used to remove solid particles from liquids.

filter funnel A cone-shaped funnel used to hold a piece of FILTER PAPER.

filter paper POROUS paper used as a FILTER in chemistry.

filtration The act of passing a material through a FILTER.

fine structure The splitting of a line in the absorption or emission SPECTRUM of an element into a number of closely spaced lines as a result of electron SPIN or by the vibrational or rotational energy levels of a molecule.

first law of thermodynamics When heat is supplied to an isolated system, the amount of heat energy, ΔQ, equals the increase in INTERNAL ENERGY, ΔU, plus the mechanical WORK, ΔW, done by the system:

$$\Delta Q = \Delta U + \Delta W$$

For a gas at constant pressure p, the work is done by a change in the volume of the gas, ΔV, so

$$\Delta Q = \Delta U + p\Delta V$$

This statement is equivalent to the LAW OF CONSERVATION OF ENERGY.

Fischer–Tropsch process A process for the manufacture of HYDROCARBONS from carbon monoxide and hydrogen. Hydrogen and carbon monoxide (*see* WATER GAS) are passed over a nickel or platinum catalyst at about 200°C, and the resulting hydrocarbon mixture is separated by FRACTIONAL DISTILLATION. The process was used in Germany towards the end of the Second World War when little crude oil was available, but it is not currently an economical alternative to the distillation of crude oil.

flame A luminous region of hot gas around a COMBUSTION PROCESS. The material in the flame is IONIZED and the colour of the flame is characteristic of the ions present.

flame-front The edge of a FLAME. In the case where the substances reacting to produce the flame are both gases, the flame front marks the boundary of the region where the reaction is taking place. If the flame front reaches the speed of sound in the gas, the COMBUSTION process becomes an EXPLOSION.

flame test A simple test for certain CATIONS. The material to be tested is dipped in concentrated hydrochloric acid, to form small quantities of the ionic chloride, which is usually the most VOLATILE salt. This is then vaporized in a BUNSEN BURNER flame. The electron arrangements of certain metals cause certain colours to be produced in the flame when they are ionized. Thus sodium produces an orange flame, whilst potassium is purple and copper blue-green.

flammable (*adj.*) Able to burn.

flash photolysis A technique for studying fast reactions involving FREE RADICALS in a gas. The reacting gas is held in a glass tube and a flash of light produces free radicals and triggers the reaction. By measuring the strength of absorption lines of various free radicals over time, it is possible to study the KINETICS of the reaction.

flash point For a VOLATILE flammable liquid, the temperature above which the VAPOUR PRESSURE of the liquid is high enough for the vapour to burn in air if it is ignited by a spark.

flask A round or cone-shape vessel with a narrow neck, used to hold liquid REAGENTS.

flavine adenine dinucleotide *See* FAD.

flint Hard, brittle mineral found in chalk and limestone. It consists of fine-grained quartz.

flocculent (*adj.*) Describing a PRECIPITATE with the appearance of wool floating in a solution.

flue A chimney or other opening through which the products of combustion, particularly from burning gases in air, can escape into the atmosphere.

fluid Any substance that can flow: a LIQUID or a GAS.

fluidization The process of blowing a gas, often air, through a powdered solid, such as coal, so it behaves like a fluid. Fluidized beds are layers of fluidized material, such as solid catalysts for some gas phase reactions and coal burnt in large furnaces. Fluidized coal may be delivered to the furnace along pipes from a nearby crushing mill.

fluorescein ($C_{20}H_{12}O_5$) An orange powder; melting point 315°C; decomposes on further heating. It dissolves in water to a yellow solution with a green fluorescence. It is used as a dye and as an ABSORPTION INDICATOR.

fluorescence An effect in which ultraviolet light is absorbed and then re-emitted immediately as visible light. This is used in whitening agents that are added to paper and some washing powders to produce a brighter appearance. *Compare* PHOSPHORESCENCE, where the effect takes place over a longer period of time.

fluoride Any BINARY COMPOUND containing fluorine. All elements except nitrogen and the NOBLE GASES readily form fluorides, due to the highly reactive nature of fluorine. Non-metallic fluorides are covalently bonded, but highly POLAR, due to the extreme ELECTRONEGATIVITY of fluorine. The metallic fluorides contain the F^- ion. In many countries, small quantities of the fluoride ion are added to drinking water to combat tooth decay.

fluorine (F) The element with atomic number 9; relative atomic mass 19.0; melting point –220°C; boiling point –188°C. Fluorine can be extracted from its main ore, fluorite, by electrolysis of the molten ore, which contains mostly calcium fluoride. The lightest and most reactive of the HALOGENS and the most electronegative (*see* ELECTRONEGATIVITY) of all elements, fluorine is too reactive to find much use as an element. However, fluorine compounds are widely used, for example PTFE (polytetrafluoroethylene) is an important PLASTIC.

fluorite A mineral form of calcium fluoride, CaF_2.

fluorocarbons Compounds containing carbon and fluorine, with the formula C_nF_{n+2}. They may be produced by replacing the hydrogen with fluorine in HYDROCARBONS. An example is fluoroethane, C_2F_6, formed by the fluorination of ethane:

$$C_2H_6 + 6F_2 \rightarrow C_2F_6 + 6HF$$

The stability of fluorocarbons makes them useful as fire-extinguishing gases and those with higher relative molecular mass form good lubricants. *See also* CHLOROFLUOROCARBONS.

flux A material added to a metal to remove impurities, for example the limestone added to iron ore in a BLAST FURNACE, or the materials used in soldering and brazing that react with metal oxides to form more volatile compounds.

fluxional molecule A molecule that undergoes rapid changes between various structural ISOMERS, so it is impossible to obtain a sample containing only a single isomer.

foam A COLLOID consisting of a gas suspended in a liquid.

folic acid A VITAMIN of the VITAMIN B COMPLEX. Its active form is tetrahydrofolic acid and it is concerned with NUCLEOPROTEIN synthesis and red blood cell formation. Lack of folic acid causes poor growth and anaemia. It occurs in liver and green leafy vegetables and is made by intestinal bacteria. It is often given to pregnant women to prevent anaemia and spina bifida.

food chain The transfer of energy and nutrients from primary producers (green plants and some bacteria), through a sequence of organisms (the consumers) in which each organism in the chain feeds on, and is dependant on, the one preceding it, which is then itself eaten by the succeeding organism. When the producers and consumers die, other organisms break down the dead material, enabling recycling of nutrients into the soil or atmosphere. The food chain is an over-simplification, since many organisms have more varied diets than it suggests.

forbidden band *See* BAND THEORY.

force constant In a model of a COVALENT BOND where the potential energy of the bond is given by

$$V = \tfrac{1}{2}k(x - x_0)^2$$

the force constant is k, where x is the separation between the two atoms at the ends of the bond and x_0 is the equilibrium value of x. In such a model, the force between the atoms is

$$F = k(x - x_0)$$

formaldehyde *See* METHANAL.

formalin A solution of METHANAL in water, used as a preservative for biological specimens.

formic acid See METHANOIC ACID.

formula A representation of the composition of a compound using symbols for the elements present. Subscripts are used to indicate the number of atoms. The way in which these symbols are used vary according to how much detail is conveyed. Thus ethane may be shown as an EMPIRICAL FORMULA, CH_3, which represents simply the ratio of carbon atoms to hydrogen; as a MOLECULAR FORMULA, C_2H_6, which shows how many atoms of each element are present in a molecule; or as a STRUCTURAL FORMULA, such as CH_3CH_3, which gives some indication of how the atoms are arranged.

fossil fuels Petroleum oil, coal and natural gas. Petroleum and natural gas are produced in SEDIMENTARY rocks from the decay of marine life. Coal forms as a result of similar geological processes compressing decayed forests. There is increasing pressure to reduce the consumption of fossil fuels as they are non-renewable. The burning of such fuels also releases carbon dioxide into the atmosphere, which is believed to be responsible for the greenhouse effect.

fractional crystallization A process for separating two substances with different solubilities in a solvent. A solution containing the two substances is cooled until crystals form, and this process is repeated several times. At each stage, the proportion of the more soluble material in the liquid phase (called the MOTHER LIQUOR) will increase whilst the crystals contain an increasing proportion of the less soluble material.

fractional distillation A process of DISTILLATION used to separate a mixture of liquids having similar boiling points, by repeated boiling and cooling. The liquid mixture is heated to boiling and the vapours rise into a fractionating column and condense to form a liquid again. The most volatile liquid rises first but then, on cooling, falls down the column and is reheated when it meets the vapours from the second most volatile liquid. This process occurs repeatedly and eventually a gradient forms in the fractionating column with the most volatile fractions at the top and the least at the bottom. The required fractions can be tapped off the column at the appropriate position. Fractional distillation is an important part in the commercial processing of crude oil.

fractionating column *See* FRACTIONAL DISTILLATION.

francium (Fr) The element with atomic number 87; melting point 27°C; boiling point 677°C. Only radioactive ISOTOPES are known, the longest lived of which (francium–223) has a HALF-LIFE of only 21 minutes. Minute amounts of francium occur in nature as a result of the decay of heavier elements.

Frasch process A technique for extracting SULPHUR from underground deposits. A hole is drilled to reach the deposit and three concentric pipes are inserted. Steam is passed down the outer pipe and compressed air down the central tube. The steam melts the sulphur, which is forced to the surface by the compressed air whilst being kept molten by the outer jacket of downward moving steam.

free electron An electron in a metal or semiconductor that is not bound to any single atom but is free to move, carrying its charge and energy through the material. The large numbers of free electrons in metals make them good conductors of electricity (charge flow) and heat (energy flow).

free energy A measure of the energy released or absorbed during a reversible process. The Gibbs free energy change, ΔG, in a reaction under constant temperature and pressure is defined as:

$$\Delta G = \Delta H - T\Delta S$$

where ΔH is the ENTHALPY change, ΔS is the change in ENTROPY and T is the ABSOLUTE TEMPERATURE. This quantity is useful since a chemical reaction will only proceed spontaneously if ΔG is negative. In other words, the heat released by the reaction, ΔH, must increase the entropy of the surroundings by a sufficient amount to compensate for any entropy decrease within the REAGENTS, or if the reaction is ENDOTHERMIC, the entropy of the reagents must increase sufficiently to compensate for the entropy decrease of the surroundings. If ΔG is positive, energy must be supplied to the reaction. In practice, reactions in which ΔG is small, either positive or negative, are neither impossible nor do they go to completion, but are regarded as REVERSIBLE REACTIONS.

The Helmholtz free energy change, ΔF, is defined by:

$$\Delta F = \Delta U - T\Delta S$$

where ΔU is the change in INTERNAL ENERGY. The Helmholtz free energy is a measure of the maximum work that may be done by a reversible process at constant temperature.

See also CHEMICAL POTENTIAL.

free expansion Any process by which a gas expands without doing any WORK against the surroundings, such as is the case where a gas expands freely into a vacuum. Most gases cool in these circumstances, as work is done against attractive INTERMOLECULAR FORCES.

Thus cooling in a free expansion is a method by which gases can be cooled sufficiently to turn them into liquids.

At higher pressures and temperatures, the repulsive part of the intermolecular forces is more effective and gases become warmer in a free expansion. With hydrogen and helium this happens at room temperature and pressure. Before liquification, such gases must be pre-cooled to a temperature called the inversion temperature, below which the gas cools in a free expansion.

free radical An atom or a group of covalently bonded atoms, with an unpaired VALENCE ELECTRON. Free radicals are highly reactive and most are capable of only a limited period of independent existence, sometimes only microseconds, before they recombine with other atoms. They are produced by HOMOLYTIC FISSION, in which a covalent bond is broken by the action of heat or light.

freeze drying A technique for removing the water from perishable products so they can be stored as a dry powder. It is widely used for food, but has also found some use with blood plasma in places where refrigerated storage is not possible. The material is cooled, typically with liquid nitrogen, and then the frozen material is placed in a vacuum so the trapped ice SUBLIMES.

freezing mixture A mixture of two or more components designed to produce a solution with a temperature below 0°C. The most common example is a mixture of common salt (sodium chloride), ice and water. The energy needed to dissolve the salt and that needed to melt the ice both come from the INTERNAL ENERGY of the molecules and the material cools. The ice continues to melt as a result of the EUTECTIC MIXTURE formed between the ice and salt, which is initially above its EUTECTIC POINT. The mixture will continue to cool until all the ice has melted, all the salt dissolved or else the eutectic temperature is reached.

Friedel–Crafts reaction In organic chemistry, a reaction used industrially in the manufacture of HYDROCARBONS and KETONES. The reaction involves the introduction of an ACYL GROUP or ALKYL GROUP to AROMATIC compounds by ELECTROPHILIC SUBSTITUTION. An example is the reaction between ACID CHLORIDES, RCOCl (where R is an alkyl group) and benzene which, in the presence of a LEWIS ACID (usually aluminium chloride, $AlCl_3$) produces ketones (phenylethanone in this example).

$$C_6H_5H + CH_3COCl \xrightarrow{AlCl_3} C_6H_5COCH_3 + HCl$$

This is Friedel–Crafts ACYLATION.

Friedel–Crafts ALKYLATION occurs with HALOGENOALKANES. For example the reaction between benzene and chloroethane in the presence of a Lewis acid produces ethylbenzene:

$$C_6H_5H + CH_3CH_2Cl \xrightarrow{AlCl_3} C_6H_5CH_2CH_3 + HCl$$

The Lewis acid is a catalyst and generates the ELECTROPHILE needed for the reaction, either an alkyl (R^+) or acyl (RCO^+) CATION.

frontier orbitals In a covalently bound molecule, the highest energy MOLECULAR ORBITAL that does contain electrons and the lowest energy molecular orbital that does not contain electrons. The energy difference between the two frontier orbitals is critically responsible for many of the spectroscopic properties of the molecule.

froth flotation A process for the separation of some ORES from waste soil by forming a froth of the mixture together with oil and water. Air is blown through the mixture and the less dense material is trapped in the bubbles of the froth whilst denser material sinks to the bottom.

fructose ($C_6H_{12}O_6$) A MONOSACCHARIDE that combines with GLUCOSE to form SUCROSE. Fructose contains a KETONE group and is therefore called a KETOSE sugar. Fructose exists in several ISOMERS, ocurring naturally in the D-form. It is sweeter than sucrose and is a component of fruit and nectar, which plants use to attract animals to assist in seed dispersal.

fuel Any material that burns to provide a source of heat or energy.

fuel cell Any device that generates electricity directly from chemical energy by the oxidation of a fuel, usually hydrogen. A number of forms of fuel cell have been used experimentally, but generally hydrogen gas is passed over one porous nickel electrode, whilst oxygen is passed over a second similar electrode, with the electrodes linked by an alkaline electrolyte. The nickel acts as a catalyst for the reactions;

the hydrogen electrode is the cathode and electrons are given up,

$$H_2 + OH^- \rightarrow 2H_2O + 2e^-$$

whilst at the anode the oxygen absorbs electrons and is reduced,

$$O_2 + 2H_2O + 4e^- \rightarrow 4OH^-$$

Whilst fuel cells offer a higher efficiency than conventional HEAT ENGINES, they are bulky and complex. They have only been used in specialist applications, such as manned space vehicles, where the production of drinking water as a by-product is an added advantage.

fullerene *See* BUCKMINSTERFULLERENE.

fuller's earth A naturally occurring porous form of CLAY that has the property of decolorizing fats and oils. It was traditionally used in the wool industry for cleaning woollen fibres, a process known as fulling.

fumaric acid ($C_4H_4O_4$) The trans form of BUTENEDIOIC ACID. It is a colourless crystalline compound that sublimes at 165°C. If heated to 230°C it is converted to maleic anhydride. Fumaric acid is an intermediate in the KREBS CYCLE. It is used in the manufacture of synthetic RESINS.

fuming sulphuric acid *See* OLEUM.

functional group A small group of atoms linked together in an arrangement that determines the chemical properties of the group. In organic compounds the functional group is attached to a carbon skeleton but it is the functional group that determines the REACTIVITY of the compound. Examples of functional groups include the CARBOXYL GROUP COOH, the AMINE group NH_2 and the HYDROXYL GROUP OH (ALCOHOLS).

fundamental constant Any quantity believed to have the same value throughout all space and time and which does not depend on the value of any other such quantity. Examples of fundamental constants include the charge and mass of an electron. *See also* BOLTZMANN CONSTANT, PLANCK'S CONSTANT, RYDBERG CONSTANT.

fundamental particle *See* ELEMENTARY PARTICLE.

furan (C_4H_4O) A colourless organic liquid; boiling point 32°C. The molecule has a five-membered ring structure consisting of four CH_2 groups and one oxygen. Furan can form RESINS under acidic conditions.

furanose A sugar with a five-sided ring structure containing four carbon atoms and one oxygen atom. Fructose is an example of a sugar that forms a furanose ring structure. It can exist as straight chains or carbon atom 2 can link with the oxygen on carbon atom 5 to form a furanose ring. Furanose rings are less stable than six-sided PYRANOSE ring structures.

G

gadolinium (Gd) The element with atomic number 64; relative atomic mass 157.3; melting point 1,312°C; boiling point 3,273°C; relative density 7.9. Gadolinium is a LANTHANIDE metal and occurs only in small quantities in the Earth's crust. It has some uses in the manufacture of semiconductors and of high temperature and FERROMAGNETIC alloys.

galactose ($C_6H_{12}O_6$) A MONOSACCHARIDE that is an ISOMER of GLUCOSE. It is an ALDOSE sugar containing an ALDEHYDE group and is chemically similar to glucose. Galactose combines with glucose to form LACTOSE and occurs naturally in the DEXTROROTATORY form as lactose in milk sugar and in the LAEVOROTATORY form as a constituent of AGAR and other seaweed POLYSACCHARIDES.

galena A mineral form of LEAD SULPHIDE, PbS. Galena occurs as grey cubic crystals and is the chief ore of lead.

gallium (Ga) The element with atomic number 31; relative atomic mass 69.7; melting point 30°C; boiling point 2,403°C; relative density 5.9. Gallium is a soft metal, widely used in the electronics industry in the manufacture of gallium arsenide, used in high speed semiconductors for its high electron mobility.

galvanizing A way of protecting steel articles from corrosion by dipping them in molten zinc, which then acts as a SACRIFICIAL CATHODE. Galvanized materials are easily recognized by the characteristic patterns formed by the crystals of zinc produced as the coating solidifies.

gamma-BHC *See* BHC.

gamma radiation, *gamma ray* High energy, short wavelength ELECTROMAGNETIC RADIATION, emitted as an atomic nucleus rearranges itself into a lower energy state after an ALPHA DECAY or BETA DECAY. Most alpha decays and many beta decays produce gamma rays.

Gamma radiation is only weakly IONIZING, so has a very long range in air, effectively falling off in an INVERSE SQUARE LAW as the radiation spreads out. A dense material such as lead will provide some reduction in the intensity of gamma rays, provided a thickness of several centimetres is used. The intensity of gamma radiation falls off exponentially in an absorbing material, at a rate that depends on the nature of the absorbing material and on the distance from the gamma ray source. The thickness of a given material needed to reduce the level of radiation by a factor of one half is called the HALF-THICKNESS.

gamma ray *See* GAMMA RADIATION.

gangue The waste rocky material that is found mixed in with a metal ore.

garnet A class of SILICATE minerals, sometimes occurring as gemstones, all of which have the form $A_3B_2(SiO4)_3$, where A is a DIVALENT metal and B is a TRIVALENT metal, usually aluminium.

gas The state of matter in which a substance will expand to fill its container. In gases, the molecules are much more widely spaced than they are in solids, so the forces between them are much weaker; thus the density of a gas is much less than for a solid.

At high temperatures and pressures, the distinction between liquid and gas can disappear. The temperature above which this happens is called the CRITICAL TEMPERATURE. Oxygen is an example of a substance that is above its critical point at room temperature and cannot be turned into a liquid simply by compressing it – such gases are called PERMANENT GASES.

See also BOYLE'S LAW, CHARLES' LAW, DALTON'S LAW OF PARTIAL PRESSURES, IDEAL GAS, IDEAL GAS EQUATION, KINETIC THEORY, PRESSURE LAW, TRANSPORT COEFFICIENT.

gas chromatography CHROMATOGRAPHY in which the moving phase is a gas. Often this is a vaporized volatile sample, carried along by hydrogen, passed over an oily HYDROCARBON on a solid support. This technique is known as gas liquid chromatography (GLC). The rate at which the materials in the sample pass along the solid phase varies, and their time of arrival at the far end of the absorbing medium is measured, typically by a hot wire that detects changes in the thermal conductivity of the gas.

gas constant *See* MOLAR GAS CONSTANT.

gas discharge The flow of electric current through a gas, often at reduced pressure. The electric field must be strong enough to accelerate ions rapidly enough for them to create further ionization when they collide with gas molecules; thus gas discharges generally occur only at relatively high voltages. As the ions recombine with electrons, light is given out, with a colour characteristic of the gas used.

gas jar A thick walled cylindrical glass container open at one end, used for the collection of gases produced in reactions. The open end of a gas jar is ground flat, so a disc of glass coated with grease can be used to cover the jar and prevent the gas escaping.

gas laws The three laws, BOYLE'S LAW, CHARLES' LAW and the PRESSURE LAW, that between them describe the properties of IDEAL GASES. They contain the same information as is contained in the IDEAL GAS EQUATION, but relate only to a fixed mass of gas, i.e. a fixed number of molecules.

gas liquid chromatography *See* GAS CHROMATOGRAPHY.

gasoline The term for PETROL in the USA.

Gatterman reaction A variation of the SANDMEYER REACTION for the synthesis of bromo and chloroarenes (for example C_6H_5Cl). A DIAZONIUM SALT is reacted with hydrogen halide and copper powder (the catalyst) and a halogen is introduced into the aromatic ring in place of an amino group. The reaction is named for the German chemist Ludwig Gatterman (1860–1920), who discovered it in 1890.

Gay–Lussac's law In any reaction where the REAGENTS and the products are all gases, the ratios of the volumes of the reagents and the products will be simple whole numbers, provided the temperature and pressure remain constant. For example, in the formation of ammonia:

$$N_2 + 3H_2 \rightarrow 2NH_3$$

1 volume of nitrogen combines with 3 of hydrogen to form 2 volumes of ammonia. Gay–Lussac's law was originally an EMPIRICAL law, but can now be understood in terms of AVOGADRO'S HYPOTHESIS and the molecular nature of chemical reactions.

gel (*n.*), ***gelatinous*** (*adj.*) A substance having some of the stiffness of a solid when subject to small deforming forces, but able to flow like a liquid under the influence of larger forces. Gels are formed as an intermediate stage in the coagulation of some COLLOIDS, with the particles in the colloids joining up to form long strands. *See also* THIXTROPIC.

gelatin(e) One of several glutinous substances obtained by boiling COLLAGEN from the skin, bones and connective tissue of animals in dilute acid. Gelatins swell in hot water to form a viscous solution, which sets on cooling to form a jelly. Gelatins are widely used in the food industry and in photography. Gelatin is rich in the amino acids GLYCINE, PROLINE and LYSINE.

gelatinous *See* GEL.

gel filtration A form of CHROMATOGRAPHY in which the mixture to be separated diffuses down a column containing a GEL. The smaller molecules are better able to penetrate the spaces between the solid particles in the gel and so move more rapidly down the column. The technique is particularly useful for large organic molecules, such as PROTEINS.

gene The basic unit of inheritance encoded by a specific length of DNA controlling one particular function or characteristic, for example eye colour. It was thought that one gene encoded one ENZYME, but it is now known that this is not always the case, since proteins are made up of several POLYPEPTIDES each encoded by a separate gene. Today, a gene is considered to be a length of DNA encoding any molecular cell product.

genetic engineering The deliberate biochemical manipulation of GENES, or DNA itself, so that they are spliced (divided), altered, added or removed or transferred from one organism to another. Such engineering has enabled large-scale production of useful chemicals such as ANTIBIOTICS and INSULIN. The manipulation of DNA involves cutting and rejoining it with special ENZYMES called RESTRICTION ENDONUCLEASES and DNA LIGASE respectively.

Plants and bacteria can be modified by genetic engineering to achieve practical ends, for example production of drugs by bacteria, better growth of plants or disease resistance. In the future it is likely that some genetic diseases will be curable by genetic engineering.

geochemistry The study of the chemistry of the Earth. It includes the study of the abundance and distribution of the naturally occurring elements and their ISOTOPES.

geology The study of the solid part of the Earth. The Earth contains a central core of molten iron and nickel, surrounded by a layer of molten rock called magma. This forms a region called the mantle. On top of this lies the Earth's crust, or lithosphere, only a few hundred kilometres thick compared to the Earth's radius of 6,400 km. The crust is made up of a number of distinct regions called tectonic plates.

In addition to the IGNEOUS rock that has solidified from the core, much of the rock found on the Earth's surface is SEDIMENTARY. Some sedimentary rocks are deposited by the processes of erosion, whereby wind or water breaks down rock in one place and deposits it elsewhere to form shale or sandstone (depending on the size of the particles involved).

Other sedimentary rocks, such as LIMESTONE, are formed from the shells of dead marine life. Pressure then converts such material into rock, though it is generally softer than igneous rocks such as granite. Folding of such rocks by plate tectonic activity can lead to them experiencing greater heat and pressure, forming harder rocks, which are described as METAMORPHIC.

geometric isomerism, *cis/trans isomerism* ISOMERS that contain the same groups joined together in the same order (unlike structural isomers) and differ only in the geometry of their bonds. This occurs in compounds with a DOUBLE BOND or with certain ring structures, where there is restricted rotation of the atoms joined by the bonds, thereby giving rise to more than one possible geometric arrangement. Geometric isomers differ in their chemical reactions.

Geometric isomerism often occurs where there is a >C=C< double bond or a carbon ring. In these cases the terms *cis* and *trans* are used to distinguish the isomers. These derive from the Latin meaning 'on this side of' and 'across' respectively, so that the prefix *cis* is applied when the groups at the end of a double bond are on the same side of the plane of the bond. The term *trans* is used when similar groups are on opposite sides of the plane of the double bond:

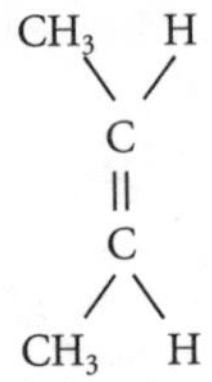

```
CH3   H        CH3   H
  \  /           \  /
   C              C
   ||             ||
   C              C
  /  \           /  \
CH3   H         H   CH3
```

cis-but-2-ene 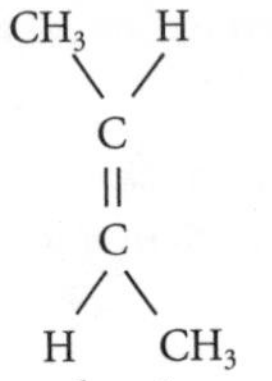   *trans*-but-2-ene

Geometric isomerism also occurs where there are >C=N– or –N=N– double bonds and then the prefixes *syn-* and *anti-* are used.

geraniol ($C_{10}H_{17}OH$) An ALCOHOL that occurs naturally in several ESSENTIAL OILS; boiling point 107°C.

germanium (Ge) The element with atomic number 32; relative atomic mass 72.6; melting point 937°C; boiling point 2,830°C; relative density 5.4. Its existence was predicted from a gap in Mendeleyev's PERIODIC TABLE, and the element was discovered in 1886.

Germanium is a semiconductor and was much used in early semiconductor devices. It has a relatively small BAND GAP, which means that a fairly large number of CHARGE CARRIERS are produced by thermal vibrations unless it is kept at a low temperature. This has lead to its replacement by silicon in many electronic applications, though it is still used for some specialized applications.

getter A device for removing small amounts of gas from vacuum vessels in order to improve the vacuum. A getter consists of a wire made of a reactive metal, such as magnesium. The vessel is evacuated and sealed and the getter vaporized by passing a current through it. The metal atoms react with any air molecules remaining in the vessel to form a non-volatile OXIDE or NITRIDE.

GFRP *See* GLASS-FIBRE REINFORCED PLASTIC.

Gibbs free energy *See* FREE ENERGY.

gibbsite A mineral form of ALUMINIUM HYDROXIDE, often found in association with BAUXITE.

glass Any of a number of transparent brittle materials. Common glass is made by melting together sand (mostly silicon dioxide) with lime (calcium oxide) and soda (mostly sodium carbonate). AMORPHOUS materials are sometimes referred to as glasses, as glass is an example of an amorphous material. In some ways glasses are more like liquids than solids, they have no melting point, but simply flow more readily as they are heated. *See also* DISORDERED SOLID.

glass electrode A device used in the measurement of pH. It comprises a thin glass membrane surrounding a platinum electrode in an acidic electrolyte. When placed in a solution, hydrogen ions will diffuse through the glass, producing an ELECTRODE POTENTIAL between the electrode inside the glass and a second electrode placed in the solution. This potential depends on the relative concentration of hydrogen ions in the two liquids, enabling the pH of the solution to be measured.

glass-fibre reinforced plastic (GFRP) A COMPOSITE MATERIAL consisting of glass fibres embedded in a plastic RESIN. Glass is very strong for its density, but cannot be used alone as a structural material since it is brittle – cracks travel rapidly through the glass resulting in a catastrophic failure. The plastic resin is tough and prevents the spread of cracks from one fibre to the next.

Glauber's salt An old name for the crystalline form of SODIUM SULPHATE, $Na_2SO_4.10H_2O$, particularly when taken medicinally as a laxative.

global warming *See* GREENHOUSE EFFECT.

globular protein Any of a range of PROTEINS that are compact and rounded in structure and usually soluble in water. These proteins provide a metabolic role rather than the structural role provided by the FIBROUS PROTEINS. ENZYMES are globular proteins, as are certain HORMONES such as INSULIN. Other examples include CASEIN, HAEMOGLOBIN and the ANTIBODIES.

globulin Any one of a group of GLOBULAR PROTEINS that are insoluble in water, soluble in weak salt solutions and have a metabolic rather than structural function. In animals, globulins are found in blood, eggs and milk. Globulins are also found in plant seeds.

glove box A device for the manipulation of dangerous materials, particularly those that are radioactive or biologically hazardous. A sealed box is maintained at a pressure slightly below atmospheric, to prevent any leakage of material out of the box, and the material inside the box is handled by use of rubber gloves that are sealed in holes in one wall of the box.

glucagon A POLYPEPTIDE hormone (29 AMINO ACIDS) produced by the pancreas and involved in the regulation of blood sugar levels. Glucagon converts GLYCOGEN in the liver to GLUCOSE in response to a drop in the concentration of glucose in the blood. This process is called GLYCOGENOLYSIS. Glucagon acts antagonistically with INSULIN.

glucocorticoid Any one of a group of CORTICOSTEROID hormones secreted by the adrenal gland and concerned with GLUCOSE metabolism. They include CORTISOL and CORTISONE. Glucocorticoids raise blood pressure and blood sugar levels, the latter by increasing formation of glucose from fat and proteins. They also inhibit INSULIN by increasing the rate of GLYCOGEN formation in the liver. Glucocorticoids are effective in reducing inflammation and are thus used in the treatment of certain conditions.

gluconeogenesis The conversion of PROTEINS and FATS into GLUCOSE by the liver. *See also* GLYCOGENESIS, GLYCOGENOLYSIS.

gluconic acid ($C_6H_{12}O_7$) The CARBOXYLIC ACID corresponding to the sugar GLUCOSE. It forms colourless crystals, melting point 125°C, and is soluble in water and alcohol. It is made by certain moulds and bacteria and can be manufactured by the oxidation of glucose.

glucose, *dextrose* ($C_6H_{12}O_6$) A MONOSACCHARIDE, the most common HEXOSE sugar. Glucose is found in plants and in the blood of animals, where it is a source of energy for the body, being used to generate ATP. Glucose is optically active (*see* OPTICAL ACTIVITY) and exists in several forms. Naturally occurring glucose is DEXTROROTATORY. It is a constituent of many POLYSACCHARIDES, including STARCH, CELLULOSE and GLYCOGEN, and is made by the HYDROLYSIS of starch or SUCROSE. Levels of glucose in the blood are regulated by the liver and the pancreatic hormones GLUCAGON and INSULIN. *See also* GLUCONEOGENESIS, GLYCOGENESIS, GLYCOGENOLYSIS.

glucoside A GLYCOSIDE in which the sugar residue is GLUCOSE.

glucuronic acid ($C_6H_{10}O_7$) A compound derived from the oxidation of GLUCOSE. Glucuronic acid can combine with hydroxyl (–OH), amino (–NH_2) or carboxyl (–COOH) groups to form a glucuronide. In animals, glucuronides can combine with toxic substances to increase their solubility. This process, termed glucuronidation, has an important role in the excretion of toxic substances in urine. Glucuronic acid is a constituent of plant gums.

glucuronide *See* GLUCURONIC ACID.

glutamic acid ($C_5H_9NO_4$) A non-essential acidic AMINO ACID, commonly called glutamate to indicate its negative charge at physiological pH. GLUTAMINE is the uncharged derivative.

Glutamic acid is found widely in plant and animal tissue. It is important in nitrogen metabolism and also acts as a NEUROTRANSMITTER. The sodium salt, MONOSODIUM GLUTAMATE, is widely used in the food industry.

glutamine ($C_5H_{20}N_2O_3$) A non-essential AMINO ACID that is an uncharged derivative of GLUTAMIC ACID. It is found in plant and animal tissue and used commercially in medicine and research.

glycerate-3-phosphate ($CHOCH(OH)CH_2O-PO_3H_2$) A phosphorylated 3-carbon compound that is produced as an intermediate in GLYCOLYSIS and in the CALVIN CYCLE of PHOTOSYNTHESIS.

glyceride Any LIPID formed by the combination of a FATTY ACID with GLYCEROL. Glycerides can be mono, di or tri, depending on how many of the three HYDROXYL GROUPS on glycerol have combined with the fatty acids. If one of the hydroxyl groups on glycerol combines with a PHOSPHATE group then a PHOSPHOLIPID is formed. Combination with a sugar results in a GLYCOLIPID. TRIGLYCERIDES are found naturally as the main constituents of fats and oils. *See also* ESTER, LIPID.

glycine ($C_2H_5NO_2$) The simplest AMINO ACID and the main one in sugar cane. Glycine is a sweet, crystalline amino acid, prepared commercially from GELATIN for use in medicine and research.

glycerine *See* GLYCEROL.

glycerol, *glycerine, propan-1,2,3-triol* ($HOCH_2CH(OH)CH_2OH$) A trihydric ALCOHOL that is a sweet, colourless liquid extracted from animal and vegetable oils. Glycerol reacts with FATTY ACIDS to form LIPIDS. It is used in ANTIFREEZE solutions, explosives and cosmetics.

glycogen A POLYSACCHARIDE (made of branched chains) of the sugar GLUCOSE. It is stored in the liver as a CARBOHYDRATE source until needed. Glycogen is the animal equivalent of STARCH. It is converted back to glucose by the pancreatic hormone GLUCAGON (a process known as GLYCOGENOLYSIS), and glucose is converted to glycogen by INSULIN (GLYCOGENESIS). These two hormones therefore act antagonistically to regulate blood sugar levels.

glycogenesis The conversion of GLUCOSE in the blood to GLYCOGEN by the liver. Glycogen is stored until needed (when blood glucose levels fall). *See also* GLUCONEOGENESIS, GLYCOGENOLYSIS.

glycogenolysis The breakdown of GLYCOGEN in the liver to GLUCOSE by GLUCAGON. *See also* GLUCONEOGENESIS, GLYCOGENESIS.

glycol *See* ETHANE-1,2-DIOL.

glycolipid A LIPID that is covalently linked to a CARBOHYDRATE. Glycolipids are found in biological cell membranes. There is a wide variation in the composition and complexity of glycolipids.

glycolysis (*Glycol* = sugar, *Lysol* = breakdown) The series of reactions resulting in the breakdown of the six carbon sugar GLUCOSE to two molecules of the three carbon compound pyruvate, with the release of energy in the form of ATP. The process requires no oxygen and is the only form of respiration, and therefore ATP synthesis, in anaerobic organisms. In aerobic organisms, glycolysis is the first stage of respiration and the pyruvate formed then enters the KREBS CYCLE, which does require oxygen. Glycolysis occurs in living cells. There is an overall gain of two molecules of ATP for each glucose molecule broken down as well as two pairs of hydrogen atoms. The hydrogen atoms go into the ELECTRON TRANSPORT SYSTEM to generate six further ATPs if oxygen is present, or are removed by the FERMENTATION process if no oxygen is present.

glycoprotein A protein covalently linked to a CARBOHYDRATE. Glycoproteins are important constituents of biological cell membranes and of body fluids. Glycoproteins occur on the surface of cells, acting as receptors, for example, for hormones. The protein can form the bulk of the molecule, as in cell-surface glycoproteins, or the carbohydrate can represent the major part. The addition of sugar residues to proteins is called glycosylation.

glycosaminoglycan, *mucopolysaccharide* Any of a group of POLYSACCHARIDES containing amino sugars (such as glucosamine). They have gel-like properties since their structure allows them to trap water. Examples include HYALURONIDASE and CHONDROITIN.

glycoside Any one of a group of sugar derivatives that are linked to a non-sugar residue (R) by a GLYCOSIDIC BOND. The hydroxyl group (–OH) on carbon 1 of the sugar is replaced by (–OR). Glycosides in which the sugar is GLUCOSE are termed GLUCOSIDES. The simplest is methylglucoside $C_6H_{11}O_5–O–CH_3$. Glycosides are commonly found in plants and have various functions. Examples include the group of plant pigments ANTHOCYANIN and several compounds, termed cardiac glycosides, that are used as heart stimulants.

glycosidic bond The bond that forms between two MONOSACCHARIDE units when they join to give a DISACCHARIDE or POLYSACCHARIDE. The bond usually forms between carbon atom 1 of one monosaccharide and carbon atom 4 of the other monosaccharide and is therefore called a 1-4 glycosidic bond. A glycosidic bond is formed by the removal of a water molecule and is therefore a CONDENSATION REACTION. Any two monosaccharides can link in this way but MALTOSE, LACTOSE and SUCROSE are the most common.

An α-glycosidic bond forms when the –OH group on carbon atom 1 is below the plane of the monosaccharide ring and a β-glycosidic bond when the –OH group is above the plane of the ring. For example, CELLULOSE consists of glucose molecules linked by 1-4 β-glycosidic bonds, whereas STARCH is glucose molecules linked by 1-4 α-glycosidic bonds.

glycosylation The addition of sugar residues to protein. *See* GLYCOPROTEIN.

gold (Au) The element with atomic number 79; relative atomic mass 197.0; melting point 1,064°C; boiling point 2,807°C; relative density 19.3. Gold is a TRANSITION METAL with a characteristic yellow colour. The low reactivity of gold means that it does not tarnish, so it is widely used in jewellery and for electrical contacts. For jewellery it is alloyed with other metals such as silver and copper, the purity of the gold being measured in carats on a scale from 0 (no gold) to 24 (pure gold).

Goldschmidt process The extraction of metals from their oxides by using powdered aluminium as a REDUCING AGENT, for example:

$$Cr_2O_3 + 2Al \rightarrow Al_2O_3 + 2Cr$$

Such reactions are highly EXOTHERMIC and produce the metal in its molten form. When carried out with iron, this reaction is called the THERMIT PROCESS.

Graham's law of diffusion The rate of DIFFUSION of a gas is inversely proportional to the molecular mass. Thus light gases, such as hydrogen, diffuse more quickly than heavier ones, such as carbon dioxide.

grain boundary In a POLYCRYSTALLINE material, the boundary between one orientation of the crystal lattice and another.

gram A unit of mass, nowadays defined as one thousandth of a KILOGRAM.

granite A hard, IGNEOUS rock.

graphite An ALLOTROPE of carbon in which carbon atoms are arranged in a hexagonal pattern held together by COVALENT BONDS, similar to those found in AROMATIC compounds. The individual layers are held together by much weaker VAN DER WAALS' bonds, which accounts for the flaky nature of graphite. The delocalized nature of the covalent bonding means that graphite conducts electricity along the layers of atoms, but not from one layer to the next. Graphite is also a good conductor of heat.

Graphite is used as the 'lead' in pencils, in electrical contacts, as a solid lubricant and as the moderator in some nuclear power stations.

gravimetric analysis Any technique of QUANTITATIVE ANALYSIS that involves measuring the mass of substance in a reaction. For example, the amount of barium in a sample could be measured by dissolving the sample in hydrochloric acid, and then adding sulphuric acid to precipitate the barium as barium sulphate, which is dried and weighed.

gray (Gy) The SI UNIT of absorbed IONIZING RADIATION. One gray is equal to an energy of one JOULE absorbed from the radiation. *See also* DOSE.

greenhouse effect A term used to describe the trapping of solar radiation in the Earth's atmosphere by various gases, creating a rise in the Earth's temperature, similar to the heat reflection in greenhouse glass. If it were not for the greenhouse effect, the surface temperature of the Earth could not sustain life. However, in recent years the greenhouse effect has become more marked, leading to an increase in the average global temperature. The main greenhouse gases building up and causing this so-called global warming are carbon dioxide (due to FOSSIL FUEL consumption and destruction of forests), methane (a by-product from agriculture), water vapour and CHLOROFLUOROCARBONS (CFCs; from refrigerators, aerosol sprays and POLYSTYRENE). Sustained global warming could eventually cause melting of the polar ice caps, a rise in sea levels and consequent flooding of low-lying land. A change in the climate would also affect crop growth and animal population. It is not clear exactly what consequences an increase in the greenhouse effect will have in the future since many effects oppose one another but the issue causes great concern to the future of humankind.

greenhouse gas Any gas, such as carbon dioxide or methane, which contributes to the GREENHOUSE EFFECT in the Earth's atmosphere by absorbing infrared radiation.

greenockite A mineral form of cadmium sulphide, CdS.

Grignard reagents A group of ORGANOMETALLIC COMPOUNDS of the type RMgHal where R is an ALKYL GROUP and Hal a HALIDE. Grignard reagents are formed by the reduction of HALOGENOALKANES with magnesium in the presence of dry ETHER. They are used in organic synthesis to form carbon-carbon bonds, thereby building up carbon skeletons. They add to the CARBONYL GROUP of ALDEHYDES and KETONES to give secondary and tertiary ALCOHOLS respectively. Their reaction with carbon dioxide yields CARBOXYLIC ACID and AMIDES and NITRILES give ketones. An example of a Grignard reagent is ethylmagnesium bromide, C_2H_5MgBr. Grignard reagents are named for the French chemist Victor Grignard (1871–1935).

Grotthus–Draper Law For reactions influenced by light, only light of wavelengths absorbed by one of the reagents can have any effect on the reaction. In modern chemistry, this is understood in terms of the photons of light being absorbed and creating an excited state, or dissociating a molecule, and this influencing the reaction.

ground state The lowest energy state of a system, such as an atom, from which it can be excited to higher energy states. *See* ENERGY LEVEL. *See also* EXCITED STATE.

Group A column within the PERIODIC TABLE, containing elements with similar chemical properties.

group displacement law *See* FAJAN AND SODDY'S GROUP DISPLACEMENT LAW.

growth hormone, *somatotrophin* A protein HORMONE produced by the pituitary gland that controls body metabolism and growth generally. Low levels in humans results in dwarfism and high levels in gigantism.

guanine, *6-hydroxy-2-aminopurine* ($C_5H_5N_5O$) An organic base called a PURINE that occurs in NUCLEOTIDES. *See also* DNA, RNA.

gum Any of a variety of sticky substances produced from a plant. Typically, gums are obtained from a cut in the bark of certain trees. Gums are generally insoluble in organic solvents, and form COLLOIDS with water. The term 'gum' is also used more generally of other sticky materials, such as those produced by the POLYMERIZATION of some partially oxidized HYDROCARBONS, which can cause problems in internal combustion engines.

gunpowder An EXPLOSIVE made by grinding together 75 per cent potassium nitrate, 15 per cent carbon, in the form of charcoal, and 10 per cent sulphur. The potassium nitrate acts as an OXIDIZING AGENT and large amounts of gas are produced as it oxidizes the carbon and sulphur.

gypsum A mineral form of calcium sulphate, consisting mostly of the DIHYDRATE $CaSO_4.2H_2O$.

H

Haber–Bosch process *See* HABER PROCESS.

Haber process, *Haber–Bosch process* An industrial process for the manufacture of AMMONIA from nitrogen and hydrogen:

$$3H_2 + N_2 \Leftrightarrow 2NH_3$$

The reaction is highly reversible and under the conditions generally used the yield is only a few per cent.

A balance has to be struck between the higher yield at low temperatures and the unacceptably long time taken to approach this yield even in the presence of a CATALYST. In practice, a temperature of about 450°C is used, at a pressure of 250 atmospheres in the presence of an iron catalyst. The ammonia is extracted from the unreacted gases by cooling it below its boiling point and the unreacted nitrogen and hydrogen are recycled.

haem ($C_{34}H_{32}FeN_4O_4$) A complex organic molecule containing iron, which combines with proteins to form HAEMOGLOBIN, MYOGLOBIN and CYTOCHROME. The haem group serves as a PROSTHETIC GROUP. Haem is a PORPHYRIN consisting of four nitrogen-containing rings chelated (*see* CHELATING AGENT) to a central iron(II) atom. The iron can bind oxygen (as it does in the respiratory pigments haemoglobin and myoglobin) or conduct electrons (as it does in the series of CYTOCHROMES in the ELECTRON TRANSPORT SYSTEM).

haematite An iron ore containing iron(III) oxide, Fe_2O_3.

haemoglobin In vertebrates and some invertebrates, a GLOBULAR PROTEIN found in the red blood cells that is used to transport oxygen from the lungs to the body tissues. It comprises a HAEM group, which contains iron, and GLOBULIN (a protein). In humans it has a RELATIVE MOLECULAR MASS of 68,000. In other species, the relative molecular mass ranges from 16,000 to 3 million. The haem group is always the same, but the globulin varies from species to species. In humans, haemoglobin consists of four polypeptide chains each with a haem group (this also varies between species), each able to carry one oxygen molecule, so the amount of oxygen that can be carried in the blood is increased compared to species with fewer haem groups.

Haemoglobin combines easily with oxygen in regions of high oxygen concentration (e.g. in the lungs or gills) to form OXYHAEMOGLOBIN (which is bright red compared with the red colour of haemoglobin alone) and is transported in the blood to the body tissues, where it is easily released where oxygen is at low concentration.

Foetal haemoglobin differs slightly from that of an adult and combines with oxygen even more easily, which allows it to obtain oxygen from the mother's haemoglobin in the placenta.

The carbon dioxide concentration affects the release and uptake of oxygen by haemoglobin. Some carbon dioxide is carried by haemoglobin as CARBAMINOHAEMOGLOBIN. Haemoglobin can also combine with carbon monoxide to form CARBOXYHAEMOGLOBIN, but this reaction is irreversible and results in death if too much carbon monoxide is inhaled.

See also MYOGLOBIN.

hafnium (Hf) The element with atomic number 72; relative atomic mass 178.5; melting point 2,230°C; boiling point 4,602°C; relative density 13.3. Hafnium is a TRANSITION METAL. It has found some use in the nuclear industry and in some high temperature alloys.

hahnium The name proposed for the element with atomic number 105. Only very short-lived ISOTOPES exist and very little is know about the properties of this element.

half-cell A single ELECTRODE immersed in an ELECTROLYTE. Two half-cells, connected together by a salt bridge can be used for the comparison of ELECTRODE POTENTIALS. *See also* REDOX HALF-CELL.

half-life The time taken for one half of the radioactive nuclei originally present in a sample to decay (*see* RADIOACTIVITY). The half-life τ is related to the DECAY CONSTANT λ by the formula

$$\tau = \ln 2/\lambda$$

half-thickness The thickness of a given material needed to reduce the intensity of GAMMA RADIATION by one half. The half-thickness of a material depends on the energy SPECTRUM of the gamma radiation concerned.

halide Any BINARY COMPOUND containing a HALOGEN; that is, a FLUORIDE, CHLORIDE, BROMIDE or IODIDE.

haloform One of four compounds with the general formula CHX_3, where X is a HALOGEN. They are chloroform ($CHCl_3$), fluoroform (CHF_3), bromoform ($CHBr_3$) and iodoform (CHI_3).

haloform reaction The reaction between the CH_3CO– group, found in carbonyl compounds (*see* CARBONYL GROUP), and a HALOGEN in alkaline solution, to yield $CHHal_3$ (where Hal stands for halogen). If the halogen is iodine the reaction is termed the TRIIODOMETHANE TEST.

halogen Any of the elements from group 17 (formerly group VII) of the PERIODIC TABLE: FLUORINE, CHLORINE, BROMINE, IODINE and ASTATINE. They all exhibit similar chemical properties, forming salts containing an ion with a single negative charge. They combine with hydrogen to form POLAR MOLECULES, which dissociate in water to form acids.

The REACTIVITY of the halogen decreases going down the series: fluorine is the most electronegative element (*see* ELECTRONEGATIVITY) and is highly reactive, while iodine is relatively unreactive and forms mostly IONIC compounds. Melting and boiling points increase with increasing atomic mass: fluorine and chlorine are gases at room temperature, whilst bromine is a volatile liquid, and iodine and astatine are solids.

halogenation The incorporation of a HALOGEN into an organic compound, by ADDITION or SUBSTITUTION REACTIONS. *See also* MARKOWNIKOFF'S RULE.

halogenoalkane, *alkyl halide* An ALKANE in which a hydrogen atom has been replaced by a HALOGEN atom. A halogenoalkane has the general formula $C_nH_{2n+1}Hal$, where Hal is a halogen. The terms primary, secondary and tertiary used to describe ALKYL GROUPS are applied to halogenoalkanes in the same way, although these terms do not strictly refer to the halogenoalkane itself. The chemistry of the halogenoalkane is determined by the halogen FUNCTIONAL GROUP rather than the alkane skeleton to which it is attached. The melting and boiling points of halogenoalkanes are higher than those of their corresponding alkane.

Halogenoalkanes can be manufactured by reacting ALCOHOLS with phosphorus halides or sulphur dichloride oxide. For example, for chloroalkanes:

$$ROH + PCl_5 \rightarrow RCl + HCl + POCl_3$$

$$ROH + SCl_2O \rightarrow RCl + HCl + SO_2$$

where R is an alkyl group.

Bromoalkanes and iodoalkanes can be manufactured by reacting alcohols with concentrated aqueous hydrogen bromide or hydrogen iodide respectively. For example:

$$C_2H_5OH + HBr \rightarrow C_2H_5Br + H_2O$$

Halogenoalkanes undergo ELIMINATION and SUBSTITUTION REACTIONS (*see* BROMOETHANE).

Useful halogenoalkanes include dichloromethane as a solvent, chlorofluoromethane as a refrigerant and aerosol propellant, bromochlorodifluoromethane as a fire extinguisher and HALOTHANE as an anaesthetic. CHLOROFORM (trichloromethane) was used as an anaesthetic and CARBON TETRACHLORIDE (tetrachloromethane) as a dry cleaning agent although these latter two are no longer in use as they are thought to induce cancer.

halothane, *2-bromo-2-chloro-1,1,1-trifluoroethane* ($CF_3CHBrCl$) A general ANAESTHETIC, now widely used since it is relatively nontoxic. It is a colourless liquid with an odour similar to CHLOROFORM.

hard (*adj.*) Describing any material that is not easily scratched.

hardness 1. Resistance to scratching.

2. The presence of dissolved calcium and magnesium salts in water. *See* HARD WATER.

hard water Water containing dissolved salts of calcium and magnesium. These react with soap to form insoluble compounds, which produce a scum on the water surface and prevent the soap from forming a lather. The main cause of hardness is the presence of carbonate salts, which dissolve in the water when it passes through porous rocks such as limestone. These are precipitated when the water is boiled (in kettles for example), forming limescale. These deposits are poor conductors of heat and deposits on heating elements lead to reduced efficiency of heating and failure of heaters, due to the high temperatures reached. *See also* ION EXCHANGE, PERMANENT HARDNESS, SOFT WATER, TEMPORARY HARDNESS, WATER SOFTENING.

heat The energy that is transferred from one body or system to another as a result of a difference in temperature. In SI UNITS, heat, like all other forms of energy, is measured in JOULES. *See also* HEAT CAPACITY, LATENT HEAT, LAW OF CONSERVATION OF ENERGY, THERMODYNAMICS.

heat capacity The amount of heat needed to change the temperature of an object by one degree CELSIUS (or one KELVIN, which is the same size). If the temperature is increasing, this much heat energy will have been taken in by the object; if it is decreasing, the energy will have been given out. *See also* MOLAR HEAT CAPACITY, RATIO OF SPECIFIC HEATS, SPECIFIC HEAT CAPACITY.

heat engine Any machine for converting HEAT energy to mechanical WORK. *See* CARNOT CYCLE, KELVIN STATEMENT OF THE SECOND LAW OF THERMODYNAMICS.

heat exchanger A device for transferring heat energy from one fluid to another, without contact between the two fluids. In many applications, a hot liquid, often water, needs to be cooled, giving up its heat to the surrounding air. The liquid flows through pipes to which thin metal plates are attached. Air is then forced over the plates carrying away heat.

heat of atomization The energy needed to convert one mole of a given substance from its normal form at the temperature specified to atoms infinitely far apart from one another. For example, the heat of atomization for hydrogen is 433 kJ mol^{-1}:

$$H_2 \rightarrow 2H, \quad \Delta H = 433 \text{ kJ mol}^{-1}$$

heat of combustion The energy of reaction when one MOLE of a given substance is completely combined with oxygen. For example, the heat of combustion for ethane is –1,556 kJ mol^{-1}:

$$C_2H_6 + 3\tfrac{1}{2}O_2 \rightarrow 2CO_2 + 3H_2O \quad \Delta H = -1{,}566 \text{ kJ mol}^{-1}$$

See also CALORIFIC VALUE.

heat of crystallization The energy released when one MOLE of a specified material is formed as a solid from a SATURATED AQUEOUS SOLUTION.

heat of dilution The energy change when one mole of a given substance is diluted (see DILUTE) from one specified CONCENTRATION to another. At low concentrations, the ions or molecules in a solution no longer interact with one another and the heat of dilution falls to zero.

heat of dissociation The energy required to convert one MOLE of a given compound into its constituent elements.

heat of formation The energy released or the ENTHALPY change when one MOLE of a given compound is formed from its elements in their usual states at the temperature specified. For example, the heat of formation of ammonia, NH_3, is –46 kJ mol^{-1}:

$$\tfrac{1}{2}N_2 + 1\tfrac{1}{2}H_2 \rightarrow NH_3, \quad \Delta H = -46 \text{ kJ mol}^{-1}$$

The minus sign indicates that energy is given off in this process. *See also* STANDARD HEAT OF FORMATION.

heat of ionization The energy needed to completely ionize (*see* IONIZATION) one MOLE of a given compound with the ions being formed in an infinitely dilute aqueous solution.

heat of neutralization The energy of reaction when one MOLE of hydrogen ions are completely reacted with a BASE. In the case of a neutralization of a strong acid by a strong base, the heat of reaction is essentially the heat of reaction for the formation of water from aqueous ions:

$$H + (aq) + OH^- (aq) \rightarrow H_2O\ (l), \quad \Delta H = 57.3 \text{ kJ mol}^{-1}$$

For weak acids or weak bases, the heat of neutralization is different by an amount needed to ionize the weak acid or base, which depends on the HEAT OF IONIZATION of these compounds.

heat of reaction The amount of energy given out, or taken in, in a given chemical reaction. It is usually quoted in terms of the number of joules of energy per MOLE of reacting substances. The symbol used for a heat of reaction is ΔH, and this is taken as negative if heat is given out (that is, lost from the reaction).

Most reactions that take place spontaneously are EXOTHERMIC, with ΔH negative. However, ENDOTHERMIC reactions may take place provided they result in an increase in ENTROPY – the breaking up of a crystal for example, or an increase in the number of molecules.

Heats of reaction are sometimes referred to as enthalpies of reaction. The difference between the two terms is only significant if gases are involved, as the ENTHALPY includes a term relating to the energy needed to push back the surrounding atmosphere if a gas is given off, keeping the reacting vessel at a constant pressure, usually specified as ATMOSPHERIC PRESSURE.

See also HESS'S LAW

heat of solution The energy change when one mole of a given substance is dissolved in a large volume of a specified SOLVENT, usually water. The volume of solvent must be such that the dissolved substance is at such a low concentration that the addition of more solvent produces no further change.

heat reservoir A hypothetical object of infinite HEAT CAPACITY, which heat can enter or leave without producing a change in temperature.

heat treatment The heating and cooling of a material, usually a metal or alloy, under controlled conditions to produce changes in the mechanical properties of the material which remain after it has returned to room temperature. *See* ANNEALING, QUENCHING.

heavy metal The collective term for metals of high atomic mass, particularly those TRANSITION METALS that are toxic and cannot be processed by living organisms, such as lead, mercury and cadmium.

heavy water Water in which both hydrogen atoms have been replaced with the DEUTERIUM ISOTOPE. Heavy water is used as a moderator in some forms of nuclear reactor.

helium (He) The element with atomic number 2; relative atomic mass 4.0; boiling point –269°C; does not solidify at ATMOSPHERIC PRESSURE. Helium is the second most abundant element in the universe, making up almost 25 per cent of the atoms. It is scarce on Earth as it occurs only as a gas and its low mass means that it easily escapes from the Earth's gravitational pull. Helium was first discovered from its ABSORPTION SPECTRUM in sunlight.

Helium is found in some porous rocks in association with the production of ALPHA PARTICLES from radioactive ores. It is used as a filling in airships and high-altitude balloons, where it has replaced inflammable hydrogen. The gas is also used to dilute oxygen when fed to deep-sea divers under high pressure. Helium has a very low boiling point, about 4 K (–269°C), which has also led to the widespread use of liquid helium in low temperature refrigeration systems, particularly for superconducting magnets (*see* SUPERCONDUCTIVITY).

See also SUPERFLUIDITY.

Helmholtz free energy *See* FREE ENERGY.

hemicellulose A POLYSACCHARIDE found in plant cell walls. It is the easily hydrolysed portion of CELLULOSE.

hemihydrate A CRYSTALLINE form of a SALT that contains one molecule of WATER OF CRYSTALLIZATION for every two molecules of the salt. An example is calcium sulphate hemihydrate, $2CaSO_4.H_2O$.

henry (H) The SI UNIT of magnetic inductance. An inductance of one henry will result in an induced ELECTROMOTIVE FORCE (e.m.f.) of one VOLT when the current producing the induced e.m.f. is changing at the rate of one AMPERE per second.

Henry's law The ratio of the CONCENTRATIONS of a given molecule in two phases is always constant at a given temperature. This accounts for the constant nature of the PARTITION CONSTANT, where the two phases are the two SOLVENTS. This also accounts for the fact that the SOLUBILITY of a gas is proportional to its pressure – here the two phases are the gas itself and the gas molecules in solution. This law does not apply to any case where the molecules change their nature in the two phases, for example in the case of hydrogen chloride or ammonia, which both form ions in aqueous solution.

heparin A GLYCOSAMINOGLYCAN that is an anticoagulant used in treating thrombosis. It is found in the liver, lungs and blood vessels and inhibits conversion of prothrombin to thrombin during the blood clotting process.

heptane (C_7H_{16}) The seventh member of the ALKANE series. There are nine possible ISOMERS. *n*-Heptane ($CH_3[CH_2]_5CH_3$) is a colourless, inflammable liquid; boiling point of 98°C. Heptane is obtained from PETROLEUM and is used in defining the OCTANE NUMBER of PETROL.

herbicide A poisonous chemical that kills plants considered to be weeds. Examples include PARAQUAT, MCPA, MCPB, 2,4-D and 2,4,5-T.

hertz (Hz) The SI UNIT of frequency. One hertz is a frequency of one oscillation per second.

Hess's Law An application of the LAW OF CONSERVATION OF ENERGY. This law states that the overall energy change involved in any chemical process is independent of the route by which that process takes place. Thus if the reaction A→B can proceed via two possible intermediates, X and Y, the HEATS OF REACTION for A→X and X→B will add to the same total as those for A→Y and Y→B, this total being the heat of reaction for A→B. The law is useful since it allows heats of reaction to be calculated for reactions that have not been observed directly.

heteroatom An odd atom in the ring of a HETEROCYCLIC compound, for example nitrogen in PYRIDINE.

heterocyclic (*adj.*) Describing any organic compound that is CYCLIC but which contains different atoms within a closed ring structure. *See also* ALICYCLIC, AROMATIC.

heterogeneous (*adj.*) Describing a substance whose properties vary from one place to another. *Compare* HOMOGENEOUS.

heterogeneous catalyst A CATALYST that is in a different phase from the reaction it catalyses. Usually a solid catalyst, such as finely divided platinum, is used to catalyse gaseous reactions.

heterolytic fission The breaking of a COVALENT BOND to give two oppositely charged fragments. The electrons forming the bond are thus shared unequally between the fragments. This is particularly common in compounds with polar bonding (*see* POLAR BOND). An example is the dissociation of hydrogen chloride in an aqueous solution:

$$HCl \rightarrow H^+ + :Cl^-$$

In contrast to HOMOLYTIC FISSION the electrons in a heterolytic reaction remain paired. Such reactions involve IONS rather than RADICALS (as in homolytic fission). Examples of heterolytic reactions are those of HALOGENOALKANES.

heteronuclear (*adj.*) Describing a COVALENT BOND that links two different nuclei, or a diatomic molecule in which the two atoms are of different types, for example, HCl. *Compare* HOMONUCLEAR.

1,2,3,4,5,6-hexachlorocyclohexane *See* BHC.

hexacyanoferrate(II) Any salt containing the complex hexacyanoferrate(II) ion $Fe\,(CN)_6^{4-}$. The salts are generally yellow in colour.

hexacyanoferrate(III) Any salt containing the complex hexacyanoferrate(III) ion $Fe(CN)_6^{3-}$. These salts are usually blood red in colour, though an important blue pigment, Prussian blue, is the DOUBLE SALT $KFe[Fe(CN)_6]$.

hexagonal close packed (*adj.*) Describing a crystalline structure in which each layer of atoms is CLOSE PACKED, with each atom surrounded by 6 others in that layer. The second layer of atoms lies above the gaps in the first, whilst the atoms in the third layer lie above gaps in the second and directly above the atoms in the first. If the layers of atoms are labelled *A* and *B*, the structure can be described as *ABAB*... Many metals occur as hexagonal close packed structures, including magnesium and zinc, though the CUBIC CLOSE PACKED structure is also common.

hexamethylene diamine *See* 1,6-diaminohexane.

hexane (C_6H_{14}) The sixth member of the ALKANE series. There are five ISOMERS of hexane. *n*-Hexane, $CH_3[CH_2]_4CH_3$, is a colourless liquid; boiling point of 69°C. It is used as a solvent.

hexose A MONOSACCHARIDE containing six carbon atoms in the molecule.

hexyl group The group

$$CH_3CH_2CH_2CH_2CH_2CH_2-$$

($-C_6H_{13}$) derived from HEXANE.

histamine, *4-(2-aminoethyl)-imidizole* ($C_5H_9N_3$) A derivative of the amino acid HISTIDINE, released in animal tissues in response to injury or stimulation by the appropriate ANTIGEN during an inflammatory or allergic reaction. Histamine causes dilation of local blood vessels and an increase in their permeability to allow, for example, ANTIBODIES through, which are needed for repair or recovery of the site. During an allergic reaction, such as hayfever, the release of histamine causes some inflammation and is responsible for the characteristic itching and sneezing.

histidine An AMINO ACID, the precursor of HISTAMINE.

histochemistry The study of the distribution of molecules within the cells and matrices of tissues using staining techniques and MICROSCOPY.

histone One of a group of basic PROTEINS containing a high level of AMINO ACIDS. Histones are found in association with NUCLEIC ACIDS in higher organisms and are a major component of chromosomes.

Hoffman clip A metal clip used to squash a rubber tube to stop or control the flow of material through the tube.

Hofmann degradation The transformation of an AMIDE to a primary AMINE containing one carbon less than the original amide. This occurs when amides react with chlorine or bromine solutions in the presence of excess alkali. For example:

$$RCONH_2 \xrightarrow[NaOH]{Br_2} RNH_2 + CO_2 + HBr + NaBr$$

In this reaction the HALOGEN first replaces a hydrogen atom of the NH_2^- group to form a chloro- or bromo-amide. This reacts with alkali to give an isocyanate (compound with an –N=C=O group), which then decomposes to give an amine and carbon dioxide.

The Hofmann degradation is used in the preparation of the chemical 2-aminobenzoic acid, which is important in the DYESTUFFS industry.

Hofmann voltammeter A VOLTAMMETER designed to collect gases given off in ELECTROLYSIS.

hole A vacant space in the electron structure of a SEMICONDUCTOR. A neighbouring electron may move to fill this hole, which then appears to move through the semiconductor in the opposite direction to the electron flow. Holes can be thought of as behaving like positive CHARGE CARRIERS.

holmium (Ho) The element with atomic number 67; relative atomic mass 164.9; melting point 1,472°C; boiling point 2,700°C; relative density 8.8. Holmium is a LANTHANIDE metal, known to be magnetic, but too difficult to separate from other elements to be commercially useful.

homogeneous (*adj.*) Describing a substance that is the same throughout, such as a material made from a single compound. *Compare* HETEROGENEOUS.

homogeneous catalyst A CATALYST that is in the same phase as the reaction it catalyses.

homologous (*adj.*) Describing a similarity in some aspect, such as structure, position or functional properties. In chemistry, organic chemicals form HOMOLOGOUS SERIES.

homologous series A group of organic compounds with similar chemical properties, forming a series in which the members differ from each other by a constant RELATIVE MOLECULAR MASS. ALKANES form such a series where methane is the first member and the series progresses by the addition of $-CH_2-$ through to ethane, propane, butane, pentane. ALKENES, ALKYNES, ALCOHOLS, ALDEHYDES, KETONES, and CARBOXYLIC ACIDS all form similar homologous series. The physical properties of members of such a series show a steady change along the series.

homolytic fission The breaking of a COVALENT BOND to give two neutral atoms or FREE RADICALS (fragments with unpaired electrons). An example is the splitting of chlorine into free radicals by the action of ultraviolet light:

$$Cl_2 \rightarrow 2Cl^-$$

Many of the reactions of SATURATED HYDROCARBONS involve homolytic fission since the ELECTRONEGATIVITIES of carbon and hydrogen are similar and carbon-carbon and carbon-hydrogen bonds exert an equal attraction on the bonding electrons. Such reactions are usually initiated by other radicals, for example fluorine atoms released by the splitting of the weak bond in molecular fluorine.

The term homolytic reaction is also used to refer to the reaction in which radicals react with each other to form a covalent bond. Examples of homolytic reactions are those between ALKANES and chlorine or bromine, which are important in the manufacture of solvents, anaesthetics and refrigerants.

Compare HETEROLYTIC FISSION.

homonuclear (*adj.*) Describing a COVALENT BOND that links two nuclei of the same type, or a diatomic molecule where both atoms are of the same type, for example, H_2. *Compare* HETERONUCLEAR.

hormone Any molecule (usually of small molecular mass) secreted directly into the blood by certain organs, called endocrine glands, and transported to a target cell or organ causing a specific response, which is not just local. Hormones regulate a number of body functions, such as general metabolism and growth (thyroid hormones), response to stress or danger (adrenal hormones), blood sugar levels (pancreatic hormones) and reproductive functions (hormones from testes and ovary). Hormones are not all of one chemical class; some (for example PROLACTIN, GROWTH HORMONE) are proteins, some (for example INSULIN, GLUCAGON) are smaller POLYPEPTIDES of less than 100 AMINO ACIDS, whilst others (for example ADRENALINE, NORADRENALINE, THYROXINE) consist of only a few amino acids.

Some hormones are derived from LIPIDS and are termed STEROID HORMONES. These are lipid-soluble and can pass through the cell membrane. When a lipid-soluble hormone reaches its target cell it forms a complex with a receptor molecule (hormone-receptor complex), passes through the cell membrane and influences some activity within the cell. The other protein hormones are water-soluble and cannot pass through the cell membrane. Water-soluble hormones act by binding to a receptor molecule on the target cell membrane, which then activates a second messenger that initiates a specific chemical change within the cell. Water-soluble hormones generally mediate short-term effects, whilst lipid-soluble steroids mediate longer-term effects. Hormones do not differ much from one species to another but their effects may differ.

Substances resembling the animal hormones are also produced in plants, such as AUXINS. These are termed plant growth substances. *See also* ENCEPHALIN, ENDORPHIN, PHEROMONE.

Hund's rule of maximum multiplicity A consequence of the repulsion between electrons in a given ORBITAL. The rule states that the electrons in p-, d- or f-orbitals tend to arrange themselves with one electron in each available orbital before a second electron enters any of the orbitals. Thus iron, for example, which has six 3d electrons, will have one D-ORBITAL with two electrons and four with a single electron.

hyaluronic acid A GLYCOSAMINOGLYCAN found in connective tissue. It is composed of units of D-glucuronic acid and N-acetyl-D-glucosamine. It lubricates joints and binds cells together, aiding wound repair. It is broken down by the enzyme hyaluronidase.

hyaluronidase An enzyme that breaks down HYALURONIC ACID and CHONDROITIN.

hybridization The process of forming a HYBRID ORBITAL.

hybrid orbital A superposition of the WAVEFUNCTIONS of a number of ORBITALS with the same energy to form a composite orbital. This is the way orbitals sometimes behave in the formation of COVALENT BONDS. Thus carbon, containing two electrons in a 2s orbital and two in 2p orbitals, forms four sp^3 hybrid orbitals, each containing one electron. These orbitals have the shape of a lobe directed towards the corners of a TETRAHEDRON centred on the carbon atom, and it is these orbitals that produce the well-known TETRAHEDRAL structure of carbon's covalent bonds.

hydrate A substance containing a particular compound in association with water, particularly WATER OF CRYSTALLIZATION.

hydrated (*adj.*) Describing a compound that contains water, particularly WATER OF CRYSTALLIZATION, or an ion that has water molecules bound to it by hydrogen bonding (*see* HYDROGEN BOND).

hydration The combination of a substance with water to produce a single product. It is the opposite of dehydration. *See also* HYDRATE.

hydrazine (N_2H_4) A colourless liquid; melting point 1.4°C; boiling point 114°C; relative density 1.0. Hydrazine is prepared by the reaction between ammonia and sodium(I) chlorate:

$$2NH_3 + NaClO \rightarrow N_2H_4 + H_2O + NaCl$$

Hydrazine is a powerful REDUCING AGENT and is used as such in some rocket fuels.

hydrazoic acid, *hydrazoic acid, azoimide* (N_3H) A highly toxic colourless liquid, melting point –80°C, boiling point 37°C. It is a strong reducing agent that explodes in the presence of oxygen. Heavy metal salts are AZIDES and these are used as detonators and in organic synthesis.

hydrazones Organic compounds formed by the condensation between an ALDEHYDE or KETONE and HYDRAZINE. They contain the group

$$\begin{matrix} R \diagdown \\ \quad C = NNH_2 \\ H \diagup \end{matrix}$$

In phenylhydrazones, one of the hydrogen atoms is replaced by a C_6H_5 group.

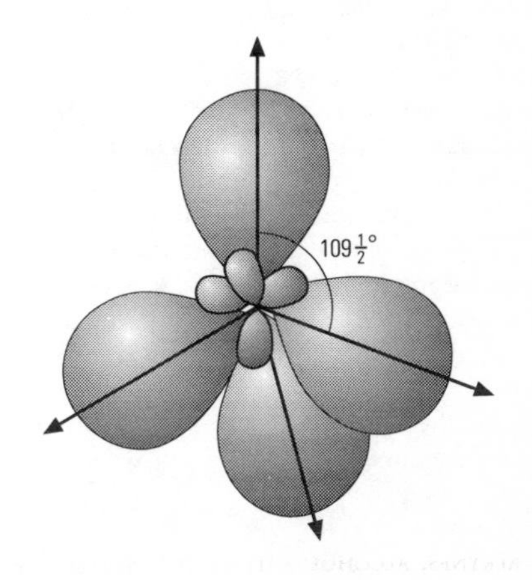

Hybrid orbital.

hydride Any BINARY COMPOUND containing a metal and hydrogen. They are unusual in containing the H^- ion. They are formed only by the more reactive metals and decompose on contact with water, for example:

$$KH + H_2O \rightarrow KOH + H_2$$

hydrocarbon A chemical compound consisting of hydrogen and carbon only. The carbon atoms can be linked by COVALENT BONDS in a long chain (ALIPHATIC) or in a ring (CYCLIC). Examples of hydrocarbons are the ALKANES, ALKENES and ALKYNES.

hydrochloric acid (HCl) A solution of HYDROGEN CHLORIDE gas in water. It is an important MINERAL ACID, widely used in industry.

hydrocortisone *See* CORTISOL.

hydrogen (H) The element with atomic number 1; relative atomic mass 1.0; melting point –259°C; boiling point –253°C. By far the most abundant element in the universe, making up about 75 per cent of the atoms. Hydrogen is also widespread on Earth, forming many compounds, including water, from which it can be extracted by ELECTROLYSIS. In its elemental form, hydrogen normally occurs as a gas. However, it is believed to have a metallic phase at extreme temperatures and pressures.

Hydrogen is present in all ACIDS and these will react with the more reactive metals to release hydrogen gas. The presence of hydrogen in the laboratory is confirmed by lighting a sample of the gas: it burns with a characteristic squeaky pop, caused by the high speed of sound in the gas. The heavy ISOTOPES of hydrogen, DEUTERIUM and TRITIUM, are important in the production of energy by nuclear fusion.

See also BOSCH PROCESS.

hydrogenation The addition of hydrogen to a substance, usually in the presence of a catalyst and at high pressures. Hydrogenation of UNSATURATED COMPOUNDS is of great industrial use, for example, in the manufacture of margarine and in the petroleum industry.

hydrogen azide (HN_3) A colourless liquid; melting point –80°C; boiling point 37°C; relative density 1.1. Hydrogen azide can be prepared from the reaction between sodium amide and sodium nitrate:

$$2NaNH_2 + NaNO_3 \rightarrow HN_3 + 3NaOH$$

Hydrogen azide is an extremely strong REDUCING AGENT.

hydrogen bond A weak bond formed between molecules that have a hydrogen atom in a POLAR COVALENT BOND. The positive charge on the hydrogen atom produces an attraction for the electronegative atom (*see* ELECTRONEGATIVITY) in a neighbouring atom.

Hydrogen bonding is particularly noticeable in hydrogen fluoride, HF, water, H_2O, and ammonia, NH_3. All three compounds have melting and boiling points much higher than methane, CH_4, which does not contain a POLAR BOND. Hydrogen bonding is also responsible for the unusual behaviour of water, which expands on cooling below 4°C, and further expands on freezing. Individual water molecules are believed to form groups such as $(H_2O)_3$.

Hydrogen bonds are much weaker than covalent bonds, having energies of typically a few tens of kilojoules per MOLE. They are easily broken by heating to temperatures much greater than room temperature, but are extremely important in holding together the structure of many biological molecules.

hydrogen bromide (HBr) A colourless gas; melting point –86°C; boiling point –66°C. Hydrogen bromide can be made by heating bromine in hydrogen in the presence of a platinum catalyst:

$$H_2 + Br_2 \rightarrow 2HBr$$

Hydrogen bromide is highly soluble in water, dissociating to form a strong acid.

hydrogencarbonate Any salt containing the hydrogencarbonate ion, HCO_3^-. Hydrogencarbonates are ACIDIC SALTS, the best known being sodium hydrogencarbonate (sodium bicarbonate), used in indigestion remedies.

hydrogen chloride (HCl) A colourless fuming gas; melting point –115°C; boiling point –85°C. Hydrogen chloride gas can be made by burning hydrogen (obtained from ELECTROLYSIS of water) with chlorine (from the electrolysis of BRINE) or by heating sodium chloride with concentrated sulphuric acid:

$$NaCl + H_2SO_4 \rightarrow Na_2SO_4 + HCl$$

Hydrogen chloride is a covalent gas (*see* COVALENT BOND), but is highly soluble in water, fully dissociating to give a strong acid known as hydrochloric acid.

hydrogen cyanide (HCN) A colourless gas; melting point –14°C, boiling point 26°C. It is a weak acid, formed by the action of strong acids on metal cyanides, for example:

$$NaCN + HCl \rightarrow HCN + NaCl$$

Hydrogen cyanide is extremely toxic. It is used in the manufacture of some plastics and is also produced when they are burnt in a limited oxygen supply, as often happens in fires involving soft furnishings.

hydrogen electrode A platinum ELECTRODE, coated with finely divided platinum (platinum black) over which hydrogen is passed at a pressure of one ATMOSPHERE, in contact with hydrogen ions at a concentration of one MOLE per decimetre cubed. *See also* STANDARD ELECTRODE POTENTIAL.

hydrogen fluoride (HF) A colourless liquid; melting point –83°C; boiling point 20°C; relative density 1.0. Hydrogen fluoride can be made by the action of sulphuric acid on calcium fluoride:

$$H_2SO_4 + CaF_2 \rightarrow 2HF + CaSO_4$$

HF is unusual in being a liquid and only forming a weak acid when dissolved in water (hydrofluoric acid). Both these features are a consequence of the highly reactive nature of fluorine, which forms a strong polar bond.

hydrogen iodide (HI) A colourless gas, melting point –51°C, boiling point –36°C. Hydrogen iodide can be formed by heating iodine vapour with hydrogen in the presence of a platinum catalyst:

$$H_2 + I_2 \rightarrow 2HI$$

Hydrogen iodide dissolves in water to form a moderately strong acid.

hydrogen peroxide (H_2O_2) A colourless liquid; melting point –1°C; boiling point 150°C; relative density 1.4. Hydrogen peroxide can be manufactured by adding acid to barium peroxide, for example:

$$H_2SO_4 + BaO_2 \rightarrow H_2O_2 + BaSO_4$$

Hydrogen peroxide decomposes readily to give oxygen and water, a reaction accelerated by the presence of light or metal ions, and which can be explosive in the case of pure hydrogen peroxide:

$$2H_2O_2 \rightarrow 2H_2O + O_2$$

Hydrogen peroxide is often supplied as an aqueous solution with its strength measured in 'volumes'. Thus one litre of '50 volume' hydrogen peroxide can be oxidized to give 50 litres of oxygen at STANDARD TEMPERATURE AND PRESSURE.

Hydrogen peroxide is an important oxidizing and bleaching agent and has been used as an oxidizer in some rocket fuels, usually with hydrogen:

$$H_2O_2 + H_2 \rightarrow 2H_2O$$

hydrogen spectrum The EMISSION SPECTRUM of HYDROGEN. As hydrogen has only a single electron its spectrum is fairly simple. It comprises a number of series of SPECTRAL LINES – the Lyman series in the ultraviolet; the BALMER SERIES in visible light; and the Paschen, Ritz and Brackett–Pfund series in infrared.

Each series corresponds to the electron moving to a particular ENERGY LEVEL from one of the higher levels. In the Lyman series, the electron transitions are all between the GROUND STATE and higher levels, in the Balmer series they all connect with the energy level above the ground state, in the Paschen series with the next higher energy level and so on. This pattern was first described empirically in the RYDBERG EQUATION, and the successful explanation of this pattern was an early triumph for QUANTUM MECHANICS.

hydrogen sulphide (H_2S) A colourless gas with a distinctive smell of rotten eggs; melting point –86°C; boiling point –61°C. Hydrogen sulphide is generally produced in the laboratory in KIPP'S APPARATUS, usually by the reaction between an acid and iron(II) sulphide:

$$FeS + 2HCl \rightarrow H_2S + FeCl_2$$

Hydrogen sulphide is highly toxic, though its characteristic smell makes it more detectable and therefore less dangerous than some toxic gases. It is a mild REDUCING AGENT and forms a black PRECIPITATE in the presence of many metal ions that form insoluble sulphides, for example:

$$H_2S + PbSO_4 \rightarrow PbS + H_2SO_4$$

hydrogensulphite Any compound containing the hydrogensulphite ion, HSO_3^-. Hydrogensulphites are ACIDIC SALTS and mild REDUCING AGENTS.

hydrolase Any one of a group of ENZYMES involved in HYDROLYSIS reactions, for example peptidases, lipases.

hydrolysis A chemical DECOMPOSITION brought about by the presence of water. For example, iron(III) chloride is hydrolysed to form iron(III) hydroxide and hydrochloric acid:

$$FeCl_3 + 3H_2O \rightarrow Fe(OH)_3 + 3HCl$$

Other examples include ATP breakdown and digestion.

hydrometer A device for measuring the DENSITY of a fluid by means of a weighted float with a long stem. The level to which the float sinks in the liquid is a measure of the liquid's density. Hydrometers are often used to determine the concentrations of certain solutions, such as sugar and ethanol in water when brewing beer, or sulphuric acid in a LEAD-ACID CELL.

hydrophilic (*adj.*) (Greek = water-loving) Describing a molecule that exerts a strong INTERMOLECULAR FORCE towards water molecules (and other POLAR solvents), usually as the result of ionic or highly polar bonding (*see* IONIC BOND, POLAR BOND). Such materials are generally soluble in water or form LYOPHILIC SOLS. DETERGENT molecules function by having a polar end, which has a strong affinity for water molecules, and a non-polar end, which has a strong affinity for non-polar materials such as grease. *Compare* HYDROPHOBIC.

hydrophobic (*adj.*) (Greek = water-hating) Describing a molecule that does not exert a strong INTERMOLECULAR FORCE on water molecules. Such materials are generally IMMISCIBLE with water or form LYOPHOBIC SOLS. HYDROCARBONS are an example. *Compare* HYDROPHILIC.

hydrosol A SOL in which solid particles are suspended in water.

hydroxide Any compound containing a metal and the hydroxide ion, OH^-. Hydroxides are formed by the reaction between a metal and water, though this only happens at high temperatures with the less reactive metals, for example:

$$Ca + 2H_2O \rightarrow Ca(OH)_2 + H_2$$

The hydroxides of the ALKALI METALS are all soluble and form alkali solutions. The ALKALINE EARTHS form hydroxides that are basic but only slightly soluble. TRANSITION and P-BLOCK metal hydroxides are generally insoluble and are AMPHOTERIC, reacting with strong bases to form negative ions. For example, aluminium hydroxide will react with a base to form the aluminate ion:

$$Al(OH)_3 + OH^- \rightarrow Al\,(OH)_4^-$$

hydroxonium ion The ion H_3O^+ formed in the dissociation of water and present in acidic solutions. They are formed by the association of a proton (hydrogen ion) with a water molecule.

hydroxyaldehyde, *aldol* The product obtained from a reaction between two ALIPHATIC ALDEHYDES. The term aldol is derived from *ald-* for –CHO and *-ol* for –OH. The simplest hydroxyaldehyde is 3-hydroxybutanal ($CH_3CHOHCH_2CHO$), which is formed from the reaction of ethanal in dilute alkali solution. The name aldol is often used to refer to 3-hydroxybutanal. *See* ALDOL CONDENSATION.

6-hydroxy-2-aminopurine *See* GUANINE.

2-hydroxybenzoic acid *See* SALICYLIC ACID.

hydroxyl group (–OH) An atom of hydrogen and an atom of oxygen bonded together in a covalently bonded compound (*see* COVALENT BOND). Hydroxyl groups are found in ALCOHOLS and PHENOLS, where they react as a single entity. The presence of a single hydroxyl group is termed monohydric, two is dihydric, three trihydric and more is polyhydric. GLYCOL is a dihydric alcohol and GLYCEROL is a trihydric alcohol.

hydroxyproline ($C_5H_9NO_3$) A hydroxylated derivative of the amino acid proline. It occurs in COLLAGEN.

2-hydroxypropanoic acid *See* LACTIC ACID.

hydroxysuccinic acid *See* MALIC ACID.

5-hydroxytryptamine *See* SEROTONIN.

hygroscopic (*adj.*) Describing a compound that can absorb water vapour from the atmosphere.

hypertonic solution A SOLUTION with a higher OSMOTIC PRESSURE compared to some other solution. In a pair of solutions, the hypertonic solution is the one with the greater osmotic pressure. *Compare* HYPOTONIC SOLUTION, ISOTONIC.

hypo *See* SODIUM THIOSULPHATE.

hypobromous acid *See* BROMIC ACID.

hypochlorite Any salt containing the chlorate(I) ion, ClO^-.

hypochlorous acid *See* CHLORIC ACID.

hypothesis A statement that has not been proved but which is used as a starting point for a logical or mathematical argument. If this argument leads to an obviously false conclusion, the hypothesis or the argument must be incorrect, a procedure called *reductio ad absurdum*, literally 'reduction to the absurd'.

hypotonic solution A SOLUTION with a lower OSMOTIC PRESSURE compared to some other solution. *Compare* HYPERTONIC SOLUTION, ISOTONIC.

I

icosahedral (*adj.*) Describing a shape that has twenty faces, each in the shape of an equilateral triangle.

ideal gas A hypothetical gas that obeys the GAS LAWS perfectly at all temperatures and pressures. For this to happen, the forces between the molecules of the gas must be negligible except during collisions, and the volume of the gas molecules must be negligible compared to the volume occupied by the gas itself. REAL GASES behave in approximately this way provided the temperature is not too low, so the KINETIC ENERGY of the molecules is large compared to the POTENTIAL ENERGY of the INTERMOLECULAR FORCES, and provided that the density is not too high so that the volume of the molecules (the EXCLUDED VOLUME) is not too large a fraction of the total volume of the gas. *See also* BOYLE'S LAW, CHARLES' LAW, IDEAL GAS EQUATION, JOULE'S LAW, KINETIC THEORY, PRESSURE LAW.

ideal gas equation, *universal gas equation* The EQUATION OF STATE for an IDEAL GAS. One particular consequence of the ideal gas equation is that the volume taken up by a gas at a given temperature and pressure depends only on the number of molecules present and not on the nature of those molecules. In particular, at STANDARD TEMPERATURE AND PRESSURE (0°C and one ATMOSPHERE), one MOLE of molecules of any gas will take up a volume of 22.4 dm^{-3}.

For a gas containing n moles of molecules, at a pressure p in a volume V:

$$pV = nRT$$

where T is the ABSOLUTE TEMPERATURE and R is the MOLAR GAS CONSTANT, $R = 8.31\ \mathrm{J\ K^{-1}\ mol^{-1}}$, or if the gas contains N molecules:

$$pV = NkT$$

where k is the BOLTZMANN CONSTANT, equal to $1.38 \times 10^{-23}\ \mathrm{J\ mol^{-1}}$.

ideal gas temperature scale A temperature scale that defines ABSOLUTE TEMPERATURE as being proportional to the pressure exerted by a fixed mass of IDEAL GAS held in a constant volume. The size of the temperature unit, the KELVIN, is fixed by defining the TRIPLE POINT of water to have a temperature of 273.15 K.

ideal solution A solution that obeys RAOULT'S LAW perfectly.

igneous (*adj.*) Describing a rock that has solidified from lava or magma.

ignition temperature The minimum temperature at which a substance will burn in air.

ignition tube A disposable glass tube, used in the laboratory for melting or boiling REAGENTS.

imides Organic nitrogen-containing ring compounds containing the group (imido group)

```
–C–NH–C–
 ||    ||
 O     O
```

They are formed by heating dibasic acids or their anhydrides with ammonia.

imines Organic compounds containing an –NH– group, in which the nitrogen atom is part of a ring structure, or the =NH group, in which the nitrogen atom is linked to a carbon atom by a double bond. Either group is called the imino group. Imines can be cyclic or linear.

immiscible (*adj.*) Describing two liquids that will not mix together, such as hexane and water. When stirred, an EMULSION is formed, which will separate if left to stand, with the less dense material floating to the surface. *See also* MISCIBLE.

immunoglobulin (Ig) Human GLYCOPROTEIN that is the ANTIBODY in an immune reaction to combat a foreign substance. There are 5 classes of immunoglobulin (IgG, IgM, IgA, IgE, IgD) differing in their structure, the degree to which they polymerize and when they are produced. IgG represents 70 per cent of human immunoglobulin and is produced in response to an infection. IgE is associated with allergic responses.

The basic immunoglobulin unit consists of two heavy and two light POLYPEPTIDE chains which form two Fab regions containing the ANTIGEN-binding site and an Fc region that determines the biological properties of the immunoglobulin, for example where the antibody binds. Specificity of the antibody is determined by a sequence of AMINO ACIDS at the amino-terminal of the Fab region – termed the hypervariable region.

Normal immunoglobulin, containing most of the body's antibodies, can be obtained from pooled blood plasma and given to patients at high risk of contracting a disease for a short period of time. A specific immunoglobulin can also be recovered from convalescing patients for administration to people at risk from a fatal disease, for example rabies.

impermeable (*adj.*) Unable to be penetrated by water. The term is used to describe some kinds of rock, such as granite, compared to limestone, for example, through which water can penetrate.

inactive electrode An ELECTRODE used in ELECTROLYSIS made from a material such as platinum or graphite that does not play any chemical role in the electrolysis.

indene (C_9H_8) A colourless liquid obtained from PETROLEUM fractions; melting point –2°C; boiling point 188°C. It consists of a five-membered ring joined to a benzene ring and is used as a solvent and in organic synthesis.

indicator A substance that produces a colour change in response to some chemical change, such as an increase in acidity. The best known indicator is LITMUS. Indicators can be used to detect the END-POINT of a reaction in a TITRATION. *See also* UNIVERSAL INDICATOR.

indigo ($C_{16}H_{10}N_2O_2$) An important blue dye occurring in the leaves of a plant as a GLUCOSIDE called indican. It was extracted from this plant but is now made synthetically.

indium (In) The element with atomic number 49; relative atomic mass 114.8; melting point 157°C; boiling point 2,080°C; relative density 7.3. Indium is a soft silvery metal. It is used as an abrasion-resistant coating on some metal parts and in the manufacture of some doped SEMICONDUCTORS.

induced fit *See* LOCK-AND-KEY MECHANISM.

inducible An ENZYME that is synthesized only when the SUBSTRATE is present. *Compare* CONSTITUTIVE.

inductive effect The effect on an organic compound of a group or atom pulling or pushing electrons towards or away from itself. Inductive effects can explain some organic reactions. Groups such as $-NO_2$ and –COOH are electron-pulling so reduce the electron density on a ring, decreasing its susceptibility to further SUBSTITUTION REACTIONS. Groups such as –OH and $-CH_3$ have the opposite effect.

inelastic neutron scattering A technique used to study the energy spectrum of vibrations with a crystal lattice. A beam of neutrons of a single energy is obtained by applying the technique of NEUTRON DIFFRACTION to neutrons from a nuclear reactor. These neutrons are then scattered off a crystal. When a neutron is scattered inelastically (i.e. with a loss of energy), that energy is carried away from the target nucleus as a vibration in the crystal lattice. Such vibrations can be studied by comparing the energy and momentum of the neutrons before and after collision.

inert gas *See* NOBLE GAS.

inert pair effect The tendency of the heavier elements in groups 3 and 4 to form compounds with VALENCIES two lower than the valencies that would normally be expected. This effect is shown by the lighter elements in the groups. For example, lead is in group 4 of the periodic table along with carbon, germanium and silicon, but unlike these elements, which usually have a valency of 4, lead usually shows a valency of 2. This is explained by the fact that forming four bonds requires the removal of electrons from the outer S-ORBITAL, which requires substantially more energy than the loss of electrons from the P-ORBITAL. In the lighter elements this is offset by the energy released by forming extra bonds, but in heavier elements balance tends to favour the formation of lower valency compounds. Similarly, thallium in group 3 usually shows a valency of 1.

inflammable (*adj.*) Able to burn.

infrared ELECTROMAGNETIC WAVES with wavelengths from about 1 mm to 7×10^{-7} m. Infrared radiation is emitted by all hot objects and can be detected by the heating effect it produces when absorbed by a blackened surface. Infrared with a wavelength close to the visible part of the ELECTROMAGNETIC SPECTRUM (called near infrared) can also be detected by modified versions of photographic film and electronic devices used to detect visible light. The fact that warm objects produce more infrared than cold objects has led to the development of many military applications based on infrared cameras for night-time surveillance and heat-seeking missiles.

infrared spectroscopy The study of the infrared ABSORPTION SPECTRUM of organic molecules. This is a useful technique for determining molecular structure, as many COVALENT BONDS have resonant frequencies corresponding to the stretching or bending of the bonds that lie in the infrared region of the electromagnetic spectrum.

inhibitor 1. A substance that slows down the rate of a chemical reaction or blocks it. *See also* ENZYME INHIBITION.

2. A material added to some glues, such as epoxy resins, to slow down the rate at which they harden.

inorganic (*adj.*) Describing compounds that contain no carbon or that are carbides or carbonates, or oxides or sulphides of carbon. Such substances are generally obtained from the Earth's crust or atmosphere. Inorganic chemistry is the study of the chemical reactions of the elements and the inorganic compounds they form.

insecticide A poisonous chemical that kills hazardous or unwanted insects. *See also* DDT.

insoluble (*adj.*) Describing a compound that will not dissolve (normally in water).

insulator A material through which current or heat cannot flow. Except for graphite, all non-metals are electrical insulators in their solid form. Organic liquids are also electrical insulators, as are all gases unless ionized. *See also* BAND THEORY, CONDUCTOR, SEMICONDUCTOR.

insulin A protein HORMONE that is responsible for the regulation of blood sugar levels. Insulin is a 51 amino acid POLYPEPTIDE, made in the pancreas. It is made up of two polypeptide chains (of 21 and 30 amino acids) linked by two disulphide bridges.

Insulin is produced when blood sugar levels are high, for example after a meal, and has the effect of reducing those levels. Insulin promotes uptake of free GLUCOSE by body cells (e.g. muscle and adipose cells) by altering the permeability of the cell membrane to glucose. It also promotes uptake of amino acids by muscles and converts excess glucose (formed as a result of CARBOHYDRATE breakdown following a meal) into GLYCOGEN for storage in the liver. This is called GLYCOGENESIS. Deficiency in the production of insulin leads to a condition called diabetes. Human insulin can now be synthetically produced from bacteria. *See also* GLUCAGON.

intercalation compound A compound in which one type of atom or ion occurs in the gaps within the regular crystal lattice of another element or compound. Such materials are often non-STOICHIOMETRIC. *See also* INTERSTITIAL.

intermetallic compound A compound formed between metals. Such compounds are ALLOYS in which the atoms of the metals are present in a simple ratio and form a regular lattice structure.

intermolecular bonding Weak INTERMOLECULAR FORCES, such as HYDROGEN BONDS, holding together molecules that are themselves covalently bound. At relatively low temperatures these bound structures can behave like single molecules, but at high temperatures they fall apart into the constituent molecules. Intermolecular bonding is particularly important in many biological processes. *See also* VAN DER WAALS' BOND.

intermolecular forces The forces that act between one molecule and another in a substance. If these forces are strong enough compared to the energy of any thermal vibrations – in other words, if the substance is cold enough – the intermolecular forces will hold the substance together as a solid or liquid. At higher temperatures, or with weaker forces, the material will behave as a gas. In an IDEAL GAS there are no intermolecular forces.

All intermolecular forces are ELECTROSTATIC in origin and are the result of forces between the electrons and the nuclei of the molecules involved, governed by the rules of QUANTUM MECHANICS. In all cases, there is a link between the strength of the intermolecular forces and the melting point and also the amount of energy needed to melt the material (the LATENT HEAT). The way in which the intermolecular forces change with distance is linked to the stiffness of the resulting material, though most materials are much weaker than might be expected from the strengths of the individual intermolecular forces, owing to the existence of defects in the lattice structure, which also account for plastic behaviour.

See also HYDROGEN BONDING, IONIC SOLID, MACROMOLECULE, METAL, VAN DER WAALS' FORCE.

internal energy The energy that atoms or molecules in a substance possess as a result of forces between themselves and the KINETIC ENERGY of their random thermal motion, as opposed to any bulk motion of the object, or external forces acting on the object. When an object is heated, provided no external WORK is done, the increase in the internal energy of the object is equal to the amount of heat energy supplied. In an IDEAL GAS, the internal energy is purely due to the kinetic energy of the molecules. *See also* FIRST LAW OF THERMODYNAMICS, JOULE'S LAW.

International Practical Temperature Scale (IPTS) A temperature scale designed to conform as close as possible to thermodynamic temperature. The unit of temperature is the KELVIN (K). The latest version, devised in 1990, has 16 fixed points. The IPTS supersedes all other temperature scales for scientific purposes.

interstitial (*adj.*) Describing an atom located in the interstices, or spaces, between atoms in a regular CRYSTAL lattice. For example, in steel, carbon atoms occupy the interstices between the much larger iron atoms.

Invar (*Trade name*) An alloy of iron (64 per cent) and nickel (36 per cent) with small amounts of carbon and manganese. The alloy has an extremely small expansion over normal temperatures. It is used where it is important that a change in temperature does not produce a change in size.

inverse square law The behaviour of any quantity that radiates from a point source or the surface of a sphere with its strength falling in such a way that it is reduced by a factor of four when the distance from the source is doubled.

inversion A reaction in which there is a change from one optically active configuration of a compound to the opposite configuration (*see* OPTICAL ACTIVITY).

inversion temperature The temperature below which a gas will cool in a FREE EXPANSION.

iodic acid Iodic(V) acid, HIO_3, and iodic(VII) acid or periodic acid, H_5IO_6. Iodic(V) acid is a pale yellow liquid, which decomposes on heating; relative density 4.6. It can be made by oxidizing iodine with hydrogen peroxide:

$$4H_2O_2 + I_2 \rightarrow 2HIO_3 + 4H_2O$$

Iodic(V) acid dissolves in water and is a strong acid and an OXIDIZING AGENT. Iodic(VII) acid is a white solid that decomposes on heating. It forms a weak acid in an aqueous solution.

iodide Any BINARY COMPOUND containing IODINE.

iodine (I) The element with atomic number 53; relative atomic mass 126.9; sublimes at 183°C; relative density 4.9. Iodine is a HALOGEN and occurs as deep purple crystals. It is insoluble in water but soluble in organic solvents such as ethanol or tetrachloromethane. Iodine is extracted as sodium iodide from sea water and used in the manufacture of silver iodide, an important light-sensitive chemical in the photographic industry. Iodine is concentrated by the body in the thyroid gland, which makes its radioactive ISOTOPES a dangerous part of nuclear fallout.

iodine oxide Any of the BINARY COMPOUNDS of iodine and oxygen. They are: iodine(IV) oxide, I_2O_4, iodine(V) oxide, I_2O_5, and iodine(IV) iodine(V) oxide, I_4O_9. Iodine (IV) oxide is a yellow powder, obtained by heating IODIC ACID with concentrated sulphuric acid and treating the resulting mixture with water. On heating, it decomposes to iodine and iodine (V) oxide. It is believed to have the structure $O{=}I{-}IO_3$.

Iodine(IV) iodine(IV) oxide is a yellow solid, formed by warming iodine in ozone,

$$4I + 3O_3 \rightarrow I_4O_9$$

It is believed to have the structure $I(IO_3)_3$. It decomposes on heating to give iodine, oxygen and iodine(V) oxide.

Iodine(V) oxide is a white solid; decomposes on heating; relative density 4.9. It is the most stable of the iodine oxides. It is formed by heating iodic acid,

$$2HIO_3 \rightarrow I_2O_5 + H_2O$$

It is a powerful OXIDIZING AGENT, but decomposes on heating to give iodine and oxygen.

iodoform Common name for TRIIODOMETHANE.

iodoform test *See* TRIIODOMETHANE TEST.

ion An ATOM or MOLECULE that is not electrically neutral, having gained or lost ELECTRONS. An ion that has lost one or more electrons (cation) is positively charged, whereas one that has gained electrons (anion) is negatively charged. *See also* IONIC BOND, IONIC SOLID, IONIZATION, IONIZATION ENERGY, ZWITTERION.

ion exchange The exchange of ions between a solution and a solid material. When dissolved ions are washed past the solid, they are removed and replaced by a different ion. This process occurs in nature, particularly in the uptake of fertilizer by plants, and it is also used to remove magnesium and calcium ions from HARD WATER, replacing them with sodium ions. To reverse the process, and regenerate the ion exchange material, a concentrated solution of sodium ions, from dissolved sodium chloride, is passed over the exchange material. Similar techniques, based on synthetic POLYMERS can be used to remove all ANIONS and replace them with HYDROXIDE ions (OH^-) or to replace CATIONS with hydrogen ions (H^+). By using these together, water can be purified, producing deionized water. Ion exchange is a very powerful technique and can be used to separate, for example, two proteins differing in a single AMINO ACID only.

ion exchange chromatography CHROMATOGRAPHY in which an ION EXCHANGE resin is used as the stationary phase. Competition between absorption rates for different ions means that different ions migrate through the chromatography column at different rates.

ionic (*adj.*) Describing a material, particularly a solid or a solution, that contains IONS or is held together by IONIC BONDS.

ionic bond A CHEMICAL BOND in which one or more electrons are transferred from one atom to another, creating a positive and negative ION. The attractive forces between these ions then holds the material together as a crystalline solid. These solids are generally hard, brittle materials with high melting points. They are usually soluble in water and insoluble in organic compounds. They always conduct electricity when molten or in solution.

Ionic bonds are generally formed between materials of highly differing ELECTRONEGATIVITY, thus the most electronegative and electropositive (*see* ELECTROPOSITIVITY) materials generally form ionic bonds. Although the removal of an electron requires an input of energy (*see* IONIZATION ENERGY), as may the formation of a negative ion (*see* ELECTRON AFFINITY), the energy released by the formation of a crystal from these ions more than offsets these requirements in substances that form ionic crystals. For example, when sodium reacts with chlorine to form sodium chloride, the formation of sodium ions from solid sodium requires 610 kJ mol^{-1}, whilst the formation of chlorine ions from molecular chlorine releases 223 kJ mol^{-1}. Thus the formation of the ions requires a net energy input of 387 kJ mol^{-1}. However, the formation of the crystal releases an energy of 779 kJ mol^{-1}, so the process as a whole is EXOTHERMIC.

Compare COVALENT BOND. *See also* FAJAN'S RULES, IONIC SOLID, SHELL.

ionic product of water The product of the CONCENTRATION of hydrogen and hydroxide ions in water, $[H^+][OH^-]$. This is a constant over all conditions of acidity and alkalinity and is equal to $10^{-14}\,mol^2\,dm^{-6}$.

ionic radius A measure of the effective size of an ION in an IONIC SOLID. The ionic radii of all ions increases with increasing ATOMIC NUMBER and is larger for ANIONS than for CATIONS in the same PERIOD. Typical ionic radii vary between about 0.03 nm (Be^{2+}) and 0.22 nm (Te^{2-}).

Ionic radii may be determined by finding the interatomic spacing in the crystal from its density and the masses of the atoms involved, or by X-RAY CRYSTALLOGRAPHY. This spacing is then allocated to the two types of ion in the crystal in a way that aims to give consistent results for all combinations of ions. The results are good enough to suggest that this technique is a valid one. However, it clearly contains some approximations arising from the different lattice structure of different crystals, the non-spherical distribution of charge in an ion, and the partially covalent nature of some bonds.

ionic solid A crystalline solid in which adjacent atoms gain and lose one or more electrons to form a regular lattice of IONS. Each ion is attracted to its nearest neighbours with the opposite charge. There is also a repulsive element in the interatomic force at short distance, caused by the effect of the PAULI EXCLUSION PRINCIPLE, which begins to promote electrons to higher ENERGY LEVELS as the atomic electron clouds of neighbouring atoms start to overlap. As a result of the strong nature of the ionic attractions, ionic materials tend to have melting and boiling points above room temperature and are usually fairly hard materials. *See also* COVALENT CRYSTAL.

ionic strength A measure of the effect of IONS in a solution on other ions in the solution. The ionic strength of an ion in solution is equal to its molar concentration multiplied by the square of its charge. The ionic strength of the solution overall is the sum of the ionic strength of the ions it contains.

ionization The process of creating IONS in a substance that previously contained neutral atoms or molecules, such as by a chemical reaction or IONIZING RADIATION. *See also* HEAT OF IONIZATION.

ionization energy, *ionization potential* The minimum energy needed to remove an electron infinitely far away from an atom, often specified either in ELECTRON-VOLTS or per mole of atoms. The first ionization energy relates to the removal of the first electron, the second ionization energy is the additional energy needed to remove a second electron and so on.

ionization potential *See* IONIZATION ENERGY.

ionized (*adj.*) Describing a material that contains IONS – for example as a result of chemical action as in acidified water, or IONIZING RADIATION.

ionizing radiation Any RADIATION that creates IONS in any matter through which it passes. Ionizing radiation may be a stream of high-energy particles (such as ELECTRONS, PROTONS or ALPHA PARTICLES) or short-wavelength ELECTROMAGNETIC RADIATION (ULTRAVIOLET radiation, X-RAYS or GAMMA RADIATION).

For all types of ionizing radiation, the level of radiation can be reduced by passing the radiation through a material containing as many electrons as possible, to provide the greatest number of opportunities for the radiation to lose energy by IONIZATION. Thus lead, which has a large number of electrons per unit volume, so is a dense material, is a much better absorber of ionizing radiation than aluminium.

See also BACKGROUND RADIATION, BECQUEREL, DECAY CONSTANT, RADIOCARBON DATING.

ion pair A pair of ions with equal and opposite charges. The energy involved in forming an ion pair from a pair of neutral atoms depends on the separation of the ions in the pair, and is an important contribution to the total energy involved in forming an ionic compound from its elements.

iridium (Ir) The element with atomic number 77; relative atomic mass 192.2; melting point 2,410°C; boiling point 4,130°C; relative density 22.4. Iridium is a silvery TRANSITION METAL. It is used in the manufacture of hard alloys for wear-resistant surfaces, though its high cost limits this to fairly small bearings in precision machinery.

iron (Fe) The element with atomic number 26; relative atomic mass 55.8; melting point 1,535°C; boiling point 2,750°C; relative density 7.9. Iron is the fourth most abundant element in the Earth's crust. It is extracted from its ores (mostly iron oxides) by reduction by carbon in a BLAST FURNACE, for example:

$$2Fe_2O_3 + 3C \rightarrow 4Fe + 3CO_2$$

The resulting PIG IRON contains too much carbon to be used directly and is subsequently reacted, in an ELECTRIC ARC FURNACE or BESSEMER CONVERTER, with oxygen to remove excess carbon, to make steel.

Iron is the most widely used metal. Its tendency to rust is offset by its high strength and low cost. Iron is a TRANSITION METAL that forms compounds of valency 2 (traditionally called ferric) and 3 (traditionally called ferrous).

iron chloride Iron(II) chloride, or ferric chloride, $FeCl_2$, and iron(III) chloride, or ferrous chloride, $FeCl_3$.

Iron(II) chloride is a greenish-yellow salt; melting point 670°C; decomposes on further heating; relative density 3.2. It can be made by heating iron in hydrogen chloride:

$$Fe + 2HCl \rightarrow FeCl_2 + H_2$$

Iron(III) chloride is a dark brown solid; melting point 306°C; decomposes on further heating; relative density 2.9. It can be formed by burning iron in chlorine:

$$2Fe + 3Cl_2 \rightarrow 2FeCl_3$$

Iron(III) chloride is soluble but hydrolyses partially to iron(III) hydroxide:

$$FeCl_3 + H_2O \rightarrow Fe(OH)_3 + 3HCl$$

iron oxide Iron(II) oxide, FeO, iron(III) oxide, Fe_2O_3, and the mixed oxide, iron(II)–iron(III) oxide, Fe_3O_4.

Iron(II) oxide is a black powder; melting point 369°C; decomposes on further heating; relative density 5.7. It can be produced by reacting iron with ethanedioic acid to form iron(II) oxalate, which then decomposes to iron(II) oxide when heated in the absence of air:

$$Fe + (COOH)_2 \rightarrow Fe(COO)_2 + H_2$$

$$Fe(COO)_2 \rightarrow FeO + CO + CO_2$$

Iron(II) oxide is a base, dissolving in strong acids to form iron(II) salts, for example:

$$FeO + H_2SO_4 \rightarrow FeSO_4 + H_2O$$

Iron(III) oxide is a reddish brown solid; melting point 1,564°C; decomposes on further heating; relative density 5.2. It is formed naturally in rusting. It can also be formed by roasting iron in oxygen:

$$4Fe + 3O_2 \rightarrow 2Fe_2O_3$$

Iron(III) oxide is readily reduced, by carbon monoxide for example:

$$Fe_2O_3 + 3CO \rightarrow 2Fe + 3CO_2$$

Iron(II)–iron(III) oxide, which occurs naturally as the mineral magnetite, is a black solid. It contains iron atoms alternately in the +2 and +3 OXIDATION STATES.

iron sulphate Iron(II) sulphate $FeSO_4$ and iron(III) sulphate $Fe_2(SO_4)_3$.

Iron(II) sulphate generally occurs as the heptahydrate, $FeSO_4.7H_2O$, a bluish-green solid; relative density 1.9; melting point 64°C; decomposes on further heating. Iron(II) sulphate can be formed by the action of sulphuric acid on iron:

$$Fe + H_2SO_4 \rightarrow FeSO_4 + H_2$$

On heating, it decomposes to give iron(III) oxide:

$$2FeSO_4 \rightarrow Fe_2O_3 + SO_2 + SO_3$$

Iron(III) sulphate is a brownish yellow solid, which decomposes on heating; relative density 3.1. It can be formed by the OXIDATION of iron(III) sulphate, for example by the action of hydrogen peroxide:

$$2FeSO_4 + H_2SO_4 + H_2O_2 \rightarrow Fe_2(SO_4)3 + 2H_2O$$

irreversible (*adj.*) Describing a change that cannot happen in reverse, usually because to do so would violate the SECOND LAW OF THERMODYNAMICS.

isobar 1. Any one of a number of NUCLEI with the same MASS NUMBER (hence the same number of NUCLEONS) but different ATOMIC NUMBERS, and so are different elements.

2. A line on a graph of volume against temperature linking points at which the pressure of a fixed mass of material will be the same. For an IDEAL GAS, the isobars are straight lines passing through the origin.

isobutane *See* BUTANE.

isoelectric point The state of particles in a COLLOID that have no net charge as a result of being equally likely to absorb positive and negative ions from the dispersed phase. Since the particles in many colloids rely on their ELECTROSTATIC repulsion to keep them separate, when a colloid is at its isoelectric point it is most likely to coagulate (*see* COAGULATION) into larger particles, forming a PRECIPITATE.

isoelectronic (*adj.*) Describing a pair or group of compounds that have the same total number of VALENCE ELECTRONS. For example, nitrogen and carbon monoxide are isoelectronic as they both have a total of 10 valence electrons.

isoleucine An ESSENTIAL AMINO ACID, present in CASEIN and body tissue.

isomer A chemical compound that possesses the same composition and MOLECULAR MASS as another but differs in its physical or chemical properties. The differences are due to the structural arrangement of the constituent atoms. For example, there are two organic compounds with the formula C_4H_{10} – butane $CH_3(CH_2)_2CH_3$ and methyl propane $CH_3CH(CH_3)CH_3$. These have the same MOLECULAR FORMULA but differ in their STRUCTURAL FORMULA and are therefore structural isomers:

$$CH_3 - CH_2 - CH_2 - CH_3 \quad \text{butane}$$

$$\begin{array}{c} CH_3 - CH - CH_3 \\ | \\ CH_3 \end{array} \quad \text{methyl propane}$$

Structural isomers are distinct compounds with differing properties and can even belong to different classes of compounds.

Structural isomers differ in the order in which the atoms are joined together but some isomers are less obvious and are due to differences in their three-dimensional structure, seen only when models are constructed. Optical isomers are mirror images of each other and geometric isomers arise as a result of the different spatial arrangement of atoms around a central plane of symmetry.

See also GEOMETRIC ISOMERISM, OPTICAL ISOMERISM, TAUTOMERISM.

isomerase One of a group of ENZYMES that causes the rearrangement of groups within a molecule.

isomorphic (*adj.*) Describing two crystalline materials with the same lattice structure. Thus sodium chloride and potassium bromide are isomorphic, as they each form a crystalline structure in which each ANION is surrounded symmetrically by six CATIONS, resulting in a cubic structure described as face-centred cubic.

isonitrile, *isocyanide, carbylamine* Organic compounds with the general formula R–NC. They are colourless liquids with an unpleasant odour and are toxic.

iso-octane *See* OCTANE.

isoprene *See* 2-METHYLBUTADIENE.

isotactic polymer A POLYMER with a regular structure where all the substituted carbons have the same stereo-configuration and are on the same side of the carbon chain, if it is considered that the carbon atoms lie in the same plane. This arrangement gives a polymer great strength, for example in POLYPROPYLENE. *Compare* ATACTIC POLYMERS, SYNDIOTACTIC POLYMER.

isothermal (*adj.*) Describing a change that takes place at constant temperature.

isotonic (*adj.*) Describing a solution that has the same OSMOTIC PRESSURE as some reference solution.

isotope Any one of a series of atomic nuclei of the same element, each with the same number of PROTONS but different numbers of NEUTRONS, hence different masses. *See also* NUCLEUS, RELATIVE ATOMIC MASS.

isotropic (*adj.*) Describing a material, usually a crystalline solid, that has the same physical properties, such as thermal or electrical conductivity, regardless of the direction in which these properties are measured. *Compare* ANISOTROPIC.

JK

joule (J) The SI UNIT of WORK and ENERGY. One joule of work is done when a force of one NEWTON moves through a distance of one metre in the direction of the force.

Joule's law For an IDEAL GAS, the INTERNAL ENERGY depends only on temperature, not on pressure or volume. REAL GASES show slight departures from Joule's law, most cooling as they expand, to compensate for work done against INTERMOLECULAR FORCES, though some gases, such as hydrogen, show an increase in temperature.

kainite A mineral DOUBLE SALT, $MgSO_4.KCl.3H_2O$. It is mined as a source of potassium.

kaolin, *China clay* A naturally occurring CLAY containing small particles of a pure white colour. It is formed by the erosion and weathering of FELDSPAR-rich rocks, such as granite, and is widely used in the manufacture of ceramic objects and as a filler in paints.

Kastner-Kellner process The ELECTROLYSIS of BRINE in a MERCURY-CATHODE CELL to produce chlorine and sodium hydroxide.

kelvin (K) The SI UNIT of temperature. The size of the kelvin is the same as the degree CELSIUS. The TRIPLE POINT of water is fixed at exactly 273.16 K: the kelvin is defined as the fraction 1/273.16 of the temperature above ABSOLUTE ZERO of the triple point of water. *See also* ABSOLUTE TEMPERATURE.

Kelvin statement of the second law of thermodynamics No system can convert heat energy entirely to mechanical WORK – a certain amount of heat must be given out to the cooler surroundings.

keratin A fibrous sulphur-rich PROTEIN found in the skin of vertebrates which toughens the outer protective layer. It is also found in hair, nails, hooves, feathers and horns. Keratin is generally insoluble and is resistant to attack by ENZYMES.

kerosene A mixture of HYDROCARBONS of the ALKANE series obtained from PETROLEUM. It is a thin oil sold as a fuel under the common name of paraffin.

ketals Organic compounds with the general formula

$$\mathrm{R(R')C(OR'')_2}$$

Ketals are formed by the addition of an ALCOHOL to a KETONE. If one molecule of alcohol reacts with one molecule of ketone a hemiketal is formed. Further reaction of this with another alcohol produces a full ketal. Ketals are colourless liquids with characteristic odours. *See also* ACETALS.

ketone, *alkanone* One of a group of organic compounds containing the CARBONYL GROUP (C=O) attached to two ALKYL GROUPS (R and R′) R-CO-R′ or RR′CO. The general formula of a ketone is $C_nH_{2n}O$, the same as an ALDEHYDE, but in aldehydes there is only one alkyl group and one hydrogen atom. The nomenclature of ketones follows that of the ALKANE with the same carbon skeleton, with the ending *-ane* being replaced by *-anone.* A number is placed before the *-one* in the name to indicate the position of the carbonyl carbon, for example pentan-3-one. Other examples include propanone (acetone) and phenylethanone. Ketones can be ALIPHATIC (usually liquids), AROMATIC (usually solids), CYCLIC or mixed. Ketones are formed by the OXIDATION of secondary ALCOHOLS:

$$\mathrm{RR'CHOH} \xrightarrow{\text{oxidized}} \mathrm{RR'C{=}O}$$

Ketones can be reduced back to secondary alcohols but are not themselves easily oxidized, unlike aldehydes, and this feature is used to distinguish them from aldehydes in FEHLING'S TEST or using TOLLEN'S REAGENT. *See also* FRIEDEL–CRAFTS REACTION, SCHIFF'S REAGENT.

ketose, *keto-sugar* A sugar having a KETONE group (C=O). *See also* MONOSACCHARIDE.

keto-sugar *See* KETOSE.

kieselguhr A SILICATE mineral formed from the fossilized remains of a class of single-celled creatures called diatoms. Kieselguhr is widely used for its porous, absorbent nature, for example in the manufacture of dynamite, and as a thermal insulator.

kilo- (k) A prefix before a unit indicating that the size of the unit is to be multiplied by 10^3, for example kilowatt (kW) is equivalent to one thousand watts.

kilogram The SI UNIT of mass. One kilogram is defined as the mass of the International prototype kilogram, a cylinder of platinum-iridium alloy kept at Sèvres, near Paris, France.

kinetic energy The energy of movement. The kinetic energy, E, of a mass m moving at a speed v is

$$E = {}^1/_2 mv^2$$

kinetics The measurement and study of the rates of CHEMICAL REACTIONS and biological processes. In particular, the determination of the MECHANISMS of such reactions, by studying the effects that factors such as temperature, pressure and CATALYSTS have on the rate. Kinetics looks at the sequence in which the events leading to a chemical reaction take place, and the intermediate products that are created. *See also* RATE OF REACTION.

kinetic theory The explanation of the properties of matter in terms of the motion of its component atoms and molecules. The temperature of a body is dependent on the INTERNAL ENERGY, and therefore the velocity, of its molecules. If heat is supplied to a substance, such as a gas, the velocity of the particles increases and the temperature rises.

Kinetic theory explains the pressure of a gas in terms of the impacts of its molecules on to the walls of its container. If the temperature rises, the number of impacts per second increases, and the pressure increases. By making various assumptions about the molecules of an IDEAL GAS (such as negligible forces between molecules, the molecules themselves take up a negligible volume, etc.), kinetic theory gives rise to the GAS LAWS. Kinetic theory also extends to liquids and solids and explains such physical phenomenon as thermal expansion and CHANGES OF STATE in terms of the thermal motion of the particles. *See also* BROWNIAN MOTION, STATISTICAL MECHANICS.

Kipp's apparatus A vessel used in laboratories for the production of gases from the reaction of a liquid (held in the upper part of the vessel) and a solid, held in the lower part. As gas pressure builds up in the apparatus, the liquid is forced back into the upper chamber, preventing further reaction. An example of the use of Kipp's apparatus is the preparation of hydrogen sulphide from the reaction of hydrochloric acid and iron sulphide.

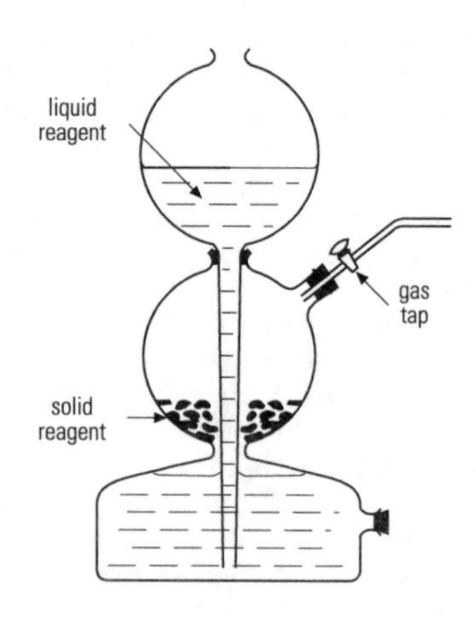

Kipp's apparatus.

Kjeldahl's method A method devised by Johan Kjeldahl (1849–1900) for determining the amount of nitrogen in an organic compound. The nitrogen is converted to ammonium sulphate (NH^{4+}) using concentrated sulphuric acid and a copper(II) sulphate catalyst. Alkali is added to neutralize the mixture and it is heated to distil off ammonia, NH_3, the amount of which can then be estimated by TITRATION. The amount of nitrogen in the original sample can then be calculated.

knocking A phenomenon occurring in petrol engines that reduces the power output. Knocking sounds result from explosions of unburned fuel-air mixture before it is ignited. The extent of knocking depends on the fuel composition and can be reduced by using fuel with a higher OCTANE NUMBER. Lead is used in petrol to reduce knocking but is being phased out due to its association with mental retardation in children.

Kohlrausch's equation An equation describing how the MOLAR CONDUCTIVITY of a strong ELECTROLYTE varies with concentration.

$$\Lambda = \Lambda_\infty - kC^{1/2}$$

where Λ is the molar conductivity at concentration C, Λ_∞ is the molar conductivity in the limit of infinite dilution and k is a constant.

Kohlrausch's law The MOLAR CONDUCTIVITY of any IONIC solution at infinite dilution is the sum of the molar conductivity of the ions it contains. This law simply expresses the idea that at low concentration, the ions in a solution carry charge independently of one another.

Kolbe reaction A method of preparing saturated or unsaturated hydrocarbons by the ELECTROLYSIS of a solution of a CARBOXYLIC ACID salt. The reaction was discovered by Herman Kolbe (1818–84).

An example is the electrolysis of ethanoic acid to give ethane.

$$2CH_3CO_2^- \longrightarrow CH_3\text{–}CH_3 + 2CO_2$$

Krebs cycle, *TCA cycle, tricarboxylic acid cycle, citric acid cycle* A series of biochemical reactions that constitute the second stage of respiration in aerobic (oxygen-requiring) organisms. In the Krebs cycle, the three-carbon compound pyruvate, formed from GLUCOSE during the first stage of GLYCOLYSIS, is converted, in the presence of oxygen, to carbon dioxide and hydrogen atoms. The Krebs cycle is named after its discoverer Hans Krebs (1900–81).

Before entering the cycle, pyruvate combines with a compound called COENZYME A to form acetyl coenzyme A, which then enters the Krebs cycle to combine with OXALOACETIC ACID. The cycle itself generates very little energy for the cell but the hydrogen atoms are carried to the ELECTRON TRANSPORT SYSTEM, by the carriers NAD or FAD, where they later generate energy in the form of ATP. The carbon dioxide is removed as a waste product. The process is cyclic, with the end-product, oxaloacetic acid, being the substrate for the next round.

The reactions of the Krebs cycle occur in the living cells. In addition to degrading pyruvate to carbon dioxide and providing hydrogen atoms for the electron transport system, the intermediate compounds of the Krebs cycle are needed to make other substances, such as FATTY ACIDS, AMINO ACIDS and CHLOROPHYLL, and so this cycle is considered to play a key role in a cell's biochemistry.

krypton (Kr) The element with atomic number 36; relative atomic mass 83.8, melting point –157°C; boiling point –152°C. Krypton is a NOBLE GAS. Its low abundance in the atmosphere, compared to argon, means it is only used in special gas discharge tubes.

kurchatovium A name proposed for element number 104. This element has been manufactured in small amounts, but all ISOTOPES are unstable, with short HALF-LIVES.

L

labile (*adj.*) Describing a compound in which certain atoms or groups of atoms, such as LIGANDS, are easily replaced by other atoms or groups.

lactams A group of colourless, poisonous solids formed when AMINO ACIDS are heated. This causes a reaction between the CARBOXYL GROUP of a molecule and the AMINO GROUP with the elimination of water, and a cyclic AMIDE is formed. The cyclic compound then has the group –NH.CO.– as part of the ring. URACIL is an example of a lactam.

lactate fermentation *See* FERMENTATION.

lactic acid, ***2-hydroxypropanoic acid*** ($CH_3CHOHCOOH$) A colourless, odourless, syrupy liquid organic acid with a sour taste; melting point 18°C; boiling point 122°C. It is produced naturally by certain bacteria during FERMENTATION and is responsible for the taste of sour milk. Lactic acid is also produced by muscle cells during vigorous exercise, when oxygen is limited. The build up of lactic acid in muscles causes the cramp pains experienced at such times. Lactic acid is manufactured by the fermentation of various sugars, including LACTOSE, and is used in the food industry and in the dyeing and tanning industries.

lactone Any one of a group of organic ANHYDRIDES formed by the reaction between a hydroxyl group on a molecule of a hydroxy-acid and a carboxyl group on the same molecule. The reaction is accompanied by the elimination of a molecule of water. The lactone formed contains a ring of atoms with –CO.O– forming part of the ring structure.

γ-lactones have five-membered rings, δ-lactones have six-membered rings and β–lactones have four-membered rings. The γ- and δ-lactones form relatively easily; they form spontaneously in concentrated solutions of the acids.

lactose A DISACCHARIDE made from the combination of GLUCOSE and GALACTOSE. Lactose is found in mammalian milk (5 per cent in cows' milk) and is important to the suckling young.

laevorotatory (*adj.*) Describing a form of OPTICAL ACTIVITY in which the plane of POLARIZATION of PLANE-POLARIZED light is rotated in an anticlockwise direction. It is also used to describe the ENANTIOMER of a compound exhibiting OPTICAL ISOMERISM that rotates plane-polarized light anticlockwise. This used to be denoted by the prefix *l*- but (–) is now used. This is not to be confused with the prefix L- used to indicate the configuration of CARBOHYDRATES and AMINO ACIDS.

lake One of several insoluble pigments made by precipitating natural or artificial organic dyes with an inorganic compound, usually HYDROXIDES, OXIDES or SALTS. A coloured complex results, which is used in paints and printing inks.

lamellar solids Solids in which the crystalline structure occurs only in a single plane, so that the material tends to form as sheets or flakes. MICA and GRAPHITE are examples. In graphite, the carbon atoms within a plane form covalent bonds with one another, to form an interlinked plane of hexagons, but there is only weak bonding between the atoms in one layer and another.

lance A water-cooled pipe used to carry a gas, usually oxygen, into a furnace. *See* OPEN HEARTH FURNACE, OXYGEN FURNACE.

lanolin A pale yellow substance obtained from wool fat; melting point 37°C. It consists of an emulsion of CHOLESTEROL and its ESTERS in water. It is used in cosmetics and medicine, as it makes a good base for creams and ointments.

lanthanide, ***lanthanoid, rare earth element*** Any one of the 15 elements with atomic numbers from 57 (LANTHANUM) to 71 (LUTETIUM). Each has its outer electrons in 4f ORBITALS (except for lanthanum). They are grouped together in the PERIODIC TABLE because they have very similar chemical properties, which differ only slightly with atomic number.

lanthanoid *See* LANTHANIDE.

lanthanum (La) The element with atomic number 57; relative atomic mass 138.9; melting point 918°C; boiling point 3,464°C; relative density 6.1. The first member of the LANTHANIDE series, it is a soft metallic element used in some alloys.

lapis lazuli A blue mineral widely used for jewellery, particularly by the ancient Egyptians. The blue colour arises from sulphur impurities in sodium aluminium silicate.

latent heat The heat energy taken in or given out when a material changes its physical state at a constant temperature. Energy is taken in when a solid turns to a liquid, or a liquid to a gas, as energy is needed to overcome the forces holding the molecules together in a solid or liquid. When a gas condenses, or a liquid turns into a solid, energy is released as chemical bonds form holding the molecules together. *See also* CLAPEYRON CLAUSIUS EQUATION.

lattice An array of points arranged in space in some symmetrical pattern, such as the positions occupied by atoms in a crystal. *See also* ISOMORPHIC.

lattice energy The energy required to separate the molecules or ions of a crystal so they are infinitely far apart from one another. Often quoted for one mole of the material. *See also* BORN–HABER CYCLE.

lava Semi-molten rock, such as that ejected from a volcano during an eruption. It is derived from magma, and when reaching the surface it cools to form igneous rocks.

law of chemical equilibrium *See* EQUILIBRIUM CONSTANT.

law of conservation of energy In any closed system, where no energy can enter or leave, the total amount of energy remains constant. Thus energy can never be created or destroyed, though energy may be transferred from one form into another.

law of constant composition *See* LAW OF CONSTANT PROPORTIONS.

law of constant proportions, *law of constant composition* A pure sample of any compound will always contain the same proportions of each of the constituent elements. If these masses are divided by the RELATIVE ATOMIC MASSES of the elements, the proportions will reduce to simple ratios. Thus sodium chloride contains 39 per cent sodium by mass and 61 per cent chlorine, and this is true for any pure sample of the material no matter how it is obtained. Once it is realized that the chlorine atoms are heavier than the sodium atoms, it can be calculated that this means that there is one atom of sodium for each atom of chlorine, and the formula NaCl can be deduced. This law, first put forward by John Dalton (1766–1844), was an important contribution to the fundamental ideas of chemistry (*see* DALTON'S ATOMIC THEORY).

law of equivalent proportions, *law of reciprocal proportions* If two elements, A and B, both form a compound with a third element C, then when A and B combine together they will do so in the same proportions in which they reacted with C. If A and B are oxygen and hydrogen for example, 12 g of carbon will combine with 4 g of hydrogen (to form methane) or 32 g of oxygen (to form carbon dioxide). When hydrogen and oxygen combine together to form water, 32 g of oxygen will combine with 4 g of hydrogen.

law of mass action The rate at which a CHEMICAL REACTION takes place is dependent on the CONCENTRATION of the reagents. In a reaction where one MOLE of A reacts with one mole of B, the rate at which the reaction takes place is proportional to the concentrations of A and B, represented in equations by [A] and [B] respectively, so

$$\text{rate of reaction} = k[\mathrm{A}][\mathrm{B}]$$

where k is a constant, called the RATE CONSTANT. For a more complex reaction, involving different numbers of moles of the reagents, but still reacting in a single step,

$$x\mathrm{A} + y\mathrm{B} + z\mathrm{C} \rightarrow \text{products}$$

where x, y and z are the numbers of moles involved,

$$\text{rate of reaction} = k[\mathrm{A}]^x\,[\mathrm{B}]^y\,[\mathrm{C}]^z$$

The law of mass action applies only if the reacting molecules have free access to one another, so is not applicable to reactions involving solids, where only the atoms on the surface of the solid are available to react. It is for this reason that finely divided solids react more rapidly. *See also* RATE OF REACTION.

law of multiple proportions If two materials, A and B, react together to form more than one compound, the mass of A that combines with a fixed mass of B in the two compounds will be a simple ratio. For example, in water, H_2O, 2 g of hydrogen combines with 16 g of oxygen. In hydrogen peroxide, H_2O_2, the same 16 g of oxygen will combine with just 1 g of hydrogen – a simple ratio of 2:1.

law of octaves An early attempt at grouping the chemical elements into families with similar properties, based on an analogy with musical notes. Whilst some elements, notably the HALOGENS, were correctly classified, the idea was superseded by the more sophisticated classification of the PERIODIC TABLE.

law of reciprocal proportions *See* LAW OF EQUIVALENT PROPORTIONS.

lawrencium (Lr) The element with atomic number 103. It does not occur naturally, and the longest lived ISOTOPE has a HALF-LIFE of only 3 minutes. Lawrencium was first synthesized by bombarding atoms of californium with high energy beryllium nuclei.

layer lattice A crystal lattice that occupies only a single plane, leading to the formation of a LAMELLAR SOLID.

LCAO (***Linear Combination of Atomic Orbitals***) An approximation to the form of a MOLECULAR ORBITAL constructed by adding or subtracting the ORBITALS of the atoms from which the molecule is made up.

leaching The process in which the passage of water or some other solvent through a porous solid material causes some part of that solid to be dissolved in the solvent. In particular, leaching is the washing away of substances out of the soil, leading to a reduction in soil fertility. Where fertilizers leach out of the soil, this can lead to pollution of rivers, etc.

lead (Pb) The element with atomic number 82; relative atomic mass 207.2; melting point 328°C; boiling point 1,740°C; relative density 11.4. Lead is a dense, soft metal, once used for covering roofs and making water pipes. Lead occurs widely in nature, mainly as lead sulphide in the ore galena. The ore is roasted to form the oxide, which is then reduced to metallic lead with carbon:

$$2PbS + 3O_2 \rightarrow 2PbO + 2SO_2$$

$$2PbO + C \rightarrow 2Pb + CO_2$$

Lead is widely used in the ELECTRODES of LEAD-ACID CELLS and in tetraethyl lead as an additive to petrol, though concerns about exposure to lead in the environment (it is a cumulative poison) have meant tighter controls on its use. Its relatively low cost and high density also makes lead a common choice for protection against IONIZING RADIATION, such as X-rays and gamma radiation.

lead(II) acetate *See* LEAD(II) ETHANOATE.

lead-acid cell A rechargeable electrochemical CELL with ELECTRODES of lead and lead sulphate and a sulphuric acid ELECTROLYTE, often used where high currents are needed, such as in motor vehicles.

lead carbonate ($PbCO_3$) A white solid; decomposes on heating; relative density 6.1. Lead carbonate is insoluble in water and can be made as a PRECIPITATE in the reaction between aqueous solutions of lead nitrate and ammonium carbonate:

$$PbNO_3 + NH_4CO_3 \rightarrow PbCO_3 + NH_4NO_3$$

Lead carbonate decomposes on heating to give lead oxide:

$$PbCO_3 \rightarrow PbO + CO_2$$

lead carbonate hydroxide, *basic lead carbonate, white lead* ($2PbCO_3.Pb(OH)_2$) A white powder; decomposes on heating; relative density 6.1. White lead was formerly used as a white pigment, but its high toxicity means it is no longer used. It reacts with hydrogen sulphide to form black lead sulphide:

$$2PbCO_3Pb(OH)_2 + 3H_2S \rightarrow 3PbS + 4H_2O + 2CO_2$$

lead chamber process A process for the commercial manufacture of SULPHURIC ACID. It has largely been superseded by the CONTACT PROCESS. In the lead chamber process, sulphur dioxide, nitrogen dioxide, steam and oxygen are mixed in a lead chamber. The sulphur dioxide is oxidized to sulphur trioxide by the nitrogen dioxide:

$$SO_2 + NO_2 \rightarrow SO_3 + NO$$

The nitrogen oxide is then regenerated by the oxygen:

$$2NO + O_2 \rightarrow 2NO_2$$

The sulphur trioxide dissolves in the water to produce acid with a concentration up to 77 per cent of saturated. The YIELD of the reactions is increased by recycling unreacted FEEDSTOCK.

lead(II) ethanoate, *lead(II) acetate, sugar of lead* ($Pb(CH_3COO)_2$) A white crystalline solid, soluble in water and slightly soluble in alcohol. The common form is as a trihydrate,

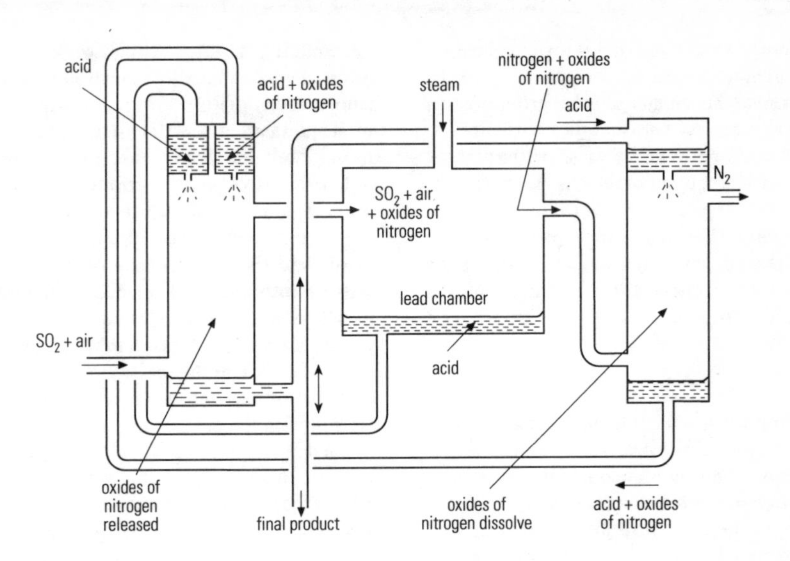

Lead chamber process.

$Pb(CH_3COO)_2.3H_2O$, but it exists as a DECAHYDRATE and in the ANHYDROUS form also. It has a sweet taste and is therefore known as the sugar of lead.

lead(IV) ethanoate, ***lead tetra-acetate*** ($Pb(CH_3COO)_4$) A colourless solid that decomposes in water and is soluble in pure ethanoic acid.

lead oxide Lead(II) oxide (or lead monoxide), PbO, or lead(IV) oxide (or lead dioxide), PbO_2.

Lead(II) oxide is a yellow solid; melting point 866°C; decomposes on further heating; relative density 9.5. Lead(II) oxide can be formed by heating lead nitrate:

$$2Pb(NO_3)_2 \rightarrow 2PbO + 4NO_2 + O_2$$

Lead oxide is AMPHOTERIC, dissolving readily in strong acids, for example:

$$PbO + 2HNO_3 \rightarrow Pb(NO_3)_2 + H_2O$$

but reacting with strong alkalis to form PLUMBATES, for example:

$$PbO + 2NaOH \rightarrow Na_2PbO_3 + H_2O$$

Lead(IV) oxide is a dark brown solid; decomposes on heating; relative density 9.4. Lead(IV) oxide can be made by the OXIDATION of lead(II) oxide, for example:

$$PbO + H_2O_2 \rightarrow PbO_2 + H_2O$$

Lead(IV) oxide is itself strongly oxidizing, and will reduce hydrochloric acid to give chlorine:

$$PbO_2 + 2HCl \rightarrow PbO + Cl_2 + H_2O$$

lead sulphate ($PbSO_4$) A white solid that is insoluble in water; melting point 1,170°C; relative density 6.2. It can be prepared as a precipitate by reacting a soluble lead compound with a soluble sulphate, for example:

$$PbCl_2 + K_2SO_4 \rightarrow PbSO_4 + 2KCl$$

Basic lead sulphate, a mixture of lead sulphate and lead hydroxide has been used as a white pigment, but concerns over the toxic nature of lead have led to the use of alternatives.

lead sulphide (PbS) A black solid; relative density 7.5; melting point 1,114°C; decomposes on further heating. Lead sulphide occurs naturally as the mineral GALENA. It is the commonest natural source of lead.

lead tetra-acetate *See* LEAD(IV) ETHANOATE.

lead tetraethyl, ***tetraethyl lead, lead(IV) tetraethyl*** ($Pb(C_2H_5)_4$) A colourless liquid; melting point −137°C; boiling point 200°C. It is used as an anti-knocking agent (*see* KNOCKING) in fuel. It is made by the reaction of CHLOROETHANE with a sodium/lead alloy, or more recently by the electrolysis of ethylmagnesium chloride (a GRIGNARD REAGENT) using a lead anode. Lead tetraethyl is toxic and causes the emission of poisonous chemicals into the atmosphere, hence its use is being reduced in favour of lead-free petrol.

Le Chatelier's principle In a REVERSIBLE REACTION, any change in conditions (such as temperature, pressure or concentration of one of the reagents) will change the equilibrium state of the reaction in such a way as to oppose the effects to the change. For example, in the formation of ammonia from nitrogen and hydrogen,

$$N_2 + 3H_2 \Leftrightarrow 2NH_3, \quad \Delta H = -92.4\ \text{kJ mol}^{-1}$$

the equilibrium will shift to the right (more ammonia) if: (i) the pressure is increased (the formation of ammonia reduces the number of molecules present, tending to reduce the pressure); (ii) the concentration of nitrogen and hydrogen is increased or the concentration of ammonia reduced; or (iii) the temperature is reduced (the reaction is EXOTHERMIC, production of ammonia tends to increase the temperature). *See also* CHEMICAL EQUILIBRIUM.

lecithin A PHOSPHOLIPID that is an important component of cell membranes of plants and animals.

Lennard–Jones 6-12 potential A form of interatomic POTENTIAL function used to model the effects of VAN DER WAALS' FORCES.

leucine, *α-aminoisocaproic acid* ($C_6H_{13}NO_2$) An AMINO ACID essential in nutrition. It is produced in several ways, in particular by the digestion of proteins by ENZYMES in the pancreas.

Lewis acid A term used to describe substances that are electron acceptors. Thus Lewis acids include conventional ACIDS but also some additional substances, such as $AlCl_3$, which can accept electrons from a chloride ion to form the $(AlCl_4)^-$ ion.

Lewis base A term used to describe substances that have available ELECTRONS and are thus electron donors. These include conventional BASES but also some additional substances, for example ammonia.

Liebig condenser A CONDENSER in which a straight glass tube is surrounded by a glass envelope through which cold water is passed.

ligand A molecule or ion that is attached to a metal ion in the formation of a complex (*see* COMPLEX ION). Water is the most common ligand, and many metal ions are HYDRATED, for example $[Cu(H_2O)]^{2+}$. The cyanide ion, CN^-, is another example; this forms complexes with iron, such as $[Fe(CN)_6]^{3-}$. Some larger ligand molecules can attach themselves to an ion at more than one point and are described as bidentate, as opposed to the more common monodentate ligands, which attach themselves to a metal ion at only one point.

ligand-field theory A version of CRYSTAL FIELD THEORY in which the detailed nature of the overlapping ORBITALS of the LIGANDS and the parent atom or ion is taken into account.

ligase One of a group of ENZYMES that forms COVALENT BONDS between two molecules using energy from the breakdown of ATP. DNA ligase in particular is very useful in genetic engineering since it joins DNA chains specifically cut by RESTRICTION ENDONUCLEASES.

light ELECTROMAGNETIC RADIATION of a wavelength to which the human eye is sensitive (visible light). The eye is sensitive to wavelengths from about 7×10^{-7} m (red) to 4×10^{-7} m (violet). The term is also commonly used for wavelengths outside this narrow range but where the properties being exploited are similar to those of visible light. Longer wavelengths are described as INFRARED and shorter wavelengths as ULTRAVIOLET. *See also* PHOTON, WAVE-PARTICLE DUALITY.

light petroleum *See* PETROLEUM ETHER.

lignin A naturally occurring organic polymer found in the cell walls of plants, composed of rings of carbon atoms joined in a chain. Lignin provides strength to the plant, and is difficult to digest so provides protection from attack by many organisms. Lignin is found in all wood and is therefore of commercial importance.

lime A term used in the common names of many calcium compounds, for example, limestone which is calcium carbonate, quicklime (calcium oxide), slaked lime (calcium hydroxide). *See also* LIME WATER.

limestone A SEDIMENTARY rock, similar to chalk but harder and more compact, formed by the deposition of the shells of marine animals, which are fused into rock by the pressure of further layers. Limestone is composed mainly of calcium carbonate and sometimes magnesium carbonate. The porous and slightly soluble nature of limestone is responsible for HARD WATER in limestone areas and for the formation of STALACTITES and STALAGMITES in limestone caves.

lime water Calcium hydroxide solution. In the presence of carbon dioxide, calcium carbonate is produced as a PRECIPITATE, though only in small quantities since calcium hydroxide is not very soluble. This can be used as a test for carbon dioxide, which turns the lime water 'milky':

$$Ca(OH)_2 + CO_2 \rightarrow CaCO_3 + H_2O$$

Linear Combination of Atomic Orbitals See LCAO

line spectrum An EMISSION SPECTRUM or ABSORPTION SPECTRUM in which the wavelengths absorbed or emitted form a number of separate very narrow ranges or lines. Each line in a spectrum represents a closely defined wavelength, corresponding to the PHOTON energy equal to the difference in energy between the ENERGY LEVELS that produced the line.

linoleic acid, *cis,cis-9,12-octadecadienoic acid* ($C_{18}H_{32}O_2$) An ESSENTIAL FATTY ACID present in many plant fats and oils, including groundnut oil, linseed oil and soya-bean oil. Linoleic acid is a polyunsaturated fatty acid with two double COVALENT BONDS. It is important in the human diet.

linolenic acid, *cis,cis,cis-9,12,15-octadecatrienoic acid* ($C_{18}H_{30}O_2$) An ESSENTIAL FATTY ACID present in some plant oils, such as linseed and soya-bean oil, and in algae. Linolenic acid is a polyunsaturated fatty acid with three double COVALENT BONDS.

linseed oil A pale yellow oil of vegetable origin, used as a DRYING OIL. It consists of a mixture of FATTY ACIDS, including LINOLEIC and LINOLENIC ACIDS.

Linz–Donawitz process The process for the production of STEEL in an OXYGEN FURNACE.

lipase One of a group of enzymes responsible for breaking down fats into FATTY ACIDS and GLYCEROL during digestion. Lipase is made by the pancreas and requires a slightly alkaline environment to function.

lipid Any one of a large group of organic compounds that are the major constituents in plant and animal fats, waxes and oils. Lipids all form ESTERS of FATTY ACIDS, and are soluble in alcohol but not water.

The most common alcohol with which fatty acids react to form lipids is GLYCEROL. The lipids formed in this way are called GLYCERIDES and can be mono, di or tri, depending on how many hydroxyl (OH) groups from glycerol have combined with the fatty acids; for example TRIGLYCERIDES have three fatty acids attached to glycerol. The properties of a lipid largely depend on the fatty acids present, because the alcohol in most lipids is glycerol. PHOSPHOLIPIDS are lipids containing glycerol in which one of the fatty acids is replaced by phosphoric acid.

In bacterial cell walls, lipids can associate with polysaccharides to form lipopolysaccharides. Some simple lipids can exist free and do not contain fatty acids. These include some STEROID HORMONES and TERPENES.

Lipids provide an energy store in plants and animals, and in the form of fat give some protection and insulation to animals and their internal organs. Lipids also provide waterproofing for plants and animals, as oily secretions and waxes formed by the combination of fatty acids with an alcohol other than glycerol. The emulsion test (which gives a cloudy solution when lipid and alcohol mix) or the Sudan III test (a red dye that detects fats and oils) detect lipids in a solution.

lipoprotein A complex of LIPIDS and protein that is a structural component of all cell membranes. CHOLESTEROL is transported in the blood within such a complex, either free or esterified (*see* ESTER) to FATTY ACIDS.

liquation A process for extracting one substance from a mixture by heating until the material with the lower melting point melts and can be poured away.

liquefaction The process of turning a gas into a liquid.

liquefied natural gas (LNG) NATURAL GAS, mostly METHANE, used as a liquid at high pressure and low temperature. Unlike propane and butane (*see* LIQUEFIED PETROLEUM GAS) methane cannot be liquefied by pressure alone, so tends to have more specialized uses. Natural gas can also be transported by ship in this form.

liquefied petroleum gas (LPG) Gaseous HYDROCARBONS, usually a mixture of PROPANE and BUTANE, stored and used as a liquid under pressure. LPG is increasingly used as an alternative to PETROL as a fuel for vehicles, with the advantages of lower cost and more complete combustion (so less of a pollutant).

liquid The state of matter in which a material is able to flow freely and take up the shape of its container, but where there is a distinct boundary between the material and its surroundings. Liquids are materials in which the molecules are closely spaced (like solids they are hard to compress, and they have similar densities to solids), but unlike most solids the molecules are randomly arranged. The forces between the molecules in a liquid are weak enough to allow liquids to flow, so they will take up the shape of any container in which they are placed, but the forces are strong enough to prevent the molecules from moving off into space, so a liquid has a fixed volume and will not expand to fill a volume.

liquid crystal Any liquid made from molecules that tend to line up under the influence of INTERMOLECULAR FORCES, producing long range ordering of a type more commonly associated with solid CRYSTALLINE materials. Under the influence of an electric field, this ordering can extend throughout the whole of the liquid.

liquid drop model A model of the atomic NUCLEUS that attempts to calculate the properties of the nucleus by treating it as a drop of liquid with a fixed DENSITY, SURFACE TENSION, specific LATENT HEAT of vaporization and so on. This model is most appropriate to nuclei containing large numbers of NUCLEONS. *See also* SHELL MODEL.

liquidus In a SOLID SOLUTION, the line on a graph of temperature against composition above which the substance is entirely liquid.

liquor A general term used in the chemical industry to describe a solution of some useful material in water, usually at a fairly high concentration.

lithium (Li) The element with atomic number 3; relative atomic mass 6.9; melting point 180°C; boiling point 1,340°C; relative density 0.5. Lithium is extracted from its ores by ELECTROLYSIS. It is the lightest of the ALKALI METALS, the small Li^+ ion making it markedly less reactive than the other alkali metals. Lithium is used increasingly in the manufacture of high energy batteries for use with cameras and similar small domestic appliances.

lithium carbonate (Li_2CO_3) A white solid; melting point 735°C; decomposes on further heating; relative density 2.1. Lithium carbonate is only slightly soluble and can be obtained as a precipitate, for example by the reaction between lithium sulphate and sodium carbonate:

$$Li_2SO_4 + Na_2CO_3 \rightarrow Li_2CO_3 + Na_2SO_4$$

Lithium carbonate is used as a drug in the treatment of some depressive illnesses.

lithium hydride (LiH) A white solid; melting point 686°C; decomposes on further heating; relative density 0.82. Lithium hydride can be formed by burning lithium in hydrogen:

$$2Li + H_2 \rightarrow 2LiH$$

Lithium hydride is a largely IONIC compound, containing the H^- ion, with the unusual property of liberating hydrogen at the ANODE in ELECTROLYSIS. Lithium hydride reacts violently with water:

$$LiH + H_2O \rightarrow LiOH + H_2$$

lithium hydroxide (LiOH) A white solid; melting point 450°C; decomposes on further heating; relative density 1.5. Lithium hydroxide can be formed by the reaction of lithium on water,

$$2Li + 2H_2O \rightarrow 2LiOH + H_2$$

Lithium hydroxide dissolves in water to form an alkaline solution, though it is less soluble than the other ALKALI METAL hydroxides. Lithium hydroxide is often used to absorb carbon dioxide, for example in manned spaceflight,

$$2LiOH + CO_2 \rightarrow Li_2CO_3 + H_2O$$

lithium oxide (Li_2O) A white solid; melting point 1,200°C; decomposes on further heating. Unlike the oxides of other ALKALI METALS, it is stable in the presence of water. It is used as a lubricant and as a flux for welding.

lithium tetrahydroaluminate ($LiAlH_4$) A pale grey powder; decomposes on heating; relative density 0.9. It is formed by the reaction of lithium hydride with aluminium chloride,

$$4LiH + AlCl_3 \rightarrow LiAlH_4 + 3LiCl$$

Lithium tetrahydroaluminate is soluble in ethoxyethane (ether) and in this form it is widely used as a powerful REDUCING AGENT for organic reactions, for example reducing CARBOXYLIC ACIDS and KETONES to ALCOHOLS. It decomposes on exposure to water to release hydrogen, so must be stored under dry conditions,

$$2LiAlH_4 + 2H_2O \rightarrow 2LiOH + 2Al + 5H_2$$

litmus A naturally occurring vegetable dye that can be used as an INDICATOR. A solution of litmus is red under ACIDIC conditions but blue in ALKALINE solutions.

litmus paper Paper soaked with LITMUS, used to test for ACIDIC or ALKALINE solutions.

litre (l) A unit of volume, correctly called the decimetre cubed (dm^3) in the SI system. One litre is equal to 10^{-3} m^3.

LNG *See* LIQUEFIED NATURAL GAS.

lock-and-key mechanism A proposed mechanism of ENZYME action. The three-dimensional structure of the enzyme provides a specific ACTIVE SITE that can be compared to a lock, while the SUBSTRATE molecule it acts upon is the key. According to this theory only the correct substrate will fit the active site of an enzyme. An enzyme–substrate complex forms that is then a different shape to the substrate alone, and it therefore falls away from the enzyme leaving it free to attach to another substrate molecule. Modern interpretations of this theory suggest that the AMINO ACIDS that form the active site alter their relative three-dimensional positions as the substrate binds. This is an induced fit.

lone pair A pair of electrons in a single ORBITAL in the VALENCE SHELL of an atom. Such a pair do not take part in the usual covalent bonding (*see* COVALENT BOND) of the atom, though they can form a CO-ORDINATE BOND. The presence of lone pairs has important implications for the STEREOCHEMISTRY of an element. Ammonia, NH_3, for example, forms a TRIGONAL PYRAMIDAL structure, which could be regarded as TETRAHEDRAL with the lone pair forming one point in the TETRAHEDRON with hydrogen atoms at the other three. On the other hand, the ammonium ion, NH_4^+, is a tetrahedral structure with the lone pair forming a co-ordinate bond with a fourth hydrogen atom.

long period A PERIOD in the PERIODIC TABLE that includes TRANSITION METALS; that is, periods 4, 5, 6, 7 and 8. *See also* SHORT PERIOD.

Loschmidt's constant The number of molecules per unit volume of an IDEAL GAS at STANDARD TEMPERATURE AND PRESSURE, 2.69×10^{23} m^{-3}.

lowering of vapour pressure The reduction in the SATURATED VAPOUR PRESSURE of a solvent, depending on the concentration of dissolved material. Provided the dissolved material does not itself have a significant vapour pressure, the reduction in vapour pressure is an approximately COLLIGATIVE PROPERTY, depending far more on the molar concentration of the dissolved material than its chemical nature.

low-grade heat Heat energy that has been used to raise the temperature of a large quantity of matter by a small amount and thus is of little use for further energy transformations. *See also* HEAT ENGINE.

Lowry–Brønsted theory *See* ACID, BASE.

LPG *See* LIQUEFIED PETROLEUM GAS.

LTH, *luteotrophic hormone* *See* PROLACTIN.

luminous intensity The amount of visible light given off per second by a light source. The SI UNIT of luminous intensity is the CANDELA.

lustre The characteristic reflective texture of metals, caused by the reflection of light by their surfaces.

luteotrophic hormone *See* PROLACTIN.

luteotrophin *See* PROLACTIN.

lutetium (Lu) The element with atomic number 71; relative atomic mass 175.0; melting point 1,663°C; boiling point 3,402°C; relative density 9.8. It is the last member of the LANTHANIDE series of metals and has some catalytic properties that are used in the petrochemical industry.

lyase One of a group of ENZYMES that cause the addition or removal of a chemical group other than by HYDROLYSIS, for example decarboxylases (*see* DECARBOXYLATION).

lycopene ($C_{40}H_{56}$) A CAROTENOID pigment; melting point 175°C. It provides the red colour to tomatoes, rose hips and many berries.

lyophilic (*adj.*) (= solvent-loving) Describing a SOL or a solid in a sol where the INTERMOLECULAR FORCES between the solid particles and the suspending liquid are strong. Such colloidal particles may break down further to form true solutions and do not coagulate readily (*see* COAGULATION). *Compare* LYOPHOBIC.

lyophobic (*adj.*) (= solvent-hating) Describing a SOL or a solid in a sol where the INTERMOLECULAR FORCES between the solid particles and the suspending particles are weak. Such materials tend not to disperse easily once coagulated (*see* COAGULATION). *Compare* LYOPHILIC.

lysergic acid ($C_{16}H_{16}N_2O_2$) A product obtained from the cereal fungus ergot; melting point 238°C. The derivative lysergic acid diethylamide (LSD) is a powerful hallucinogen.

lysine ($C_6H_{14}N_2O_2$) An AMINO ACID essential for growth. It is produced by the HYDROLYSIS of certain proteins. Lysine is found particularly in PROTAMINE and HISTONE classes of protein.

M

macromolecule A very large MOLECULE. Examples of substances with macromolecules are POLYMERS, PROTEINS and HAEMOGLOBIN.

The term macromolecule may also be applied to a crystal structure in which the atoms are held together by COVALENT BONDS, so that the whole crystal is effectively a single giant molecule. Diamond and silicon dioxide are examples of such a structure, which are both tetrahedral in shape.

See also COVALENT CRYSTAL.

macronutrient Any chemical substance needed in relatively large amounts by plants and animals for their normal growth and development. Macronutrients include nitrates, phosphates, sulphates, calcium, sodium, iron, chlorine, potassium and magnesium. Other substances are required in much smaller amounts and are called MICRONUTRIENTS. Substances usually fall into the same category for plants and animals but there are some exceptions, such as chlorine, which is a major element in animals but a trace element in plants. These essential nutrients may fulfil one or more of a variety of metabolic roles.

macroscopic (*adj.*) Visible to the naked eye. In science, a macroscopic state is a description of the behaviour of large-scale features of a system, such as temperature and pressure, as opposed to that of the individual atoms or molecules from which a system is made up.

magnesite A mineral form of MAGNESIUM CARBONATE, $MgCO_3$, often similar in appearance to chalk, and mined as a source of magnesium.

magnesium (Mg) The element with atomic number 12; relative atomic mass 24.3; melting point 651°C; boiling point 1,107°C; relative density 1.7. Magnesium is an ALKALINE EARTH metal, widespread in nature. It is extracted by the ELECTROLYSIS of fused (molten) magnesium chloride. The metal burns with an intense white flame, and is used in flares to illuminate large areas at night. Magnesium's low density makes it a useful ingredient in some high strength, low density alloys used in the aerospace industry. It is also present in CHLOROPHYLL, so magnesium is a vital element for plant growth. *See also* THERMIT PROCESS.

magnesium carbonate ($MgCo_3$) A white solid; decomposes on heating; relative density 2.6. Magnesium carbonate often occurs in nature alongside calcium carbonate, for example in the mineral dolomite, $CaCO_3.MgCO_3$. On heating, magnesium carbonate decomposes to give magnesium oxide:

$$MgCO_3 \rightarrow MgO + CO_2$$

magnesium chloride ($MgCl_2$) A white solid; melting point 714°C; boiling point 1,412°C; relative density 2.3. Magnesium chloride can be formed by burning magnesium in chlorine:

$$Mg + Cl_2 \rightarrow MgCl_2$$

ANHYDROUS magnesium chloride is DELIQUESCENT, and the salt frequently occurs as the hexahydrate $MgCl_2.6H_2O$. On heating, this hydrolyses (*see* HYDROLYSIS):

$$MgCl_2 + H_2O \rightarrow MgO + 2HCl$$

magnesium hydroxide ($Mg(OH)_2$) A white solid; decomposes on heating; relative density 2.4. Magnesium hydroxide can be formed by the reaction between magnesium and steam:

$$Mg + 2H_2O \rightarrow Mg(OH)_2 + H_2$$

It decomposes on heating:

$$Mg(OH)_2 \rightarrow MgO + H_2O$$

Magnesium hydroxide is insoluble in water, but a suspension is used in medicines, such as milk of magnesia, to neutralize excess stomach acid:

$$Mg(OH)_2 + 2HCl \rightarrow MgCl_2 + 2H_2O$$

magnesium oxide (MgO) A white solid; melting point 3,800°C; decomposes on further heating; relative density 3.6. Magnesium oxide can be made by burning magnesium in air:

$$2Mg + O_2 \rightarrow 2MgO$$

Magnesium oxide is used as a REFRACTORY material, and as a hard antireflective coating on the glass surfaces of some lenses.

magnesium sulphate ($MgSO_4$) A white solid; decomposes on heating; relative density 2.7. Magnesium sulphate is soluble and occurs naturally in a HYDRATED form as the mineral Epsom salt, widely used in the processing of cloth and leather. Magnesium sulphate decomposes on heating to give magnesium oxide:

$$2MgSO_4 \rightarrow 2MgO + 2SO_2 + O_2$$

magnet Any object that is surrounded by a MAGNETIC FIELD and attracts or repels other magnets. Magnets also attract unmagnetized pieces of iron, nickel or cobalt and some alloys containing these elements (such materials are described as FERROMAGNETIC). This is the result of the magnet temporarily magnetizing the ferromagnetic material; a phenomenon known as induced magnetism.

magnetic field The field of force surrounding a magnet or a current-carrying conductor. A magnetic field will exert a force on a moving charge. A small permanent magnet, such as a plotting compass, will tend to turn to point in the direction of the field. The SI UNIT of magnet field is the tesla.

magnetism The collective term for all the effects resulting from the presence of MAGNETIC FIELDS. Magnetic fields are produced by electric currents and by many ELEMENTARY PARTICLES, including the electron. The magnetic effects of electrons are responsible for magnetic properties of several elements.

magnetite A mineral form of iron oxide, containing the mixed oxide, iron(II)–iron(III) oxide, Fe_3O_4.

malachite A mineral form of copper carbonate hydroxide, $CuCO_3.Cu(OH)_2$. It is bright green in colour and is used for ornamental purposes. It is also occasionally mined as a source of copper.

malate A SALT or ESTER of MALIC ACID.

maleic acid ($C_4H_4O_4$) The cis form of BUTENEDIOIC ACID. It is a colourless crystalline compound; melting point 139°C. When heated it eliminates water to form maleic anhydride, which is cyclic. On prolonged heating at 150°C it converts to the trans form. It is used in making synthetic RESINS.

malic acid, *hydroxysuccinic acid* ($C_4H_6O_5$) An organic acid; melting point 100°C. It is found particularly in apples but also other fruit. Malic acid is present in all living cells and is an intermediate of the KREBS CYCLE.

malleable (*adj.*) Able to be beaten into a new shape without breaking. Malleability is an important property of many metals.

malonic acid *See* PROPANEDIOIC ACID.

maltase An enzyme that breaks down MALTOSE into its constituent GLUCOSE molecules.

maltose ($C_{12}H_{22}O_{11}$) A DISACCHARIDE; melting point 102–103°C. It is made from the combination of two GLUCOSE molecules. Maltose is a major constituent of malt and is therefore important in the manufacture of beer and whisky.

manganate(VI) Any salt containing the manganate(VI) ion, MnO_4^-. Salts containing this ion are dark green in colour. The ion can be converted to MANGANATE(VII) by manganese(IV) oxide in alkaline conditions:

$$MnO_4^- + MnO_2 + 2OH^- \rightarrow 2MnO_4^{2-} + H_2O$$

manganate(VII), *permanganate* Any salt containing the manganate(VII) ion, MnO_4^{2-}. Salts with this ion are deep purple in colour. Potassium manganate(VII), K_2MNO_4, is widely used as an OXIDIZING AGENT. Manganate(VII) ions can be converted to MANGANATE(VI) ions in acid solutions:

$$3MnO_4^{2-} + 4H^+ \rightarrow 2MnO_4^- + MnO_2 + 2H_2O$$

manganese (Mn) The element with atomic number 25; relative atomic mass 54.9; melting point 1,244°C; boiling point 2,040°C; relative density 7.4. Manganese occurs in nature as manganese oxide, and the metal can be extracted by heating the oxide with powdered magnesium:

$$MnO_2 + 2Mg \rightarrow Mn + 2MgO$$

Manganese is a TRANSITION METAL with some catalytic properties (particularly in manganese oxide) and is also used in the manufacture of some steels.

manganese-alkaline cell A common type of electrochemical CELL, non-rechargeable and more expensive than ZINC-CARBON CELLS, but longer lasting.

manganese(IV) oxide (MnO_2) A black solid; decomposes on heating. Manganese dioxide occurs naturally and is a strong OXIDIZING AGENT. It also has catalytic properties: the presence of manganese(IV) oxide will greatly increase the rate at which hydrogen peroxide, H_2O_2, decomposes. Manganese(IV) oxide is widely used in ZINC-CARBON CELLS, where it acts as a depolarizer, removing spent reagents from the ELECTROLYTE.

manganic (*adj.*) An obsolete term describing compounds containing the Mn^{3+} ion, for example manganic oxide, Mn_2O_3.

manganous (*adj.*) An obsolete term describing compounds containing the Mn^{2+} ion, for example manganous oxide, MnO.

mannans *See* MANNOSE.

mannose ($C_6H_{12}O_6$) A MONOSACCHARIDE that occurs naturally in polymerized forms known as mannans, found in plants, fungi and bacteria as food stores. It is a stereoisomer of GLUCOSE (*see* STEREOISOMERISM).

manometer An instrument for measuring pressure, comprising a U-shaped glass tube filled with a liquid. The difference in the height of the liquid in the two arms of the tube is proportional to the pressure difference between the two ends of the manometer, one of which is usually left open to ATMOSPHERIC PRESSURE.

marble A METAMORPHIC rock formed by the action of heat and pressure, on LIMESTONE. Small amounts of impurities, such as iron and copper carbonates, give some marbles their characteristic colours and patterning.

Markownikoff's rule During the ELECTROPHILIC ADDITION of hydrogen HALIDES (such as HBr) to an unsymmetrical ALKENE, the hydrogen goes to the end of the DOUBLE BOND that already has the greatest number of hydrogen atoms and the halide to the end with the least hydrogen atoms. For example, the addition of hydrogen bromide to propene ($CH_3CH{=}CH_2$) yields $CH_3CHBrCH_3$ instead of the alternative $CH_3CH_2CH_2Br$. This occurs because of the greater stability of the intermediate CARBONIUM ION formed during the production of the former. Sometimes the alternative product is formed, which is termed anti-Markownikoff addition.

Marsh's test A chemical test for ARSENIC in a compound. The substance to be tested is treated with hydrochloric acid and zinc powder. The hydrogen released combines with any arsenic to produce ARSINE. The gases released are passed along a heated tube and any arsine present decomposes to give a brown layer of metallic arsenic on the wall of the tube. To distinguish arsenic from antimony, which also produces a brown layer, the deposit is treated with sodium chlorate(I), which will dissolve arsenic but not antimony.

mass A measure of the total amount of MATTER in an object, expressed either in terms of the resistance of an object to having its motion changed (inertial mass) or the effect of a gravitational field on the object (gravitational mass). The SI UNIT of mass is the KILOGRAM.

mass defect The difference between the mass of an atomic NUCLEUS and the mass of the NEUTRONS and PROTONS from which it is made.

mass number, *nucleon number* The total number of NUCLEONS (NEUTRONS and PROTONS) in a particular atomic NUCLEUS.

mass spectrometer An instrument for the measurement of the mass of atoms, molecules or fragments of a molecule, and the relative abundance of each mass present. A sample of material is IONIZED by bombardment with an electron beam. If the sample is molecular, the molecule will also be broken into fragments. The charged particles are accelerated in an electric field. The ions then enter a velocity selector, a region in which electric and magnetic fields are applied at right angles to one another to produce opposing forces on the ions. These forces balance out only for particles moving at one particular speed, which then enter the next region of the device, where there is a magnetic field that deflects particles according to their charges and masses. A detector produces a reading of abundance against mass/charge ratio.

As well as measuring the relative abundance of ISOTOPES, the mass spectrometer is an important tool in determining the structure of organic molecules. In a KETONE or ALDEHYDE, for example, a fragment corresponding to the OH grouping (17 amu) would not be found, but in ALCOHOLS it will be present.

mass spectroscopy The use of a MASS SPECTROMETER to determine the structure of a compound or as a tool in QUALITATIVE ANALYSIS.

matter The collective term for all ATOMS.

Maxwell–Boltzmann distribution A description of the range of energies possessed by molecules in a system in which those molecules are able to exchange energy freely with one another to maintain thermal equilibrium. In the Maxwell–Boltzmann distribution the number of particles with an energy greater than E is proportional to $e^{-E/kT}$, where T is the ABSOLUTE TEMPERATURE and k is the BOLTZMANN CONSTANT. This quantity is called the BOLTZMANN FACTOR, and if E is the ACTIVATION ENERGY, the rate at which an ACTIVATION PROCESS takes place will depend on this factor. The Maxwell–Boltzmann distribution can also be used to describe range of speeds of molecules in a gas.

MCPA, ***2-methyl-4-chlorophenoxyacetic acid, methoxone*** ($C_9H_9ClO_3$) A white crystalline solid; melting point 118–119°C. It is a selective weedkiller.

MCPB, ***4-(4-chloro-2-methylphenoxy)-butyric acid*** ($C_{11}H_{13}ClO_3$) An organic chemical that is used as a selective weedkiller. In itself it is harmless to plants, but once absorbed by plant cells it is converted to a toxic chemical that kills the plant.

mean free path The average distance that a molecule in a fluid, usually a gas, travels before colliding with another molecule.

mechanism A description of the process by which a reaction takes place, particularly when relating to organic reactions. A reaction mechanism provides a step-by-step account of how the products of a reaction are produced from the reagents. Common mechanisms are given special names, such as NUCLEOPHILIC SUBSTITUTION. The different mechanisms of many reactions are known, and techniques such as FLASH PHOTOLYSIS can be used to investigate intermediate states in some reactions, but the mechanism of other reactions are still disputed.

mega- (M) A prefix indicating that the size of a unit is to be multiplied by 10^6. For instance, one megawatt (MW) is equal to one million watts.

M.E.K. *See* BUTANONE.

melamine ($C_3N_6H_6$) A THERMOSETTING PLASTIC that is heat resistant, scratch resistant and difficult to break, so has many household uses. Melamine consists of a six-membered ring structure of alternating carbon and nitrogen atoms and three NH_2 groups. This is polymerized with METHANAL to give the thermosetting plastic.

melting point The temperature at which a solid turns into a liquid, or vice versa. More technically, the melting point is the one temperature for a given pressure at which the solid and liquid can exist in equilibrium together. Most materials increase in volume, though only slightly, when they melt, and the melting point increases with increasing pressure. An important exception to this is water, which takes up less volume as a liquid than as a solid, thus water tends to melt under pressure.

membrane potential *See* DONNAN EQUILIBRIUM.

mendelevium (Md) The element with atomic number 101. Several isotopes are known, but the longest lived (mendelevium–256) has a HALF-LIFE of just 1.3 hours, too short for the material to be of any practical use.

menthol ($C_{10}H_{19}OH$) A TERPENE alcohol with a minty flavour; melting point 42°C. It is used in peppermint flavouring and oils and also for the alleviation of nasal congestion and rheumatic pain.

mercaptan *See* THIOL.

mercuric (*adj.*) An obsolete term for any compound containing the Hg^{2+} ion, for example mercuric oxide, HgO.

mercurous (*adj.*) An obsolete term for any compound containing mercury in its +1 oxidation compound, for example mercurous chloride, Hg_2Cl_2. Mercurous compounds usually contain the ion Hg_2^{2+}.

mercury (Hg) The element with atomic number 80; relative atomic mass 200.6; melting point –39°C; boiling point 357°C; relative density 13.6. Mercury is a metal, the only one that is liquid at room temperature. Formerly known as quicksilver, it is used in thermometers and in many applications that require a liquid that is electrically conducting or has a high density (10 times that of most liquids). Mercury occurs in nature as mercury sulphide, in the ore cinnabar. Mercury sulphide decomposes to mercury on heating:

$$HgS \rightarrow Hg + S$$

Mercury exhibits two valencies (*see* VALENCY). MONOVALENT compounds contain the mercury(I) ion, Hg_2^{2+}, while mercury(II) compounds containing the Hg^{2+} ion are more common. Mercury is a cumulative poison and concern has been expressed about high levels of mercury in the food chains of some industrialized areas.

mercury-cathode cell A device for the extraction of chlorine and sodium hydroxide from BRINE. The brine is electrolysed in a vessel with graphite ANODES at the top (from which chlorine gas is released) and a stream of mercury acting as the CATHODE at the base of the cell. Sodium released in the ELECTROLYSIS forms an AMALGAM with the cathode and is carried out of the cell. Water is then passed over the amalgam, and the sodium reacts, producing sodium hydroxide solution and hydrogen. The mercury is then recirculated to the electrolytic cell. This process is called the Kastner–Kellner process.

mercury chloride Mercury(I) chloride, Hg_2Cl_2, and mercury(II) chloride, $HgCl_2$. Mercury(I) chloride is a white powder; melting point 302°C; boiling point 384°C; relative density 7.0. Mercury(I) chloride can be made by heating mercury(II) chloride with an excess of mercury:

$$HgCl_2 + Hg \rightarrow Hg_2Cl_2$$

Mercury(II) chloride is a white powder; melting point 276°C; boiling point 303°C; relative density 5.4. Mercury(II) chloride can be made by heating mercury in chlorine:

$$Hg + Cl_2 \rightarrow HgCl_2$$

mercury oxide (HgO) A red or orange powder, which decomposes on heating. Mercury oxide can be formed by heating mercury in oxygen:

$$2Hg + O_2 \rightarrow 2HgO$$

At higher temperatures, this reaction is reversed and mercury oxide decomposes to mercury and oxygen.

mercury sulphide (HgS) A red or black solid. Mercury sulphide occurs naturally as the ore cinnabar, an important source of mercury. It decomposes on heating:

$$HgS \rightarrow Hg + S$$

mesomerism An obsolete term for the occurrence of RESONANCE HYBRIDS within the structure of a covalently bonded molecule.

meta- In organic chemistry, a prefix used in disubstituted derivatives of BENZENE which indicates that the substituted groups are at positions 1 and 3 in the ring, for example 1,3-dichlorobenzene (*meta*-dichlorobenzene):

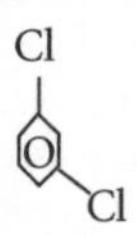

The use of the term *meta* in other situations, for example metaphosphoric acid, does not have the same meaning and is always written in full. *See also* ORTHO-, PARA-.

metabolism The chemical processes occurring within a living organism. Metabolism is a continual process of building up of body tissue (anabolism) and breaking down of living tissue into energy and waste products (catabolism). The control of metabolism is complex, involving HORMONES and ENZYMES.

metabolite A substance required for or produced by METABOLISM.

metal Any of a class of elements that are typically lustrous, MALLEABLE, DUCTILE solids (mercury is a liquid) that are good conductors of heat and electricity. Metal ions replace the hydrogen in an ACID to form a SALT, and combine with the HYDROXIDE (OH^-) ion to form a BASE. Metals form ALLOYS with each other.

About 75 per cent of the known elements are classed as metals. The ALKALI METALS and ALKALINE EARTHS are generally soft silvery reactive metals that form positive ions. The TRANSITION METALS are harder, less reactive metals, and they form COMPLEX IONS.

The structure of a metal is typically a rigid lattice of positive ions, through which the outermost electrons, the VALENCE ELECTRONS, are free to move. The forces between the ions in the metal lattice are repulsive at short distances, but at larger separations the electrons screen the repulsion between ions, resulting in an attractive force that is strong enough to give most metals melting points well above room temperature.

The presence of FREE ELECTRONS in a metallic lattice make metals good conductors of heat and electricity. The electrical conductivity allows metals to reflect electromagnetic radiation, including light, resulting in their characteristic shiny appearance. The interatomic forces in metals are generally weaker then those in IONIC SOLIDS, so metals are relatively soft, and in their pure forms are usually malleable and ductile.

See also METALLIC BONDING.

metaldehyde ($C_4O_4H_4(CH_3)_4$) A solid polymer of ethanal. It is used in slug pellets and as a fuel for portable stoves.

metallic bonding The bonding that holds together atoms in a METAL. The overlapping of the ORBITALS in a metallic solid produces an energy band that is effectively a DELOCALIZED

ORBITAL, so that VALENCE ELECTRONS can move freely from one atom to the next, allowing the metal to conduct heat and electricity. A simple model of a metallic structure is of a lattice of positive ions surrounded by a 'sea' of electrons. The relatively weak nature of metallic bonding compared to ionic bonding (*see* IONIC BOND) accounts for the softness and low melting points of metals when compared to IONIC SOLIDS.

metallic radius A measure of the effective size of a metal atom, defined as half the distance between adjacent atoms in a metal. Metallic radii increase with ATOMIC NUMBER within a given GROUP in the PERIODIC TABLE. The metallic radii of the ALKALI METALS are somewhat larger than for other metals of similar atomic number as a result of the lower packing fraction achieved by the BODY CENTRED CUBIC lattice structure adopted by the alkali metals. Typical values lie between 0.1 nm and 0.3 nm.

metallocene ORGANOMETALLIC COMPOUNDS consisting of AROMATIC rings complexed to a metal ion or atom by the pi electrons of the ring. They are formed by the reaction between the CYCLOPENTADIENE ion $(C_5H_5)^-$ and a metal. In ferrocene the metal is iron; in nickleocene it is nickle.

metalloid, ***semi-metal*** Any element that has some of the properties of a METAL, but which is not completely metallic. Many of them are SEMICONDUCTORS, or form compounds that are semiconductors. Typical examples are ARSENIC, BORON, GERMANIUM, SILICON and TELLURIUM.

metallurgy The study of the properties of METALS, and more usually their ALLOYS, particularly in respect of their engineering properties.

metamorphic (*adj.*) Describing a rock that was originally SEDIMENTARY in nature but has had its structure altered by subsequent heat and pressure.

metastable state A state of some physical system that is unstable, but in which the system will remain for an unusually long time, or a time sufficiently long for the state to be regarded as stable. SUPERHEATED water is an example of a metastable state, and some ENERGY LEVELS in atomic nuclei are also metastable.

methacrylate resins ACRYLIC RESINS made by polymerizing 2-methylpropenoic acid or its ester derivatives.

methacrylic acid *See* 2-METHYLPROPENOIC ACID.

methanal, ***formaldehyde*** (HCHO) An ALDEHYDE that is a gas at ordinary temperatures with a pungent odour. It is used dissolved in water as formalin (a 37 per cent solution), which is a preservative for biological specimens. It is also a strong disinfectant (as a vapour or a spray) and can be used to sterilize surgical instruments. Other uses include the manufacture of foam, dyes, plastics and resins.

methane (CH_4) A colourless, odourless gas; boiling point –164°C; melting point –184°C. It is the simplest HYDROCARBON and the main component of NATURAL GAS. It reacts explosively with air, and is odourless, which makes its presence ('firedamp') in coal mines a particular hazard. It is emitted by decaying vegetable matter and is therefore found in marshlands (as marsh gas) and is given off during sewage disposal. Methane contributes to the GREENHOUSE EFFECT.

methanoic acid, ***formic acid*** (HCOOH) A colourless liquid CARBOXYLIC ACID that fumes slightly and has an unpleasant odour; boiling point 100.5°C; melting point 8.4°C. It occurs in ants, stinging nettles, sweat and urine. It is prepared by heating ethanedioic acid (oxalic acid) with GLYCEROL. It is used in textile dyeing and leather tanning.

methanol, ***methyl alcohol*** (CH_3OH) A simple colourless liquid ALCOHOL with a pleasant odour; boiling point 64.5°C. It is highly poisonous, causing blindness. It is usually made from coal or NATURAL GAS but can also be made by the dry DISTILLATION of wood, hence it is known as wood spirit. Methanol is used in the manufacture of methanal, MTBE and many other organic compounds, and as a solvent.

methionine, ***2-amino-4-(methylthio)butanoic acid*** $(C_5H_{11}NO_2S)$ An ESSENTIAL AMINO ACID containing sulphur; melting point 183°C. It is present in many proteins. Methionine is the first amino acid to be incorporated into a POLYPEPTIDE chain during PROTEIN SYNTHESIS and is therefore crucial in this process.

methoxone See MCPA.

methoxyethane An ETHER produced by a SUBSTITUTION REACTION of BROMOETHANE. *See* WILLIAMSON ETHER SYNTHESIS.

methyl alcohol *See* METHANOL.

methylamines Derivatives of ammonia in which 1, 2 or 3 hydrogen atoms have been replaced by METHYL GROUPS. Substitution of one of the hydrogen atoms gives monomethylamine or methylamine, CH_3NH_2. This is a colourless gas with a strong odour of ammonia. It occurs naturally but can be synthesized by heating methanal with ammonia and is used in the manufacture of weedkillers and fungicides. The dimethylamine, $(CH_3)_2NH$, is a colourless liquid with a strong odour of ammonia, and used to make other chemicals such as solvents, weedkillers and fungicides. The trimethylamine, $(CH_3)_3N$ is a colourless liquid with a strong fishy odour. It occurs naturally.

methylated spirit ETHANOL containing METHANOL, making it undrinkable. It is used for industrial purposes as industrial methylated spirit.

methylation The process by which a METHYL GROUP is added to a compound. In ALIPHATIC compounds this may be the substitution of a hydrogen atom in, for example, HYDROXYL, AMINO or IMINO GROUPS. In AROMATIC compounds, substitution of one of the hydrogen atoms in the ring may also occur (*see* FRIEDEL-CRAFTS REACTION).

methylbenzene, *toluene* ($C_6H_5CH_3$) An AROMATIC compound consisting of a BENZENE RING with a CH_3 (methyl) group attached. It is a colourless liquid derived from PETROLEUM; boiling point of 111°C; melting point –95°C. It is used in the manufacture of the explosive TNT, as a solvent and aircraft fuel and in the manufacture of PHENOL.

methyl bromide *See* BROMOMETHANE.

2-methylbutadiene, *isoprene* ($CH_2{=}C(CH_3)$-$CH{=}CH_2$, C_5H_8) A liquid obtained from PETROLEUM that is used in the manufacture of synthetic RUBBERS. Its boiling point is 34°C. Isoprene is polymerized by ZIEGLER-NATTA CATALYSTS.

2-methyl-4-chlorophenoxyacetic acid See MCPA.

methylene (CH_2:) A highly reactive CARBENE. The divalent CH_2 group in a compound is called the methylene group.

methyl ethyl ketone *See* BUTANONE.

methyl group The organic group CH_3O–.

methyl methacrylate An ESTER of 2-METHYLPROPENOIC ACID used to make METHACRYLATE RESINS. *See* PERSPEX.

methyl orange, *4-dimethylamino-4′-azobenzene sodium sulphonate* ($C_{14}H_{14}N_3NaO_3S$) An organic dye used as an INDICATOR, particularly in TITRATIONS involving weak bases, and as a biological stain in the preparation of microscope slides. It is red below pH 3.1 and changes to yellow above pH 4.4. *See also* AZO COMPOUND.

2-methylpropenoic acid, *methacrylic acid* ($C_4H_6O_2$) A white crystalline unsaturated CARBOXYLIC ACID, used in making METHACRYLATE RESINS.

methyl t-butyl ether *See* MTBE.

metre The base unit of length in the SI system (*see* SI UNIT). The metre was originally defined as the length of a standard metal bar, then in terms of a certain number of wavelengths of light, but now defined as the distance travelled by light in a vacuum in 1/(299,792,458) seconds.

metric system Any system of measurements based on the METRE as the unit of length, the GRAM as the unit of MASS and the SECOND as the unit of time, or on some multiple of these units. In science, the system almost universally used is the SI system of units, which defines units for all physical quantities, derived from seven base units. *See* SI UNITS.

mica Any of a group of SILICATE minerals having a layered structure. Silicon and oxygen atoms covalently bound together form the layers, with metallic CATIONS and HYDROXIDE ions between the layers. Mica can readily be broken into large transparent flakes, which were once widely used as electrical insulation, but plastic materials are now more commonly used for such applications, although mica is still sometimes used where its resistance to high temperatures is important.

micelle A small group of molecules loosely clumped together, in a COLLOID for example.

micro- (μ) A prefix indicating that a unit is to be multiplied by 10^{-6}. For instance, the microampere (μA) is equal to one millionth of an AMPERE.

micronutrient Any chemical substance required in very small amounts by plants and animals for their normal growth and development. Micronutrients are often found in COFACTORS and COENZYMES. The micronutrients include the elements manganese, copper, iodine, cobalt, zinc, molybdenum, boron, selenium, chromium, silicon and fluorine. These are termed trace elements. VITAMINS can also be considered as micronutients. These essential nutrients may be required for one or more metabolic roles. *Compare* MACRONUTRIENT.

microscope An optical device that uses a system of lenses to magnify objects too small to be seen in fine detail with the naked eye. In 1665 Robert Hooke (1635–1703) was the first to

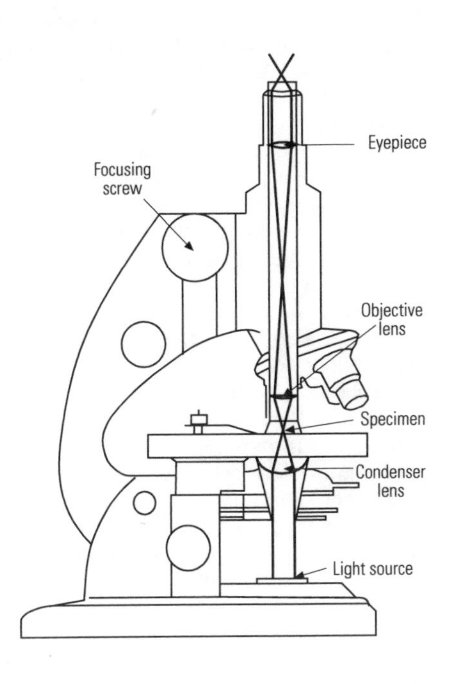

Light microscope.

record microscopic examination of cells in cork, and Anton van Leeuwenhoek (1632–1723) recorded bacteria in 1683.

A simple microscope has a single lens but limited powers of magnification, whereas a compound microscope uses two lenses and light passes from an object through the first lens (objective) to produce a magnified image that is then magnified further by the second lens (eyepiece). The total magnification is the product of the magnification of each lens and is maximally 1,500–2,000 times in a light microscope, achieved by oil-immersion objective lenses. The resolution of a light microscope is limited to two points 0.2 μm apart. The thinner the material being observed, the greater the clarity of image.

Preserved or fixed tissues are usually embedded in paraffin wax and thin sections (3–20 μm) are cut using an instrument called a microtome, but even thinner sections can be obtained by freezing the tissue in liquid nitrogen and cutting sections using a cryostat at –200°C. Various staining methods are used to enhance the image seen and to highlight specific cells or materials.

See also ELECTRON MICROSCOPE, FLUORESCENCE MICROSCOPY.

microscopy The study or use of MICROSCOPES.

microwave ELECTROMAGNETIC RADIATION with a wavelength in the range 30 cm to 1 mm. Microwave frequencies coincide with the resonant frequencies of some covalently bonded molecules. In particular, waves of 12.6 cm wavelength are resonantly absorbed by water molecules. This effect is used in microwave ovens, where microwave energy is resonantly absorbed and converted to heat by water in food.

microwave spectroscopy The study of the absorption spectra (*see* ABSORPTION SPECTRUM) of MICROWAVES by gases, typically using microwaves with a wavelength of a few millimetres. Since these waves have frequencies that excite the natural frequencies of rotation of simple covalently bonded molecules (*see* COVALENT BOND), it is possible to study certain properties of these bonds, such as their natural length, and their angle to one another.

migration 1. The movement of ions, in particular in an ELECTROLYTE or GEL, under the influence of an electric field. *See also* ELECTROPHORESIS.

2. In organic molecules, the movement of an atom, DOUBLE BOND or FUNCTIONAL GROUP from one part of the molecule to another.

mild steel An alloy of iron with a few per cent of carbon. Mild steel is cheap and easy to machine and to weld, but rusts quickly.

mill In the chemical industry, a machine for breaking large lumps of material into smaller pieces, such as coal for burning in a furnace. Also used in a wider context to describe an industrial plant where a particular material is produced by chemical processes, for example a steel mill.

milli- (m) A prefix indicating a unit is to be multiplied by 10^{-3}. For instance, one milliampere (mA) is equal to one thousandth of an AMPERE.

mineral A naturally occurring non-living substance that has a characteristic chemical composition or contains a certain compound. Rocks are made up of mixtures of minerals. Technically, a mineral is often CRYSTALLINE, having essentially a single chemical composition. More generally, a mineral is any material of commercial value that can be extracted from the Earth's crust, and under this definition includes PETROLEUM and NATURAL GAS.

mineral acid An ACID obtained from MINERALS and used in chemical processes to produce other commercially important substances. The chief mineral acids are hydrochloric acid, sulphuric acid and nitric acid.

mineralocorticoid Any one of a group of CORTICOSTEROID hormones, secreted by the adrenal gland, that is concerned with the metabolism of minerals. An example is ALDOSTERONE.

mineralogy The branch of GEOLOGY concerned with MINERALS.

mineral oil Any OIL derived from a MINERAL, particularly one derived from PETROLEUM.

miscible (*adj.*) Describing two liquids that can mix together to form a single liquid. Most organic liquids are miscible with one another, but not with water. Water and ethanol are miscible. Liquids are most likely to be miscible if the degree of polarity is similar (*see* POLARIZATION). *See also* IMMISCIBLE.

mixture A substance containing two or more elements or compounds, but where there is no chemical bonding between the constituents of the mixture. The constituents can be separated without a chemical reaction taking place. For example, in a mixture of iron and sulphur, the iron can be removed using a magnet, but once the iron and sulphur have combined to form iron sulphide (a compound) they cannot be separated by physical means. *See also* CLATHRATE, COLLOID.

m.k.s. system A system of physical units derived from the METRIC SYSTEM and based on the metre, kilogram and second. It has been replaced by SI UNITS.

mobility, ionic The contribution of one type of ion to the MOLAR CONDUCTIVITY of an electrolyte. At infinite dilution, the molar conductivity of a solution is equal to the sum of the ionic mobilities of the ions in the solution. Ionic mobility depends largely on the charge and size of an ion and on whether or not it is hydrated.

molar (*adj.*) Relating to one MOLE or to a concentration of one mole per decimetre cubed.

molality The CONCENTRATION of a solution, measured in terms of the number of MOLES of SOLUTE dissolved in one kilogram of SOLVENT.

molar conductivity The electrical conductivity of a solution divided by its CONCENTRATION, in order to allow the contribution to the conductivity of each MOLE of SOLUTE ions to be determined. In a strong ELECTROLYTE (one that is fully IONIZED) the molar conductivity is essentially independent of concentration, except at very high concentrations. In a weak (only partially ionized) electrolyte, the molar conductivity rises only gradually with dilution, the material being fully ionized at very low concentrations. *See also* KOHLRAUSCH'S LAW.

molar gas constant (R) The constant of proportionality linking the ABSOLUTE TEMPERATURE of one MOLE of gas to the product of its pressure and volume. R is equal to 8.3 J K^{-1} mol^{-1}.
See IDEAL GAS EQUATION.

molar heat capacity The HEAT CAPACITY of a piece of material containing one MOLE of molecules. For an IDEAL GAS, the molar heat capacity at constant pressure is greater than that at constant volume by R, the MOLAR GAS CONSTANT. For a material with a molar heat capacity at constant pressure of C_P and a molar heat capacity at constant volume of C_V:

$$C_P - C_V = R$$

See also DULONG AND PETIT'S LAW, RATIO OF SPECIFIC HEATS.

molarity The CONCENTRATION of a SOLUTION measured in MOLES of SOLUTE per decimetre cubed of solution. Molarity is often denoted by the letter M, so a 0.1 M solution will contain 0.1 moles of solute molecules in 1 dm^3 of solution.

molar volume The volume occupied by one MOLE of a given material. All IDEAL GASES have the same molar volume at a given temperature and pressure. At STANDARD TEMPERATURE AND PRESSURE this is 22.4 dm^3.

mole The SI UNIT of amount of substance. One mole is defined as being the amount of substance containing as many atoms (or molecules or ions or electrons) as there are carbon atoms in 12 g of carbon–12.

molecular beam A beam of atoms, ions or molecules, formed by vaporizing a small amount of a substance and allowing it to pass through a narrow hole into a vessel maintained at a very low pressure by a vacuum pump. Since there are few collisions between particles in the beam, the properties of the individual molecules can be studied without disturbance by these collisions. In particular, molecular beams are often used in high resolution SPECTROSCOPY.

molecular distillation A low pressure DISTILLATION technique in which molecules escaping from a liquid are condensed onto a cold condensing surface, after a passing through a short distance in a vacuum. This technique is used at low temperatures to distil molecules that decompose on heating, and also to distil molecules that react with oxygen in the air.

molecular formula A chemical FORMULA that gives the number of atoms present in a molecule. For example, ethane has a molecular formula of C_2H_6, though its EMPIRICAL FORMULA is CH_3. The molecular formula is determined by knowing the empirical formula and the RELATIVE MOLECULAR MASS. *See also* STRUCTURAL FORMULA.

molecular mass *See* RELATIVE MOLECULAR MASS.

molecular orbital An ORBITAL formed by the overlapping of orbitals between two atoms that are linked together by a COVALENT BOND. It is the reduction in energy brought about by the formation of one of the possible molecular orbitals, as compared to the two separate atomic orbitals, that makes the bond stable. *See also* ANTIBONDING ORBITAL, BONDING ORBITAL, DELOCALIZED ORBITAL, PI-BOND, SIGMA-BOND.

molecular sieve *See* SEMIPERMEABLE MEMBRANE.

molecular weight An obsolete term for RELATIVE MOLECULAR MASS.

molecule The smallest part of a chemical COMPOUND that can exist without it losing its chemical identity. Molecules are made of one or more ATOMS held together by IONIC or COVALENT BONDS. *See also* MACROMOLECULE.

mole fraction The proportion of a specified element or compound in a mixture, expressed in terms of the number of molecules present. Thus a mixture of 16 g of oxygen (O_2) with 4 g of helium (He) may contain 80 per cent oxygen by mass, but the mole fractions are 50 per cent each as there are two moles of gas molecules present: one mole of O_2 and one of He.

molybdenum (Mo) The element with atomic number 42; relative atomic mass 94.9; melting point 2,610°C; boiling point 5,560°C; relative density 10.2. Molybdenum is a TRANSITION METAL, used in the manufacture of hard steels.

monatomic (*adj.*) Describing a elemental substance, such as a NOBLE GAS, whose atoms behave as separate entities and do not form molecules.

Mond gas *See* SEMI-WATER GAS.

Monel metal An alloy of nickel (typically 65 per cent) with copper (approximately 30 per cent) with smaller quantities of iron, manganese, silicon and carbon. It is widely used in the chemical industry for its ability to resist attack by most acids.

monobasic acid Any ACID with a BASICITY of 1.

monoclinic (*adj.*) Describing a CRYSTAL structure where the UNIT CELL has two sets of faces at right angles to one another and the third is at some other angle. The size and shape of the unit cell are characterized by three lengths and one angle.

monoethylamine *See* ETHYLAMINE.

monohydric (*adj.*) Describing a compound with one HYDROXYL GROUP.

monomer A simple chemical compound that, under suitable conditions, can join with other identical monomers to form a long chain POLYMER.

monosaccharide A single sugar with the general formula $(CH_2O)_n$ that cannot be split into smaller CARBOHYDRATE units. When n is 3, the sugar is called a triose sugar, when n is 5 it is a pentose sugar, and when n is 6 it is hexose sugar. Monosaccharides are either aldoses (aldo-sugars), which have an ALDEHYDE group (CHO), or ketoses (keto-sugars), which have a KETONE group (C=O). Both GLUCOSE and FRUCTOSE have the formula $C_6H_{12}O_6$ but glucose is an aldose and fructose is a ketose, so their properties are different. Both can easily form ring structures: glucose usually has a six-sided pyranose ring and fructose a five-sided furanose ring, although both can form either ring structure. Most carbohydrates can form ISOMERS. Monosaccharides are sweet, soluble crystalline molecules. *See also* DISACCHARIDE, POLYSACCHARIDE.

monosodium glutamate ($C_8H_8NNaO_4.H_2O$) A white crystalline salt with a taste similar to that of meat, widely used as a food additive. It is prepared from GLUTAMIC ACID.

monotropy That form of ALLOTROPY in which there is only one stable form, with the other forms being unstable at all temperatures. Over time, the unstable form will always tend to change into the stable form. The two forms of phosphorous, red phosphorous and white phosphorous are an example of this. The white form will gradually turn into the red form. *See also* ENANTIOTROPY.

monovalent (*adj.*) Having a VALENCY of 1.

mordant A colourless chemical with which a cloth is treated before being dyed. The mordant is absorbed into the fibres of the cloth and then forms a coloured complex with the dye molecules.

morphine ($C_{17}H_{19}NO_3$) An ALKALOID of opium that is used for the relief of severe pain. It reacts to give methylmorphine (codeine) and diacetylmorphine (diamorphine or heroin). These drugs are addictive and so must be used with caution.

Mössbauer effect The absorption of the recoiling momentum of a nucleus emitting a low-energy gamma ray by the whole of a crystal lattice rather than by the single nucleus. As a result, the gamma rays concerned are emitted with a far narrower spread of energies than is normally the case. By moving a source of such gamma rays, the DOPPLER EFFECT can be used to produce a source of nuclear radiation whose frequency can be varied over a narrow range. The absorption of this energy by a target nucleus, in a technique called Mössbauer SPECTROSCOPY, can be used to measure some nuclear ENERGY LEVELS very precisely and to detect CHEMICAL SHIFTS, which provide information about the chemical environment of the nucleus concerned. Because only a few materials demonstrate the Mössbauer effect, this technique can only be used on a fairly limited range of nuclei.

mother liquor The concentrated solution from which crystals are formed or which remains after a CRYSTALLIZATION process. *See also* FRACTIONAL CRYSTALLIZATION.

MTBE, *methyl t-butyl ether* An additive in petrol, used instead of lead to improve the OCTANE NUMBER.

mucopolysaccharide *See* GLYCOSAMINOGLYCAN.

multicentre bond A COVALENT BOND in which a pair of electrons are distributed over a region containing more than two nuclei. Such bonds are found in ELECTRON-DEFICIENT compounds such as BORANE.

multiple bond A double or triple COVALENT BOND; that is, one involving the formation of both a SIGMA-BOND and one or two PI-BONDS.

multiplet A group of lines in the ABSORPTION or EMISSION SPECTRUM of an atom, ion or molecule which are separated by very small energy differences as a result of the FINE STRUCTURE of the ENERGY LEVELS.

mustard gas, *dichloryl diethyl sulphide* ($C_lCH_2CH_2SCH_2CH_2Cl$) A poisonous gas, synthesized from ethene and disulphur dichloride, S_2Cl_2. It has been used in chemical warfare.

mutarotation A change in the OPTICAL ACTIVITY of a substance as a result of a spontaneous chemical reaction, for example, dissolving in water, acids or bases.

myoglobin A GLOBULAR PROTEIN found in vertebrate muscle that is closely related to HAEMOGLOBIN and binds oxygen. It is a single polypeptide chain of relative molecular mass 17,800 and has a compact structure. Myoglobin has a single HAEM group (human haemoglobin has four) and a greater affinity for oxygen than haemoglobin. Myoglobin is therefore able to store oxygen until it is needed in situations of extreme exertion, when the blood oxygen supply from haemoglobin is not sufficient to keep up with demands of muscle cells. Myoglobin is red in colour and is responsible for the coloration of meat.

myosin A PROTEIN that is a major constituent of muscle fibres. Two classes of myosin exist and both consist of a double-headed globular region and a helical tail region. Myosin I is involved in cell locomotion and myosin II is involved in muscular contraction, forming the so-called thick filaments of muscle fibres. During muscular contraction, myosin interacts with another protein ACTIN, which binds to the head region of myosin forming a complex called ACTOMYOSIN.

N

NAD (nicotinamide adenine dinucleotide) A COENZYME derived from the vitamin NICOTINIC ACID, which is an electron carrier in the ELECTRON TRANSPORT SYSTEM and in the KREBS CYCLE in RESPIRATION. When reduced NAD receives a hydrogen atom it becomes NADH, which carries the electrons. In its phosphorylated form NAD is NADP.

NADP (nicotinamide adenine dinucleotide phosphate) The phosphorylated form of NAD (*see* PHOSPHORYLATION). NADP is important as an electron carrier in PHOTOSYNTHESIS. When reduced, NADP receives a hydrogen atom to become NADPH. NADP is not as abundant in animal cells as NAD.

nano- (n) A prefix indicating that a unit is to be multiplied by 10^{-9}. For instance, one nanometre (nm) is one billionth of a metre.

napalm A gel made by mixing liquid hydrocarbons with detergents, used in warfare for its ability to cling to materials whilst it burns.

naphthacene ring system A system of benzene rings, numbered as shown.

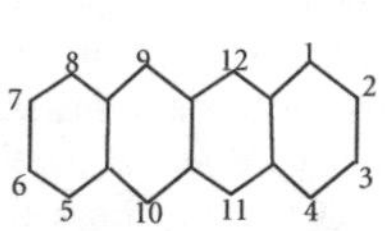

naphthalene ($C_{10}H_8$) A volatile white solid with a smell of mothballs; melting point 80°C; boiling point 218°C. It is an AROMATIC hydrocarbon, consisting of two benzene rings, and is obtained from crude oil. Naphthalene is used in the manufacture of RESINS and PLASTICIZERS.

naphthol ($C_{10}H_7OH$) One of two PHENOLS prepared from NAPHTHALENE with the same formula but differing in the position of the –OH group. β-naphthol (naphthalen-2-ol) is a white solid; melting point 123°C; boiling point 295°C. It is used as an ANTIOXIDANT in rubber and in the manufacture of dyes. α-naphthol (naphthalen-1-ol) is a colourless solid also used in dye manufacture; melting point 94°C; boiling point 280°C.

nascent hydrogen The state of HYDROGEN in which it is produced in a reaction, for example by the action of hydrochloric acid on zinc. Some reactions will take place with hydrogen produced in this way, but not with any other form of hydrogen. It is thought that such reactions produce hydrogen molecules in a more reactive EXCITED STATE.

native (*adj.*) Describing a MINERAL that contains essentially a single element not combined chemically. Only relatively unreactive materials occur in this form, such as copper, gold and sulphur.

natural gas A mixture of gases often found in pockets above PETROLEUM in the Earth's crust and which is one of the three main FOSSIL FUELS. It consists of HYDROCARBONS, mainly methane with some ethane, butane and propane.

negative feedback Where the end-product of a pathway inhibits the ENZYME at the start of the pathway, as occurs in many metabolic reactions.

neodymium (Nd) The element with atomic number 60; relative atomic mass 144.2; melting point 1,016°C; boiling point 3,068°C; relative density 7.0. Neodymium is a pale yellow metal of the LANTHANIDE series. Its SALTS are pink and sometimes used to colour glass and ceramics. Neodymium salts are also used in some lasers.

neon (Ne) The element with atomic number 10; relative atomic mass 20.2; melting point –249°C; boiling point –246°C. Neon is an inert gas that occurs in very small amounts in air. It is obtained by FRACTIONAL DISTILLATION of liquid air. The bright orange-red colour of its GAS DISCHARGE has led to its widespread use in decorative lighting and advertising signs.

neoprene, *polychloroprene* A synthetic rubber made by polymerization of 2-CHLOROBUTA-1,3-DIENE. It is very strong, resistant to chemical attack and to abrasion and so is used instead of rubber where these properties are needed. For example, it is used to coat wires and cables and in road and building construction.

neptunium (Np) The element with atomic number 93. The longest lived ISOTOPE (neptunium–237) has a HALF-LIFE of 2 million years. Neptunium is produced in many nuclear reactors from the BETA DECAY of uranium–239, which is formed when the common isotope uranium–238 is struck by a neutron. The beta decay of neptunium produces plutonium. Chemically, neptunium is a reactive member of the ACTINIDE series of metals.

Nernst equation An equation that enables a REDOX POTENTIAL under non-standard conditions to be calculated from STANDARD REDOX POTENTIALS. If the standard redox potential is $E^\ominus$, then if the ABSOLUTE TEMPERATURE is T and the number of electrons transferred per reaction is z, the new redox potential E will be

$$E = E^\ominus + (kT/ze)\ln([\text{O.A.}]/[\text{R.A.}])$$

where e is the charge on an electron, k is the BOLTZMANN CONSTANT, and [O.A.] and [R.A.] are the concentrations of the oxidized and reduced (*see* OXIDATION, REDUCTION) forms of the reagents, respectively.

Nernst heat theorem Any chemical change involving pure crystalline materials, and taking place at ABSOLUTE ZERO will involve no change in ENTROPY.

Nessler's reagent A solution of mercury(II) iodide in potassium hydroxide and potassium iodide. It is used to test for AMMONIA, with which it forms a brown precipitate of NH_2Hg_2OI.

neurotransmitter A chemical of low relative molecular mass that is released by nerve cells and transmits nerve impulses between nerve cells or between nerve and muscle cells. Neurotransmitters can be excitatory or inhibitory. About 50 neurotransmitters are known, including ACETYLCHOLINE, NORADRENALINE, ADRENALINE, ENDORPHINS, ENCEPHALINS. A number of drugs exist that can mimic neurotransmitters; for example, AMPHETAMINES mimic the action of noradrenaline and NICOTINE mimics natural neurotransmitters. Other drugs affect the release of neurotransmitters, for example caffeine increases release, whereas β-blockers inhibit release.

neutralization The process of an ACID reacting with a BASE to form a SALT plus water, for example:

$$HCl + KOH \rightarrow KCl + H_2O$$

Such reactions are all based on the formation of a water molecule from a hydrogen ion and a hydroxyl ion:

$$H^+ + OH^- \rightarrow H_2O$$

neutrino A light (probably massless) elementary particle. Neutrinos have no charge, probably zero rest mass, and move at the speed of light.

neutron The neutral particle found in the nuclei of all elements except hydrogen. It is slightly more massive than the PROTON. Outside the nucleus, the neutron has a mean life of about 12 minutes, before decaying into a proton, an ELECTRON and an antineutrino (the antiparticle of the NEUTRINO). *See also* ATOM, ISOTOPE, MASS NUMBER.

neutron diffraction A technique similar to X-RAY DIFFRACTION, but using low energy NEUTRONS, which have a DE BROGLIE WAVELENGTH similar to the X-rays they replace. As the neutrons are scattered by the nuclei rather than by the electrons of the crystal, this technique can be used to obtain further information about crystal structures.

newton The derived SI UNIT of force. Force is equal to the mass times the acceleration, so one newton is defined as the force that will make a mass of one kilogram accelerate at one metre per second per second.

niacin *See* NICOTINIC ACID.

Nicad cell *See* NICKEL-CADMIUM CELL.

Nichrome (*Trade name*) Any of a range of alloys of nickel (typically 80 per cent) and chromium. They are often used as electrical heating elements as they have a good resistance to oxidation and a high electrical resistivity.

nickel (Ni) The element with atomic number 28; relative atomic mass 58.7; melting point 1,450°C; boiling point 2,840°C; relative density 8.9. Its salts often have a characteristic pale green colour.

Nickel is a hard metal, and is fairly unreactive, which has led to its use as an electrolytically deposited coating on some steel items (*see* ELECTROPLATING). It is also used in the manufacture of NICKEL-CADMIUM CELLS. Nickel is widely used as a catalyst in a number of reactions in the petrochemical industry, such as HYDROGENATION. Along with iron, nickel is the major constituent of the Earth's core and is also found in some meteorites.

nickel-cadmium cell, *Nicad* A high capacity rechargeable electrochemical CELL, often used to power portable electrical appliances.

nickel carbonyl ($Ni(CO)_4$) A colourless liquid; melting point –25°C; boiling point 43°C. Nickel carbonyl is a CO-ORDINATION COMPOUND, formed by passing carbon monoxide gas over the powdered metal. The reaction is reversed at higher temperatures leading to the decomposition of the compound.

nickelic (*adj.*) An obsolete term describing compounds containing the Ni^{3+} ion, for example nickelic oxide, Ni_2O_3. The existence of such compounds is not well established, and nickelic oxide is believed by some to be a SOLID SOLUTION of nickel(II) oxide, NiO, and nickel peroxide, NiO_2. *See also* NICKELOUS.

nickelous (*adj.*) An obsolete term describing a compound containing the Ni^{2+} ion, particularly in those rare situations where it was necessary to distinguish these from NICKELIC compounds. An example is nickelous oxide, NiO.

nickel oxide Nickel(II) oxide, NiO, or nickel(III) oxide (nickel peroxide), Ni_2O_3.

Nickel(II) oxide is a green solid; decomposes on heating; relative density 6.6. It can be made by heating nickel nitrate in the absence of air:

$$2Ni(NO_3)_2 \rightarrow 2NiO + 4NO_2 + O_2$$

Nickel(III) oxide is a grey solid; decomposes on heating; relative density 4.8. It can be made by burning nickel in air, or by heating nickel(II) oxide in air:

$$4Ni + 3O_2 \rightarrow 2Ni_2O_3$$

$$4NiO + O_2 \rightarrow 2Ni_2O_3$$

nicotinamide adenine dinucleotide *See* NAD.

nicotinamide adenine dinucleotide phosphate *See* NADP.

nicotine, *3-(1-methyl-2-pyrrolidyl)pyridine* ($C_{10}H_{14}N_2$) An ALKALOID present in tobacco. It is a colourless liquid that darkens on exposure to air. Nicotine is poisonous and used as an insecticide.

nicotinic acid, *niacin, vitamin B₃* ($C_6H_5NO_2$) A VITAMIN of the VITAMIN B COMPLEX. Nicotinic acid is an essential component of the mammalian diet. It is found in liver and in sunflower and groundnut seeds. Deficiency causes pellagra, which results in skin lesions, diarrhoea and mental disorders. Its AMIDE derivative, nicotinamide, is a component of NAD and NADP, which are important electron carriers in metabolic pathways.

Ninhydrin, *1,2,3-triketohydrindene hydrate* ($C_9H_4O_3,H_2O$) A compound that reacts with proteins, peptides and amino acids to give a blue colour. It is used for their detection as a spray in PAPER CHROMATOGRAPHY.

niobium (Nb) The element with atomic number 41; relative atomic mass 92.9; melting point 2,468°C; boiling point 4,742°C; relative density 8.6. Niobium is a TRANSITION METAL, used in the manufacture of some special steels and in niobium-tin alloy. This alloy is the conductor in many superconducting magnets, as it retains its superconductivity up to the relatively high temperature of 22 K and is able to remain superconducting in quite strong magnetic fields.

nitrate Any salt containing the nitrate ion, NO_3^-. Nitrates form stable crystals containing WATER OF CRYSTALLIZATION. Most nitrates are soluble and many are important fertilizers since the nitrate ion is an important source of fixed nitrogen (*see* NITROGEN FIXATION). *See also* NITROGEN CYCLE.

nitration In organic chemistry, the introduction of a nitro-group (NO_2) to ARENES. This is achieved using a mixture of concentrated nitric and sulphuric acids, which produces the nitronium ion (NO_2^+), a strong ELECTROPHILE. The reactions are ELECTROPHILIC SUBSTITUTIONS. Nitration is an important step in industry in the production of a wide range of compounds. For example the nitronium ion reacts with benzene to give nitrobenzene.

nitric acid (HNO_3) An important MINERAL ACID; melting point –42°C; boiling point 83°C; relative density 1.5. Nitric acid is manufactured by the action of concentrated sulphuric acid on NITRATES, for example:

$$KNO_3 + H_2SO_4 \rightarrow KHSO_4 + HNO_3$$

This produces concentrated nitric acid as a gas. An alternative process is the OXIDATION of ammonia by excess air with a platinum-rhodium catalyst:

$$NH_3 + 2O_2 \rightarrow HNO_3 + H_2O$$

In practice this reaction proceeds via nitrogen oxide, NO, some of which is oxidized to nitrogen dioxide, NO_2, which dissolves in water to form nitric acid, with the remaining NO being recycled. The acid formed in this process is dilute, but can be concentrated by DISTILLATION up to 68.5 per cent. Nitric acid is an important FEEDSTOCK in the fertilizer, dyestuff and explosives industries.

nitric oxide *See* NITROGEN MONOXIDE.

nitride Any BINARY COMPOUND containing NITROGEN. The ALKALI METALS and ALKALINE EARTHS form nitrides on heating. These contain the ion N^{3-}, but are fairly unstable. They are hydrolysed (*see* HYDROLYSIS), for example, to form the metal HYDROXIDE and ammonia:

$$Na_3N + 3H_2O \rightarrow 3NaOH + NH_3$$

nitriding A process for producing a hard outer layer in certain special steels, especially those containing chromium, aluminium or molybdenum. The steel article is machined normally then heated for several hours in an atmosphere of ammonia. The ammonia dissociates slightly and nitrogen atoms are adsorbed into the surface of the steel (*see* ADSORPTION).

nitrification The process occurring in the soil by which ammonia (from UREA, urine and the break down of protein by AMMONIFICATION) is oxidized by bacteria to form NITRATES. The bacterium *Nitrosomonas* oxidizes ammonium ions (NH_4^+) to NITRITES (NO_2^-), which are toxic but quickly oxidized to nitrates (NO_3^-) by *Nitrobacter*. These processes release energy, which the bacteria use for their own respiratory processes. Nitrification is reduced if soil temperature and pH are low. Because nitrates are soluble, they can easily leach out of the soil, causing nitrogen deficiency, so artificial fertilizers are often added to prevent growth limitation. *See also* NITROGEN CYCLE, DENITRIFICATION.

nitrile An organic compound containing a carbon-nitrogen TRIPLE BOND, an organic CYANIDE. The general formula is RCN, where R is a HYDROCARBON group, for example CH_3CN, ethanenitrile. Nitriles are considered to be CARBOXYLIC ACID derivatives. Nitriles are formed by dehydration (*see* DEHYDRATE) when AMIDES are heated with phosphorus(V) oxide:

$$CH_3CONH_2 \xrightarrow[-H_2O]{P_2O_5} CH_3C{\equiv}N$$

or by the reaction of a HALOGENOALKANE with sodium/potassium cyanide in ethanol:

$$C_2H_5Br + CN^- \rightarrow C_2H_5CN + Br^-$$

When reacted with acid or alkali, nitriles are hydrolysed to the corresponding carboxylic acid. They are reduced to primary AMINES. Nitriles are useful as intermediates in the synthesis of certain organic compounds, particularly in providing a way of adding an extra carbon atom to a chain. They have other uses, such as in insect repellents, weed control and fuel additives.

nitrile rubber A synthetic alternative to natural RUBBER, formed by the COPOLYMERIZATION of buta-1,3-diene and PROPENONITRILE. It has the important property of being resistant to a wide range of organic solvents.

nitrite Any salt containing the nitrite ion, NO_2^-. Nitrites are easily oxidized (*see* OXIDATION) to NITRATES.

nitrobenzene ($C_6H_5NO_2$) A colourless liquid; boiling point 211°C; melting point of 6°C.

NO_2

Nitrobenzene is produced by reacting benzene with a mixture of sulphuric and nitric acids. The NITRONIUM ION is an intermediate in this process. Most of the nitrobenzene manufactured is used in the dyestuffs industry, either as nitrobenzene or reduced to PHENYLAMINE.

nitrocellulose See CELLULOSE NITRATE.

nitro compounds A large group of organic compounds containing the nitro group, NO_2, bound to a carbon atom. They are produced by NITRATION and on reduction give AROMATIC AMINES. Some nitro compounds can be themselves used as dyes; others are reduced for this purpose.

nitrogen (N) The element with ATOMIC number 7; relative atomic mass 14.0; melting point –210°C; boiling point –196°C. Nitrogen gas makes up 78 per cent of the atmosphere, but is chemically fairly unreactive due to the TRIPLE BOND in the N_2 molecule ($N{\equiv}N$).

Nitrogen is an essential element for life, being present in all PROTEINS and NUCLEIC ACIDS. Some bacteria are able to 'fix' nitrogen from the air and incorporate it into the growth of certain plants (*see* NITROGEN CYCLE, NITROGEN FIXATION). NITRATES are often used as a fertilizer since this is a form of nitrogen that can be readily used by many organisms.

Nitrogen is obtained commercially from the FRACTIONAL DISTILLATION of liquefied air

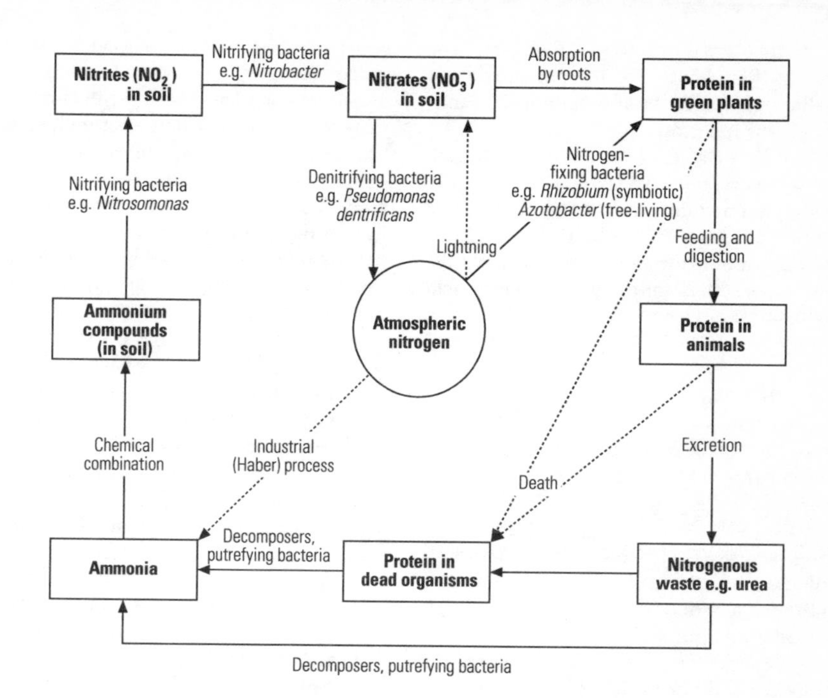

A summary of the nitrogen cycle.

and is used to manufacture AMMONIA (*see* HABER PROCESS). The gas is also used to provide an inert environment in welding and metallurgy. Liquid nitrogen is commonly used to cool other objects to low temperatures.

nitrogen cycle The circulation of nitrogen, mostly by living organisms, through the environment. Nitrogen is an essential mineral for all organisms because it is used to make proteins and other organic compounds. Although the atmosphere is 78 per cent nitrogen, this cannot be readily used by most organisms (*see* NITROGEN FIXATION). Plants obtain nitrogen from NITRATES in the soil, by absorption through the roots, and convert them to PROTEINS. The proteins are passed to plant-eating animals (herbivores) and then to meat-eating animals (carnivores) along the food chain, and nitrogen is eventually returned to the soil as excrement or when organisms die.

There are several important groups of bacteria involved in the processes of the nitrogen cycle. The bacteria *Nitrosomonas* and *Nitrobacter* are important in NITRIFICATION, which is the process by which ammonia is oxidized to form nitrates. Bacteria that can utilize atmospheric nitrogen by nitrogen fixation to make nitrogenous compounds are called nitrogen-fixing bacteria.

Anaerobic bacteria, such as *Pseudomonas denitrificans* and *Thiobacillus denitrificans*, are also important because they convert nitrates in the soil back to atmospheric nitrogen in the process of DENITRIFICATION. Decomposers (organisms capable of feeding on excrement and other dead organisms) are crucial in the nitrogen cycle for breaking down proteins, AMINO ACIDS and other nitrogenous compounds to form ammonia, ammonium ions (by AMMONIFICATION) and AMINES (by putrefaction), which are used by the nitrifying bacteria.

nitrogen dioxide *See* DINITROGEN TETROXIDE.

nitrogen fixation The process by which atmospheric nitrogen is converted to nitrogenous compounds by the action of nitrogen-fixing bacteria or cyanobacteria. These organisms convert atmospheric nitrogen to ammonia, which they use to make amino acids.

Certain types of nitrogen-fixing bacteria, such as *Rhizobium,* live in specialized root nodules on the roots of leguminous plants, such beans and peas.

See also NITROGEN CYCLE.

nitrogen hydride *See* AMMONIA.

nitrogen monoxide (NO) A colourless gas; melting point –164°C; boiling point –152°C. Nitrogen monoxide can be prepared by the reaction of sodium nitrite with sulphuric acid and iron(II) sulphate:

$$2NaNO_2 + 2H_2SO_4 + 2FeSO_4 \rightarrow 2NO + Na_2SO_4 + Fe_2(SO_4)_3 + 2H_2O$$

Nitrogen monoxide is readily oxidized to dinitrogen tetroxide:

$$2NO + O_2 \rightarrow N_2O_4$$

nitroglycerine ($C_3H_5(ONO_2)_3$) A colourless, odourless oil produced by treating GLYCEROL with a mixture of nitric and sulphuric acids. It is a powerful explosive used in the preparation of dynamite. Nitroglycerine is also used in the treatment of the chest complaint angina.

nitro-group The NO_2 group.

nitronium ion (NO_2^+) A strongly electrophilic ion (*see* ELECTROPHILE) produced by the reaction of concentrated nitric and sulphuricacids:

$$HONO_2 + 2(HO)_2SO_2 \rightarrow NO_2+ + H_3O+ + 2HOSO_3^-$$

The nitronium ion is important in NITRATION reactions.

nitrosamines Compounds with the general formula RR′NNO, where R and R′ can be any of a variety of side groups. Nitrosamines are mostly yellow oils and are carcinogenic, causing cancer particularly in liver, lungs and kidneys. They are present in cigarette smoke.

nitrous acid (HNO_2) A weak acid, formed by the action of strong acids on nitrites, for example:

$$Ba(NO_2)_2 + H_2SO_4 \rightarrow 2HNO_2 + BaSO_4$$

Nitrous acid is stable only as a gas or a dilute aqueous solution. The aqueous acid dissociates on heating to give nitric acid and nitrogen monoxide:

$$3HNO_2 \rightarrow HNO_3 + 2NO + H_2O$$

nitrous oxide *See* DINITROGEN OXIDE.

NMR An abbreviation for NUCLEAR MAGNETIC RESONANCE.

nobelium (No) The element with atomic number 102. It does not occur naturally, since the longest lived ISOTOPE (nobelium–259) has a HALF-LIFE of only 3 minutes.

noble (adj.) An archaic term describing a metal or gas that is unreactive.

noble gas, ***inert gas, rare gas*** Any of the elements HELIUM, NEON, ARGON, KRYPTON, XENON and RADON that occupy GROUP 18 (formerly group VIII or group 0) of the PERIODIC TABLE. They are characterized by having full VALENCE SHELLS of electrons. Thus they do not readily form IONS or COVALENT BONDS, though some will form compounds with fluorine.

Helium was discovered first in the spectroscopic analysis of the Sun. Apart from hydrogen, it is by far the most abundant element in the Universe. It is, however, relatively rare on Earth as it is does not combine chemically to form a solid and the gas is light enough to escape quite rapidly from the Earth's atmosphere. Some NATURAL GAS sources contain significant amounts of helium, believed to be a result of ALPHA PARTICLE production in nearby uranium deposits.

The other noble gases (except RADON) were discovered by spectroscopic analysis of the products of FRACTIONAL DISTILLATION of liquid air. Argon is by far the most abundant and is used to provide an inert atmosphere for example, in light bulbs, and in welding easily oxidized materials such as aluminium.

See also SHELL.

noble metal An archaic term for a metal that does not react with the common MINERAL ACIDS. Gold and platinum are examples. Such metals are used in jewellery as they do not TARNISH, and are also used in electrical contacts as they resist OXIDATION. *See also* BASE METAL.

nonmetal Any element that is not a METAL. Nonmetals are poor conductors of electricity and tend to have low melting points; many of them are gases at room temperature. Nonmetals are electronegative (*see* ELECTRONEGATIVITY) and tend to form negative ions or COVALENT BONDS. The OXIDES of nonmetals generally dissolve in water to form acids.

non-reducing sugar A sugar that cannot act as a REDUCING AGENT in solution, as indicated by a negative BENEDICT'S TEST or FEHLING'S TEST. *See also* CARBOHYDRATE.

non-stoichiometric compound A compound in which the elements do not combine in fixed proportions so it cannot be represented by a fixed formula such as A_xB_y, where *x* and *y* are integers. For example, many TRANSITION METALS form non-stoichiometric oxides owing to the variable VALENCY of the metal. *Compare* STOICHIOMETRIC.

noradrenaline, *norepinephrine* ($C_8H_{10}NO_3$) A HORMONE and NEUROTRANSMITTER that is, like ADRENALINE, derived from the AMINO ACID tyrosine. Noradrenaline is secreted by the adrenal gland and some nerve endings. It maintains arousal in the brain, for example in response to external stress, dreaming and emotion.

norepinephrine *See* NORADRENALINE.

normal salt A SALT formed when all the hydrogen available in an ACID (*see* ACIDIC HYDROGEN) has been replaced by metal ions. An example is sodium sulphate, Na_2SO_4, compared to sodium hydrogensulphate, $NaHSO_4$, which is an ACIDIC SALT. *See also* BASIC SALT.

normal solution An obsolete term for a solution that has a CONCENTRATION such that one decimetre cubed of the solution contains, or will react with, one MOLE of hydrogen atoms. Thus a normal solution of hydrochloric acid, H_2SO_4, or copper sulphate, $CuSO_4$ will have a concentration of 0.5 mol dm^{-3}, whilst a normal solution of hydrochloric acid, HCl, will have a concentration of 1 mol dm^{-3}. The normality of a solution is often represented by the letter N, thus 2 N indicates a solution of twice normal concentration.

nuclear magnetic resonance (NMR) A technique widely used in organic chemistry to identify organic molecules and to determine their structures. The technique can also be used to examine living organs without destroying them, which has revolutionized the diagnosis of disease.

Some of the nuclei in a molecule possess the property of SPIN, which can be in one of two orientations. If a magnetic field is applied to the molecule these differences in spin cause a splitting of the nuclear ENERGY LEVELS. The molecule is then subjected to an additional weak, oscillating magnetic field. At a precise frequency the nuclear magnets resonate and it is this that is recorded and amplified.

The resonance frequencies of a particular element depend on its environment. Thus using NMR it is possible, for example, to detect three different types of hydrogen atoms in ethanol – those in CH_3, CH_2 and OH groups. In addition, information about molecular structure can be obtained by the coupling of nuclei.

nuclease A general term for an ENZYME that degrades NUCLEIC ACID. *See* RESTRICTION ENDONUCLEASE.

nucleic acid The complex organic acid present in the cells of all organisms that is responsible for their genetic make-up. The two types of nucleic acid are DNA and RNA, and each is made of long chains of NUCLEOTIDES.

nucleon A particle found in an atomic NUCLEUS: a NEUTRON or a PROTON.

nucleophile (Greek = nucleus loving) Any molecule or ion that forms a new COVALENT BOND by donating or sharing its electrons, such as the hydroxyl ANION (OH^-).

Nucleophiles are often negatively charged but do not need to be. For example, water is a nucleophile because of the LONE PAIR of electrons on the oxygen atom. A nucleophile is a LEWIS BASE.

nucleophilic substitution A SUBSTITUTION REACTION in which an atom or group in a molecule is replaced by a NUCLEOPHILE (which provides the electrons to form the new COVALENT BOND). The general reaction can be considered to be as follows:

$$R\text{–}X + :NUC \rightarrow R\text{–}NUC + X^-$$

where R is a HYDROCARBON, metal or METALLOID group and the nucleophile, NUC, and X can be a variety of organic or inorganic ANIONS.

There are two types of nucleophilic reactions: Sn1 (substitution, nucleophilic, unimolecular) and Sn2 (substitution, nucleophilic, bimolecular) reactions. In Sn1 reactions the R–X bond is broken first and the nucleophile then enters the reaction to provide the electrons to form the R–NUC bond. In Sn2 reactions the nucleophile forms the R–NUC bond at the same time as the R–X bond breaks (hence the reaction is bimolecular).

Nucleophilic substitution reactions are used to introduce various FUNCTIONAL GROUPS into organic molecules; for example into HALOGENOALKANES. In particular, primary halogenoalkanes usually react by nucleophilic substitution reactions.

nucleoprotein A general term referring to any PROTEIN that occurs in combination with a NUCLEIC ACID in the nuclei of cells. The proteins are usually HISTONE proteins and the nucleic acid can be considered to be the PROSTHETIC GROUP. Chromosomes are largely made up of nucleoproteins.

nucleoside The organic base and PENTOSE sugar part of a NUCLEOTIDE, i.e. a nucleotide without the PHOSPHATE group. In RNA, the sugar is RIBOSE and the common nucleosides are adenosine, guanosine, cytidine and uridine. In DNA, the sugar is DEOXYRIBOSE and the nucleosides found are the same as RNA except that uridine is replaced by thymidine.

nucleotide The constituent unit of the nucleic acids DNA and RNA, which itself consists of an organic base, a PENTOSE sugar (RIBOSE, $C_5H_{10}O_5$, or DEOXYRIBOSE, $C_5H_{10}O_4$) and a PHOSPHATE group. There are five organic bases that are either PURINES (adenine and guanine), which have double rings (one with six sides, one with five), or PYRIMIDINES (cytosine and thymine or uracil), which have two single, six-sided rings. The organic bases are abbreviated to A, G, C, T or U, and the order in which they are placed in the nucleic acid strand contains the genetic code; that is, an organism's specific genetic information. The three components of a nucleotide join together in a CONDENSATION REACTION. Links then form similarly between the sugar and phosphate groups of two or more nucleotides, to form dinucleotides or polynucleotides. Although the main role of nucleotides is in the formation of nucleic acids, they are also found in other molecules, for example AMP, ADP, ATP, NAD, NADP and FAD.

nucleus (*pl.* ***nuclei***) The positively charged massive centre of an ATOM. It is made up of particles called PROTONS, which are positively charged, having a charge equal in size to the negative charge on an ELECTRON, and NEUTRONS, which have no charge and have slightly more mass than the protons. Protons and neutrons are collectively called nucleons. The neutrons and protons are both far more massive than the electrons that surround the nucleus, but the nucleus is far smaller than the atom itself. A typical atom is 10^{-10} m in diameter, whilst a nucleus is 10^{-14} m across.

The number of protons in the nucleus determines the number of electrons needed to produce a neutral atom. It is the arrangement of these electrons that determines the chemical properties of an ELEMENT. Thus different numbers of protons in the nucleus produce atoms of different elements. The number of neutrons in a nucleus has no affect on the chemical properties, but does affect the mass of the atom. The number of protons in a nucleus is called the ATOMIC NUMBER (Z). The total number of nucleons is called the MASS NUMBER (A). A NUCLIDE may therefore be represented by the notation $^{A}_{Z}X$ where X is the element.

See also LIQUID DROP MODEL, SHELL MODEL.

nuclide An atomic NUCLEUS identified as having a particular number of NEUTRONS and PROTONS. Thus nuclei of different ISOTOPES or different elements are different nuclides.

nutrient Any chemical substance required by plants and animals for their normal growth and development. In animals, these are largely obtained through the diet and include CARBOHYDRATES, PROTEINS, LIPIDS, VITAMINS and certain MINERALS. Plants obtain their nutrients through the soil and air. *See also* MACRONUTRIENT, MICRONUTRIENT.

nylon A synthetic POLYAMIDE similar in structure to protein. It was first synthesized as an alternative to silk and is more elastic and stronger than silk. There is a range of nylons, the most common being nylon–6 and nylon–6,6. Like proteins, nylon fibres contain NH and CO groups but in the latter there are different carbon chains between these groups. In nylon–6 the repeating unit is $-NH(CH_2)_5CO$, the number referring to the number of carbon atoms in this unit. Nylon–6,6 is a condensation POLYMER made from two different molecules, a diacid $HOCO(CH_2)_4COOH$ and a diamine $H_2N(CH_2)_6NH_2$. Nylon has many uses, in hosiery, carpets, textiles, and moulded plastics.

nylon-6,6	nylon-6
NH	NH
$(CH_2)_6$	$(CH_2)_5$
NH	CO
CO	NH
$(CH_2)_4$	$(CH_2)_5$
CO	CO
NH	

O

occlusion 1. The occurrence of small pockets of liquid or vapour in a crystalline material.

2. The ADSORPTION of gas molecules, especially hydrogen into INTERSTITIAL sites in a metallic LATTICE. The occlusion of hydrogen by palladium has been proposed as a safe way of storing hydrogen as a fuel for road vehicles.

ochre The yellow or red mineral form of iron(III) oxide, Fe_2O_3, often found mixed with clay. The term especially applies to this mineral when powdered and used as a pigment.

octadecanoic acid *See* STEARIC ACID.

octahedral (*adj.*) Having the shape of an octahedron; that is, a figure with eight triangular faces, each side having the same length. In chemistry, the term is used to describe a molecule or RADICAL in which a central atom is surrounded by six others, held by COVALENT BONDS or HYDROGEN BONDS. Common examples include the hexacynanoferrate(III) ion, $Fe(CN)_6^{3-}$, and many hydrated TRANSITION METAL ions, such as $Cu(H_2O)_6^{2+}$.

octane (C_8H_{18}) The eighth member of the ALKANE series, of which there are 18 possible ISOMERS. They are found in PETROLEUM and have boiling points between 99°C and 125°C. 2,2,4-trimethylpentane (iso-octane) is the most important isomer, being a colourless liquid with a boiling point of 99°C. It has anti-knock properties (*see* KNOCKING) and is used as a standard in determining the OCTANE NUMBER of petrol.

octane number, *octane rating* A numerical representation of the ability of PETROL to resist KNOCKING. This is measured by comparing the ease with which a petrol mixture burns in comparison to iso-octane (2,2,4-trimethylpentane) in a blend with heptane. Pure heptane is given a value of 0 and pure iso-octane a value of 100. Thus petrol with an octane value of 97 burns like a mixture of 97:3 iso-octane:heptane, by volume under standard conditions. Petrol with a higher value burns faster than that with a lower value.

octane rating *See* OCTANE NUMBER.

octanoic acid ($C_8H_{16}O_2$) A FATTY ACID found in sweat or as an ESTER in fusel oil, and GLYCERIDES in animal and coconut milk; melting point 16°C; boiling point 239°C.

octet A group of eight electrons in the outermost shell, the stable configuration of an atom or ion; that is, a full set of S- and P-ORBITALS.

octet rule The broad principle that elements tend to react in ways that leave them with an octet of VALENCE ELECTRONS.

oestrogen A STEROID HORMONE produced in the ovary of female mammals. The term oestrogen actually refers to a group of hormones (including synthetic ones), of which oestradiol ($C_{18}H_{24}O_2$) is the main one in humans. Oestrogens cause the development of secondary sexual characteristics, for example breast development and fat deposition, and help prepare the uterus for pregnancy and maintain the pregnancy if it follows. Oestrogens also repair the uterine lining following menstruation if pregnancy does not follow. Synthetic oestrogens are a major component of oral contraceptives and are most often thought to be responsible for the side-effects.

oil One of many types of naturally occurring HYDROCARBONS, which can be solid (FATS, WAXES) or liquid. Oils are flammable and usually insoluble in water. Mineral oils are those obtained from refining PETROLEUM and are used as fuels and lubricants. Essential oils are those obtained from plants which possess pleasant odours used in perfumes and flavourings. Fixed oils are LIPIDS found in animals and plants, such as fish and nuts, and are used, for example, in foods, soaps and paints.

olefin The common name for ALKENE.

oleic acid, *cis-9-octadecenoic acid* ($C_{18}H_{34}O_2$) An unsaturated FATTY ACID with one DOUBLE BOND; melting point 13°C. Oleic acid is a major constituent of animal and plant fats. It occurs in lard, groundnut oil, soya-bean oil and butterfat.

oleum Concentrated SULPHURIC ACID containing dissolved sulphur trioxide, SO_3.

olivine Any of a group of SILICATE minerals based on iron and magnesium, having the general formula $(Mg,Fe)_2SiO_4$. The colours vary from green to light brown.

-onium A suffix used to indicate an ion formed by the addition of a proton to a neutral molecule, particularly in hydroxonium, H_3O^+, and ammonium, NH_4^+.

open hearth furnace A device, now largely obsolete, for the manufacture of steel from PIG IRON. The pig iron is melted by burning a mixture of air, gas and oil above it in a shallow furnace (the 'hearth'). Oxygen is injected into the top of the furnace through water-cooled pipes called lances, and this OXIDIZES impurities in the steel, in particular carbon. The carbon may also be oxidized by the inclusion of HAEMATITE, an ore containing iron oxide. LIMESTONE is added to form a SLAG with the less volatile impurities. The molten steel is then removed from the furnace in large ladles, the slag is skimmed off, and the steel poured into moulds. The exhaust gas is used to heat a number of bricks arranged in an open lattice pattern. The gas flow is periodically reversed so that two sets of bricks, one at each end of the furnace, are alternately used to pre-heat the incoming gas and heated by the outgoing gases. The open hearth furnace is a slow process compared to the OXYGEN FURNACE and the ELECTRIC ARC FURNACE, which have largely replaced it. *See also* BESSEMER CONVERTER.

opiate A term referring to drugs derived from OPIUM. The term opiate is also used to refer to natural pain-relieving chemicals produced by the body, such as ENDORPHINS and ENCEPHALINS.

opium A drug that is obtained from unripe seeds of the opium poppy *Papaver somniferum*. It contains a number of ALKALOIDS, including the pain-killing substances morphine and codeine, and also a highly poisonous substance called thebaine. Morphine is a strong but addictive pain-killing drug and heroin is an even stronger synthetic derivative of morphine.

optical activity The ability of certain substances to rotate the plane of polarized light (*see* POLARIZATION). An optically active substance may be a crystal, liquid or solution. Optical activity arises from a lack of symmetry in the three-dimensional structure of the molecules (or crystal lattices) concerned. It gives rise to optical ISOMERS, which differ in the direction in which they rotate the plane of polarized light. Optically active substances are either LAEVOROTATORY or DEXTROROTATORY, according to the direction of rotation. *See also* OPTICAL ISOMERISM, POLARIMETER.

optical isomerism The property possessed by ISOMERS that are mirror images of one another. Unlike structural isomers, optical isomers are only obvious when models are constructed. They arise due to asymmetry of the molecule and differ in the direction in which they rotate the plane of polarized light (*see* POLARIZATION). The two mirror images cannot be superimposed on one another and are related in the same way as the right hand is to the left hand. This is termed chirality and the molecules are said to be CHIRAL.

The two isomers are called enantiomers. A pair of enantiomers are identical in all respects except for their orientation in space. Their physical and chemical properties are the same except for the direction in which they rotate the plane of a beam of polarized light when passed through solutions containing each enantiomer. One enantiomer rotates the plane of polarized light to the right (dextrorotatory; denoted +) and the other rotates it an equal amount to the left (laevorotatory; denoted –). Although most chemical reactions of two enantiomers are identical they do differ in their reaction with other chiral molecules.

Many naturally occurring compounds are chiral (for example AMINO ACIDS) but usually only one enantiomer is found in nature.

See also GEOMETRIC ISOMERISM, RACEMIC MIXTURE.

orbital A region of space in an ATOM or MOLECULE that can be occupied by a maximum of two electrons. An orbital is often thought of as the volume that encloses the space within which the probability of finding an electron is greater than a specified figure, since in principle, electron WAVEFUNCTIONS extend out infinitely far from the atom.

Two electrons in a single orbital often form a stable configuration called a LONE PAIR, whilst a single electron is often active in the formation of CHEMICAL BONDS. Orbitals may overlap and share electrons between atoms to form a COVALENT BOND, or an orbital may gain or lose an electron so that the atom or molecule forms an ION.

Atomic orbitals are labelled by a number in the sequence 1, 2, 3 etc., called the PRINCIPAL QUANTUM NUMBER, with higher numbers indicating higher energies. The principal quantum number indicates the size of the orbital. They are also labelled by a letter, s, p, d, f, called the SUBSIDIARY QUANTUM NUMBER. This denotes the shape of the orbital and the angular momentum of an electron in that orbital. For a principal quantum number of 1, only an S-ORBITAL exists. For a principal quantum number of 2 there is one s-orbital and three P-ORBITALS. With a principal quantum number of 3 there is a single s-orbital, three p-orbitals and five D-ORBITALS, and so forth. The energy of the orbitals, and hence the sequence in which they are filled by electrons is basically

1s, 2s, 2p, 3s, 3p, 4s, 3d, 4p, 5s, 4d, 5p, 6s, 4f, 6p, 7s, 5f

The order in which electrons go into these orbitals to form neutral atoms is responsible for the structure of the PERIODIC TABLE.

See also HUND'S RULE OF MAXIMUM MULTIPLICITY, HYBRID ORBITAL, MOLECULAR ORBITAL, PAULI EXCLUSION PRINCIPLE, SHELL.

order A measure of the extent to which the rate of a reaction depends on the CONCENTRATION of the reagents involved. In the reaction

$$A + B \rightarrow \text{products}$$

the RATE OF REACTION is proportional to $[A]^m[B]^n$, where [A] is the molar concentration of A and so on. The reaction is said to be mth order in A and nth order in B, with an overall order of $m + n$.

ore Any naturally occurring rock from which a metal can be extracted, such as iron ore, which is usually iron(III) oxide, Fe_2O_3, or iron(II) carbonate, $FeCO_3$.

organic A term that was used in the late eighteenth century to refer to compounds obtained from living material in contrast to those derived from minerals (which were termed INORGANIC). Now 'organic' refers to all compounds containing both carbon and hydrogen, except the HYDROGENCARBONATES.

Organic compounds are more numerous than inorganic compounds and form the basis of life. Although many organic compounds are made only by living organisms, such as PROTEINS and CARBOHYDRATES, it is now possible to manufacture many synthetically. Many organic compounds are derived from PETROLEUM, which is the remains of microscopic marine organisms.

Organic compounds may consist only of carbon and hydrogen – HYDROCARBONS – linked by COVALENT BONDS, although frequently they also contain oxygen, nitrogen, sulphur, phosphorus or HALOGENS similarly linked to the carbon atoms. This contrasts to the IONIC BONDING typical of many inorganic compounds.

Organic chemistry is the study of organic compounds. Examples of organic compounds include ALKANES, ALKENES, ALKYNES, ALCOHOLS, ALDEHYDES, AMIDES, AMINES, AMINO ACIDS, ESTERS, ETHERS, KETONES, proteins and carbohydrates.

organometallic compounds Organic compounds that are linked to metal or metal-like atoms, such as $NaCH_3$. CYANOCOBALAMIN (vitamin B_{12}) is the only naturally occurring

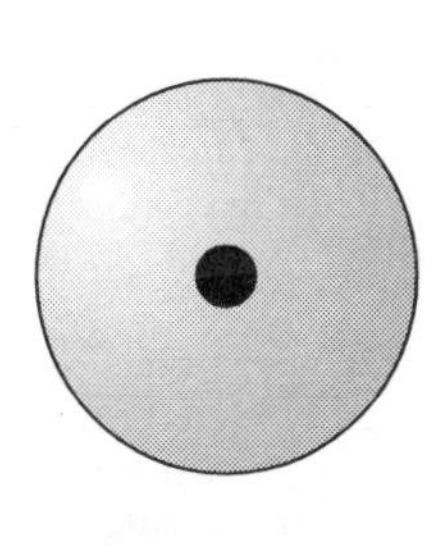

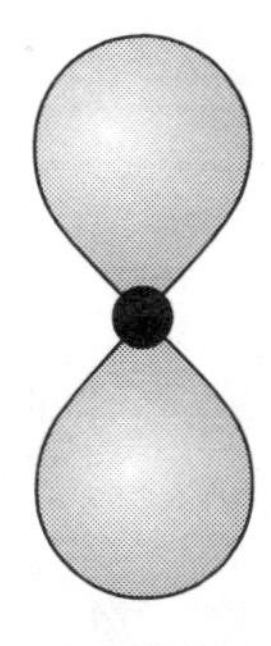

Electron orbitals.

organometallic compound. The compounds are used in organic chemistry as GRIGNARD REAGENTS and as ZIEGLER–NATTA CATALYSTS.

ornithine ($C_5H_{12}N_2O_2$) An AMINO ACID that is an intermediate in the UREA CYCLE and in the synthesis of ARGININE. It is not a constituent of proteins.

ornithine cycle *See* UREA CYCLE.

ortho- In organic chemistry, a prefix used in disubstituted derivatives of BENZENE which indicates that the substituted groups are at positions 1 and 2 in the ring, such as 2-nitrophenol (*ortho*-nitrophenol, *o*-nitrophenol).

OH, NO_2

The use of the term *ortho* in other situations, for example orthocarbonates, does not have the same meaning and is always written in full. *See also* META-, PARA-.

orthorhombic (*adj.*) Describing a crystal structure in which the UNIT CELL has all its faces at right angles to one another, with all the faces rectangular, but none square, so there are three different lengths characterizing the size of the unit cell.

ortho-sulpho benzimide *See* SACCHARIN.

osmiridium A naturally occurring alloy containing mostly osmium (typically 20 to 50 per cent) and iridium (typically 50 per cent), together with varying quantities of other elements.

osmium (Os) The element with atomic number 76; relative atomic mass 190.2; melting point 3,045°C; boiling point 5,027°C; relative density 22.6. Osmium is a transition metal and is the densest of all the elements, with a density more than three times that of iron. Osmium is a hard bluish metal used in some steels and has some catalytic properties.

osmium oxide Either of the BINARY COMPOUNDS of osmium and oxygen, osmium(IV) oxide, OsO_2, and osmium(VIII) oxide (osmium tetroxide), OsO_4. Osmium(II) oxide is a brown solid; decomposes on heating; relative density 7.7. It can be obtained as the hydrate $OsO_2.2H_2O$ as a precipitate from a solution of potassium chloro-osmate(IV) to which an alkali is added,

$$K_2OsCl_6 + 4NaOH \rightarrow OsO_2 + 4NaCl + 2KCl + 2H_2O$$

On heating, it disproportionates to metallic osmium and osmium(VIII) oxide,

$$2OsO_2 \rightarrow Os + OsO_4$$

Osmium(VIII) oxide is a yellow solid; melting point 41°C; boiling point 131°C; relative density 4.9. It can be formed by burning powdered osmium in air,

$$Os + 2O_2 \rightarrow OsO_4$$

The vapour has a pungent smell and is highly toxic. On heating it decomposes to give the metal and oxygen. Osmium(VIII) oxide dissolves in water to form a very weak acid. It is used as a catalyst and as an oxidizing agent in some organic reactions.

osmosis The movement of a liquid SOLVENT, usually water, from a less concentrated SOLUTION to a more concentrated solution through a SEMIPERMEABLE MEMBRANE (one that is permeable in both directions to water but varying in permeability to the SOLUTE) until the two concentrations are equal or isotonic.

Osmosis is a passive process requiring no energy. If external pressure is applied to the more concentrated solution, osmosis is prevented and this provides a measure of the OSMOTIC PRESSURE of the more concentrated solution, which is measured in pascals (Pa). The osmotic pressure is greater the more concentrated the solution.

Osmosis is vital in controlling the distribution of water in living organisms and in maintaining a constant water/salt balance.

osmotic potential *See* OSMOTIC PRESSURE.

osmotic pressure, *osmotic potential* The pressure difference that can occur across a SEMIPERMEABLE MEMBRANE as a result of OSMOSIS. It is defined as the pressure that needs to be applied across a semipermeable membrane to prevent osmosis. KINETIC THEORY shows that the osmotic pressure is proportional to the CONCENTRATION of the SOLUTE and to the ABSOLUTE TEMPERATURE of the liquid.

Ostwald's dilution law For a solution in which a substance AB only weakly DISSOCIATES into ions A^+ and B^-, the degree of IONIZATION, provided it is small, is inversely proportional to the square root of the concentration. This result follows from the application of CHEMICAL EQUILIBRIUM laws to the system of ions and molecules. If the degree of dissociation is α and the number of molecules present before dissociation is n in a volume V, then

$$\alpha = (KV/n)^{1/2}$$

where K is a constant.

Ostwald process An industrial process for the manufacture of NITRIC ACID from ammonia. The ammonia is first oxidized by heating with a platinum/rhodium catalyst,

$$4NH_3 + 7O_2 \rightarrow 4NO_2 + 5H_2O$$

The nitrogen(IV) oxide is then dissolved in water, and disproportionates to give nitric acid and nitrogen(II) oxide, which can be added to the ammonia entering the cycle and oxidized back to nitrogen (IV) oxide,

$$3NO_2 + H_2O \rightarrow 2HNO_3 + NO$$

overpotential *See* OVERVOLTAGE.

overvoltage, *overpotential* The amount by which the voltage across an ELECTROLYTIC CELL through which a current is being driven (in ELECTROLYSIS for example) must exceed the theoretical value derived from the ELECTRODE POTENTIALS for the cell. The size of the overvoltage depends on the nature of the electrodes, being particularly small for platinum electrodes. It appears to arise as a result of the ACTIVATION ENERGY required for the transfer of electrons through the electrode surface and is greater in any reaction in which a gas is liberated. The presence of differing overvoltages for competing processes is particularly important in electrolysis and ELECTROPLATING. This means, for example, that zinc can be deposited onto a zinc electrode in preference to the release of hydrogen, making possible the electrolytic refining of zinc.

oxalic acid *See* ETHANEDIOIC ACID.

oxaloacetic acid ($C_4H_4O_5$) A colourless, crystalline acid that combines with ACETYL COENZYME A at the beginning and end of the KREBS CYCLE. It is also produced in certain plants during PHOTOSYNTHESIS. *See also* CALVIN CYCLE.

oxidase *See* OXIDOREDUCTASE.

oxidation In general, any chemical reaction that increases the OXIDATION STATE of an element, so the element concerned is now in a state where it has lost extra electrons (or where extra electrons could be imagined to have been lost, if COVALENT BONDS are treated as IONIC). In particular, oxidation is the addition of oxygen or the removal of hydrogen from an element, or the removal of electrons to form a positive ION (or an ion more positive than was originally present).

The simplest examples of oxidation are those in which oxygen is added, for example magnesium is oxidized when it burns in air:

$$2Mg + O_2 \rightarrow 2MgO$$

This example also involves the removal of electrons from the magnesium as magnesium oxide is ionic:

$$Mg \rightarrow Mg^{2+} + 2e^-$$

and an increase of the oxidation state of the magnesium, from 0 to +2. *See also* OXIDIZING AGENT, REDOX REACTION, REDUCTION.

oxidation number *See* OXIDATION STATE.

oxidation state, *oxidation number* The number of electrons that are lost, or that can be imagined to be lost, by an element in forming a particular compound, ion or COMPLEX ION. If the compound is IONIC, the oxidation state is simply the charge on the ions, or adds up to this charge if the ion is a FREE RADICAL. In COVALENT BONDS, the electrons involved in the bond are imagined to be wholly attached to the more electronegative of the two elements (*see* ELECTRONEGATIVITY). If the two elements have equal electronegativity, the electrons are imagined to be shared equally. These rules mean that for any neutral compound, the oxidation states must add up to zero.

Fluorine always has an oxidation state of –1 and the oxidation state of oxygen is –2, except in PEROXIDES, where it is –1. The oxidation state of hydrogen is +1 except in metallic HYDRIDES, where it is –1. ALKALI METALS all have an oxidation state of +1 and ALKALINE EARTHS of +2. HALOGENS usually have oxidation states of –1, though +7 is sometimes found, such as in the ClO_3^- ion. Sulphur exhibits a wide range of oxidation states, though apart from –2 in H_2S, only +4 and +6 are common.

Many of the TRANSITION METALS can form ions or complexes with different oxidation states. For example, iron can form Fe^{2+} and Fe^{3+} ions with oxidation states of +2 and +3. In the case of the transition metals, the oxidation state is often shown by inserting the oxidation number, shown as a Roman numeral, in the chemical name.

oxidative phosphorylation The process by which ATP is formed through aerobic respiration by the transfer of electrons in the ELECTRON TRANSPORT SYSTEM, which results in the oxidation of hydrogen atoms.

oxide Any BINARY COMPOUND containing oxygen. Both metals and non metals form oxides, many by direct combustion in oxygen; for example, magnesium burns in oxygen to form magnesium oxide:

$$2Mg + O_2 \rightarrow 2MgO$$

and sulphur burns to form sulphur dioxide:

$$S + O_2 \rightarrow SO_2$$

The oxides of metals are BASIC, or AMPHOTERIC in the case of some TRANSITION METALS, and have high melting points. The oxides of non-metals have lower melting points (many are gases at room temperature) and are ACIDIC.

oxidize (*vb.*) To bring about an OXIDATION process.

oxidizing agent A material that readily brings about an OXIDATION, itself being reduced (*see* REDUCTION). Oxygen is an obvious example. Hydrogen peroxide, which is itself reduced to water, and potassium manganate(VII), reduced to manganese(II) oxide, are other examples.

oxidoreductase Any one of a group of ENZYMES that transfer O and H atoms between substances in oxidation/reduction reactions. They include dehydrogenases and oxidases.

oxoacid An acid containing a non-metallic atom bound to one or more oxygen atoms. Sulphuric acid, H_2SO_4, and nitric acid, HNO_3, are two examples.

oxonium ion An ion of the type R_3O^+, where R can be hydrogen or an organic group. An example is the hydroxonium ion, H_3O^+.

oxo process An industrial process for making ALDEHYDES.

2-oxopropanoic acid *See* PYRUVIC ACID.

2-oxy-4-amino pyramidine *See* CYTOSINE.

oxygen (O) The element with atomic number 8; relative atomic mass 16.0; melting point –214°C; boiling point –183°C. Oxygen makes up 20 per cent of the atmosphere, mostly as the molecule O_2 though some occurs as OZONE, O_3. It is the most common element in the Earth's crust, occurring in SILICATES, SILICA and CARBONATES. It is very reactive and will combine with all elements apart from the NOBLE GASES and fluorine.

Oxygen is vital to organisms that carry out aerobic RESPIRATION. These organisms absorb oxygen directly from the atmosphere or make use of dissolved oxygen in water. Oxygen also supports COMBUSTION – many materials will burn very rapidly in pure oxygen. Oxygen is extracted from liquefied air by FRACTIONAL DISTILLATION and in its liquid form is a common OXIDIZING AGENT in rocket propulsion.

oxygen furnace A device used in modern steel making. The furnace is a steel vessel with a REFRACTORY lining. Molten PIG IRON is introduced into the furnace and oxygen blown onto the surface through a water-cooled LANCE. This OXIDIZES many impurities, particularly carbon. The less volatile impurities form a SLAG with LIMESTONE, which is added to the furnace. Other elements may be added to alter the composition of the steel. Once the steel is ready, the furnace is tipped onto its side and the steel tapped off into moulds. This process is called the Linz–Donawitz process. *See also* ELECTRIC ARC FURNACE.

oxyhaemoglobin HAEMOGLOBIN combined with oxygen. Oxyhaemoglobin occurs in organisms in regions of high oxygen concentration, such as the lungs or gills. Oxygen is transported in the blood in this form to the body tissues, where it is easily released in regions of low oxygen concentration. This dissociation of oxyhaemoglobin is encouraged by the presence of carbon dioxide.

ozone (O_3) A colourless gas with a distinctive odour; melting point –192°C; boiling point –112°C. Ozone can be made by the action of an electric discharge on oxygen:

$$3O_2 \rightarrow 2O_3$$

Ozone is produced in the upper atmosphere and plays an important part in protecting the Earth's surface from ultraviolet radiation. It is a powerful OXIDIZING AGENT. *See also* OZONE LAYER.

ozone layer A protective layer consisting of the gas OZONE, O_3, 15–40 km above the Earth's surface. It is is formed by the effect of ultraviolet (UV) radiation on oxygen molecules. UV light splits oxygen (O_2) molecules into two atoms, one of which then combines with oxygen to create ozone. The ozone layer prevents harmful UV radiation reaching the Earth's surface, but in recent years it has become clear that the layer is being damaged by human activities. *See also* CHLOROFLUOROCARBON.

ozonolysis A method for adding ozone to an unsaturated organic compound, resulting in an ozonide, which is usually unstable. The method was used to determine the structure of ALKENES.

PQ

PAGE *See* POLYACRYLAMIDE GEL ELECTROPHORESIS.

palladium (Pd) The element with atomic number 46; relative atomic mass 106.4; melting point 1,551°C; boiling point 3,140°C; relative density 12.3. Palladium is a TRANSITION METAL with important catalytic properties. It is also notable in that it can absorb as much as 1,000 times its own volume of hydrogen by the process of OCCLUSION.

palmitic acid, *hexadecanoic acid* ($C_{16}H_{32}O_2$) An unsaturated FATTY ACID; melting point 63°C. It occurs in most animal and vegetable fats and in various waxes. It is used, with STEARIC ACID, for making candles and soap.

pantothenic acid, *vitamin B$_5$* ($C_9H_{17}NO_5$) A VITAMIN of the VITAMIN B COMPLEX. Pantothenic acid forms part of ACETYL COENZYME A, which is important in the KREBS CYCLE. It is found in many foodstuffs, including cereals, liver, egg yolk and yeast.

paper chromatography CHROMATOGRAPHY in which the stationary phase is absorbent paper. A spot of the mixture to be separated is placed near the base of a paper strip, the bottom edge of which is placed in a solvent. The solvent moves up the paper by capillary action, and the different components in the mixture are carried along at different rates. Colourless components may be made visible by 'developing' the CHROMATOGRAM – spraying it with some material that reacts to make the components visible. Colourless components that fluoresce (*see* FLUORESCENCE) may be viewed in ultraviolet light.

para- In organic chemistry, a prefix used in disubstituted derivatives of BENZENE which indicates that the substituted groups are at positions 1 and 4 in the ring. An example is 1,4-nitrophenol (*para*-nitrophenol, *p*-nitrophenol):

OH
|
(benzene ring)
|
NO_2

The use of the term *para* in other situations, for example, paraformaldehyde, has no structural significance and is always written in full. *See also* META-, ORTHO-.

paracetamol, *4-acetamidophenol* ($C_8H_9NO_2$) A popular pain-killing drug that also reduces fever. It is less damaging to the stomach than ASPIRIN, but liver damage can be caused by excess doses.

paraffin *See* ALKANE.

paramagnetism A weak form of magnetism found in some elements and molecules (such as O_2). A paramagnetic material will align parallel to any applied magnetic field.

paraquat, *1,1'-dimethyl-4,4'-bipyridylium dimethylsulphate* or ***dichloride*** A white solid; melting point 175–180°C. It is a contact HERBICIDE that kills all vegetation. Paraquat is poisonous if ingested, causing irreversible lung damage.

partial pressure In a mixture of gases, that part of overall pressure that can be attributed to the presence of one specified gas in the mixture; that is, the pressure that it would exert if it were alone. *See also* DALTON'S LAW OF PARTIAL PRESSURES.

partition constant A quantity used to determine the feasibility of SOLVENT EXTRACTION. For any given material, the partition constant is determined by dissolving some of the material in a mixture of SOLVENTS. The partition constant is the ratio of the concentration in solvent B to the concentration in solvent A. It is largely independent of the amount of material used and the amount of solvent present, provided neither concentration approaches saturation. If the partition constant is large, solvent extraction can be successfully used with solvent B extracting the material from solvent A. *See also* HENRY'S LAW.

pascal (Pa) The SI UNIT of PRESSURE. One pascal is equivalent to a pressure of one NEWTON per metre squared.

passive (*adj.*) Describing a substance that takes no part in a reaction. A passive metal is one, such as aluminium, that has been coated, for example by an oxide layer, so will not CORRODE as rapidly as might be expected on the basis of its REACTIVITY.

Pauli exclusion principle A consequence of QUANTUM THEORY, which states that no two electrons may occupy the same quantum state. This has applications in the arrangements of electrons in atoms, where each ORBITAL contains only two electrons, each in one of the two possible SPIN states, described as spin up and spin down, depending on the direction of the spin angular momentum relative to some reference direction.

Pauling electronegativity A commonly used scale for comparing the ELECTRONEGATIVITY of elements. Pauling measured electronegativities from differences in the energies of COVALENT BONDS and deduced a scale in which the most electronegative element, fluorine, is allocated an electronegativity of 4.0. The least electronegative elements, the ALKALI METALS, have electronegativities of about 0.9 on this scale.

p-block element Any chemical element in GROUPS 13, 14, 15, 16, 17 and 18 (formerly groups III to VIII) of the PERIODIC TABLE; that is, an element in which the outer electrons are in P-ORBITALS.

p.d. *See* POTENTIAL DIFFERENCE.

penicillin A group of ANTIBIOTICS derived from various *Penicillium* moulds, especially *P. notatum* and *P. crysogenum*, or synthetically made. All contain the basic ring structure of penicillanic acid. The first penicillin was benzylpenicillin (penicillin G). Penicillins are used in the treatment of numerous bacterial infections, where they act by disrupting synthesis of bacterial cell walls.

pentahydrate A HYDRATE containing five parts water to one part of the compound. For example, sodium thiosulphate forms the pentahydrate $Na_2S_2O_3.5H_2O$.

pentane (C_5H_{12}) The fifth member of the ALKANE series. There are three possible ISOMERS of pentane. *n*-Pentane has a boiling point of 38°C. The pentanes are all inflammable liquids extracted from petroleum.

pentanoic acid, *valeric acid* ($C_5H_{10}O_2$) A colourless liquid CARBOXYLIC ACID with an unpleasant odour; melting point –34°C; boiling point 186°C.

pentanol, *amyl alcohol* ($C_5H_{11}OH$) The fifth member in the series of ALCOHOLS. There are eight isomers of this formula. 1-Pentanol has a boiling point of 137°C. They are all oily liquids that do not mix with water.

pentose A MONOSACCHARIDE containing five carbon atoms in the molecule. Examples are RIBOSE and DEOXYRIBOSE, the sugar components in NUCLEIC ACIDS.

pentose phosphate pathway, *pentose shunt* A series of biochemical reactions in which NADPH, (*see* NADP) generated as glucose-6-phosphate, is oxidized to ribose-5-phosphate. NADPH is needed by living cells for reductive biosyntheses, such as the synthesis of FATTY ACIDS from ACETYL COENZYME A. The pentose phosphate pathway is distinct from the biochemical reactions that generate ATP; that is, GLYCOLYSIS, KREBS CYCLE and OXIDATIVE PHOSPHORYLATION. However, there are enzymes that provide a link between glycolysis and the pentose phosphate pathway when needed.

The ribulose-5-phosphate generated by the pathway (and its derivatives) are important constituents in molecules such as COENZYME A, ATP, NAD, FAD, DNA and RNA. The pathway also catalyses the interconversion of three-, four-, five-, six- and seven-carbon sugars, thereby providing living cells with a source of sugars. In animals, the pathway occurs at various sites but is more active in liver and adipose tissue than it is in skeletal tissue. This reflects sites where there is a greater need for NADPH for reductive syntheses. In plants, the pathway participates in the formation of hexoses from carbon dioxide in PHOTOSYNTHESIS.

pentose shunt *See* PENTOSE PHOSPHATE PATHWAY.

pepsin A PROTEASE enzyme found in the stomach that breaks down proteins during digestion. Pepsin is secreted as its precursor, pepsinogen, which is activated by the stomach acidity to form pepsin.

pepsinogen *See* PEPSIN.

peptidase *See* PROTEOLYTIC ENZYME.

peptide Two or more AMINO ACIDS joined together by a peptide bond (–CO-NH–), which forms between the amino group (NH_2) of one amino acid and the carboxyl group (COOH) of another amino acid with the loss of a water molecule. A long chain peptide (more than three amino acids) is called a polypeptide, which can then fold or twist to form a PROTEIN. Some peptides are HORMONES.

peptidoglycan A MACROMOLECULE that gives rigidity to the cell wall of bacteria. Peptido-

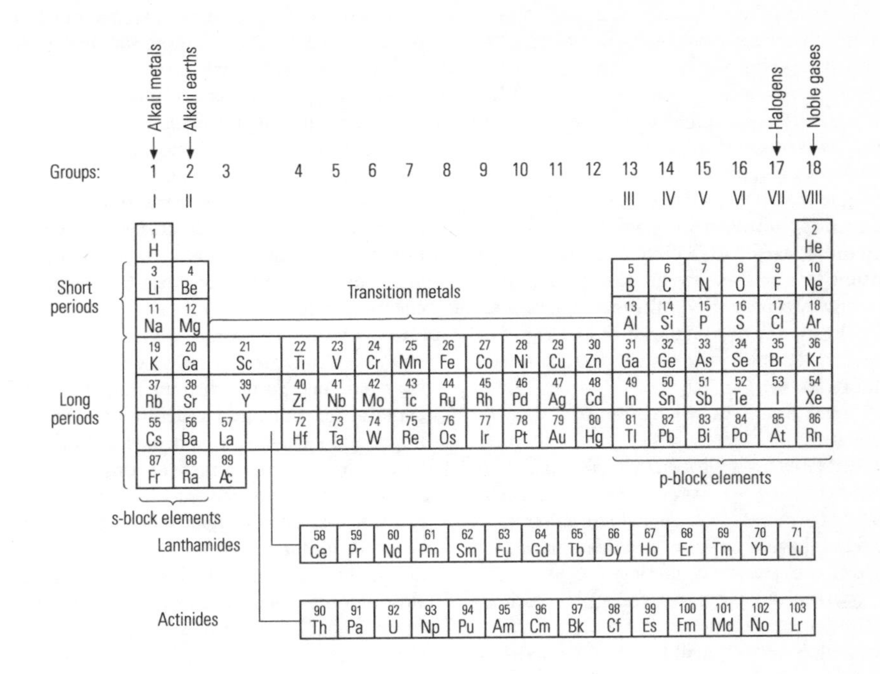

The periodic table of the elements.

glycan consists of chains of N-acetylglucosamine and N-acetylmuramic acid linked by a tripeptide of alanine, glutamic acid and lysine or diaminopimelic acid. It is not found in higher organisms.

perchlorate Any salt containing the chlorate(VII) ion, ClO_4^-.

perchlorous acid *See* CHLORIC ACID.

period A horizontal row of elements in the PERIODIC TABLE, containing a sequence of elements of consecutive ATOMIC NUMBER and steadily changing chemical properties. *See also* LONG PERIOD, SHORT PERIOD.

periodic acid *See* IODIC ACID.

periodic law The principle that certain physical and chemical properties of the elements, such as IONIZATION ENERGY, reactivity, melting point and boiling point, vary in a periodic way with ATOMIC NUMBER. It was this recognition that led to the construction of the PERIODIC TABLE.

periodic table An arrangement of the ELEMENTS in order of increasing ATOMIC NUMBER that brings out the similarities between them. Atomic number increases along the rows of the table (called periods) and the table is arranged so that elements with similar properties are placed in the same column or group. Thus the left-hand column of the table, group 1 (formerly group I), contains the ALKALI METALS, metals that reacts violently with water to form a HYDROXIDE containing a singly charged positive metal ion. The next column, group 2 (formerly group II), contains the ALKALINE EARTHS, metals that are reactive enough to form oxides on exposure to air, but which react only slowly with water. To the right of the table are group 17 (formerly group VII) and group 18 (formerly group VIII). Group 17 contains the HALOGENS, reactive elements that tend to form singly charged negative ions, whilst group 18 contains the NOBLE GASES, with a full outer SHELL of electrons, which are the least reactive of the elements.

In the middle of the table are the TRANSITION METALS, which show only slight changes with increasing atomic number, as electrons fill the D-ORBITALS.

The periodic table also contains two long series of elements, the LANTHANIDES and the ACTINIDES. Within each of these series the chemical properties are generally very similar, with each element differing only in the number of electrons in f-orbitals.

permanent gas A gas that cannot be turned into a liquid by the action of pressure alone. Oxygen and nitrogen are examples. *See* GAS.

permanent hardness HARDNESS in water that cannot be removed by boiling. It is caused principally by dissolved calcium sulphate and, to a lesser extent, magnesium sulphate. *See* WATER SOFTENING. *See also* HARD WATER, TEMPORARY HARDNESS.

permanganate The old name for MANGANATE(VII).

permutite Any naturally occurring mineral, such as ZEOLITE, used as an ION EXCHANGE medium in WATER SOFTENING processes. Modern synthetic materials have largely replaced these materials.

peroxide An OXIDE containing the O_2^{2-} ion. They are all strong OXIDIZING AGENTS, such as sodium peroxide, Na_2O_2.

peroxosulphuric(VI) acid (H_2SO_5) A colourless crystalline solid; decomposes on heating. Peroxosulphuric(VI) acid can be made by the OXIDATION of concentrated sulphuric acid by hydrogen peroxide:

$$H_2SO_4 + H_2O_2 \rightarrow H_2SO_5 + H_2O$$

The reaction is reversed on exposure to excess water or on heating. Peroxosulphuric(VI) acid is a powerful OXIDIZING AGENT.

Perspex (Trade name for ***poly(methylmethacrylate)***, ***poly(methyl 2-methylpropenoate)***) A clear, tough plastic with many uses, for example in advertising signs, protective screens and aeroplane canopies. It is an addition POLYMER containing carbon-carbon bonds similar to POLY(ETHENE) and POLYVINYL CHLORIDE but more rigid due to its structure-hindering movement of carbon chains.

pesticides Poisonous chemicals used to kill pests that may be hazardous to health, or simply unwanted by humans. Pesticides are named after the organisms they kill: insecticides kill insects, fungicides kill fungi, herbicides kill plants considered as weeds and rodenticides kill rodents. Pesticides are used in farming and gardening, but because of their toxic nature they can cause a pollution problem. A good pesticide is one that is specific to the organism to which it is directed, easily broken down to harmless substances so it does not persist and does not accumulate within an organism to be passed along the food chain.

petrochemical industry Those industries that use PETROLEUM or NATURAL GAS as the raw materials for the manufacture of a wide range of products, such as detergents, fertilizers and paints. Most petroleum is used in the manufacture of fuel and it is only the remaining 10 per cent that forms the basis of the petrochemical industry.

petrol, *gasoline* (*US*) A fuel used in motor vehicles which consists of a mixture of HYDROCARBONS obtained from PETROLEUM. Leaded petrol contains tetraethyl lead and dibromoethane to improve the engine performance, but the lead from exhaust fumes enters the atmosphere as simple lead compounds, which are thought to cause mental impairment in young children. For this reason unleaded petrol was introduced and is gradually taking over in the UK, although there is a suggestion that the fumes from this may contain some CARCINOGENS. Unleaded petrol has a lower OCTANE NUMBER than leaded petrol.

petroleum, *crude oil* A natural mineral oil, one of the three FOSSIL FUELS that have formed from the decayed remains of marine micro-organisms millions of years ago. Petroleum is found in layers of porous rock such as limestone and sandstone, often below a pocket of NATURAL GAS, which keeps it under pressure. Thus when a pipe from an oil well is forced into an oil pocket sometimes the oil rises up the borehole under natural pressure, although often pumping is necessary to extract it.

Petroleum consists of HYDROCARBONS mixed with varying amounts of oxygen, sulphur, nitrogen or other elements. It is a green/brown flammable liquid. Crude petroleum is separated by FRACTIONAL DISTILLATION into other products, including fuel oil, petrol, kerosene, diesel, lubricating oil, paraffin wax, petroleum jelly and asphalt. The larger ALKANES obtained from fractional distillation are broken down further by a process called CRACKING into lower alkanes, which are more in demand as motor or aviation fuel.

About 90 per cent of the world's oil supply is used as fuel and the rest is used in the manufacture of many everyday products, such as detergents, plastics, insecticides, fertilizers, drugs, synthetic fibres, paints, toiletries. The burning of petroleum fuel is a major cause of environmental pollution, as are the oil spillages occurring during transportation.

petroleum ether, *light petroleum* A colourless, flammable mixture of lower ALIPHATIC HYDROCARBONS, mainly pentane and hexane. It can be made in a number of grades with boiling ranges between 30–80°C. It is used as a solvent.

petrology The branch of GEOLOGY that deals with rocks, in particular their formation and chemical and physical structure.

pewter An alloy of tin (typically 65 per cent) and lead. It was traditionally used for the manufacture of plates and drinking vessels, but is little used today.

pH A measure of the acidity or alkalinity of a solution. pH 7 indicates a neutral solution, numbers smaller than this indicate ACIDIC solutions, whilst larger numbers indicate ALKALINE solutions. The pH value actually indicates the CONCENTRATION of hydrogen ions in the solution. If $[H^+]$ is the concentration of hydrogen ions in mol dm^{-3}:

$$pH = -\log[H^+]$$

phase Any one of the different arrangements in which the molecules of a certain substance may exist, as a gas, liquid, or as one or more solid ALLOTROPES.

phase diagram A graph of temperature against pressure that shows changes in melting and boiling points with pressure. Three lines on this diagram indicate the combinations of pressure and temperature at which two states can exist together in equilibrium. Where the three lines meet is the TRIPLE POINT. At pressures below that of the triple point, the liquid state does not exist and a solid when heated will sublime (turn from solid to gas). Carbon dioxide at normal (atmospheric) pressures is an example of this: solid carbon dioxide (called dry ice) will turn directly into a gas.

phase rule A rule that determines the number of DEGREES OF FREEDOM for any system containing one or more PHASES in equilibrium. If the number of phases present is P, the number of degrees of freedom is F and the number of chemically distinct components present in the system is C, then

$$F + P = C + 2$$

phenol, *carbolic acid* (C_6H_5OH) The simplest member of a group of AROMATIC chemical compounds (the phenols) characterized by a HYDROXYL GROUP (OH) being attached to an aromatic ring. Phenol itself consists of a hydroxyl group attached to a BENZENE RING:

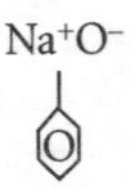

Pure phenol forms colourless crystals with a strong odour; melting point 43°C; boiling point 183°C. Phenol is insoluble in water at room temperature but MISCIBLE with water in all proportions at 84°C. Phenol is weakly acidic but is caustic and toxic by absorption through the skin. There are a number of processes by which phenol can be manufactured, including the CUMENE PROCESS. Other phenols can be made by warming aqueous solutions of DIAZONIUM SALTS.

Since phenols possess a hydroxyl group they have many reactions in common with ALCOHOLS. For example they form ESTERS with ACID CHLORIDES or ACID ANHYDRIDES. The carbon-oxygen bond is not easily broken so it is not possible to substitute a HALOGEN group in a phenol in the same way that HALOGENOALKANES are formed from alcohols. They are much more acidic than alcohols but less so than CARBOXYLIC ACIDS. A solution of phenol in alkali gives the PHENOXIDE ion. Phenol is used in the manufacture of THERMOSETTING PLASTICS, dyes, explosives and pharmaceuticals.

phenol-formaldehyde resin RESIN formed by the reaction of PHENOL with METHANAL (formaldehyde). It is a THERMOSETTING PLASTIC, used for mouldings.

phenolphthalein A dye that is colourless in its solid form, but in aqueous solution changes structure depending on the pH value of the solution. Below pH 8 it is colourless, but above pH 10 it is red. It used as an INDICATOR in the TITRATION of weak acids with strong alkalis. It is also used in medicines as a laxative.

phenoxide The ion formed when PHENOL is placed in an alkali solution.

phenoxyresins THERMOPLASTICS made by the condensation of PHENOLS, used to make containers.

phenylalanine An ESSENTIAL AMINO ACID, normally converted to the amino acid TYROSINE in the body. It is the main component of ASPARTAME (Nutrasweet).

phenylamine, *aniline* ($C_6H_5NH_2$) A colourless, oily liquid with a distinctive odour; melting point –6.2°C; boiling point 184°C. It is a simple AROMATIC compound consisting of a BENZENE RING with an NH_2 group attached to the ring. Phenylamine is highly poisonous. On contact with air it turns brown. It is made by the reduction of NITROBENZENE. Phenylamine has an important use in the rubber industry and in the manufacture of dyes and drugs.

***N*-phenylethanamide** *See* ACETANILIDE.

phenylethene, *styrene* (C_8H_8) An organic liquid made from BENZENE and ETHENE; boiling point 146°C. It is used to manufacture POLYMERS such as POLYSTYRENE and the synthetic rubber SBR.

phenyl group (C_6H_5) A group derived from BENZENE with the removal of a hydrogen atom.

3-phenylpropenoic acid *See* CINNAMIC ACID.

pheromones Chemical signals (such as odours) considered to be 'social HORMONES' that operate in a hormone-like manner between individuals of a species rather than within an individual. They are often used to attract mates, for example female silk moths can attract male mates many kilometres away by releasing the pheromone bombykol. Ants produce pheromones that warn other ants of danger and bees produce several pheromones, each with its own effects.

phosgene *See* CARBONYL CHLORIDE.

phosphagen One of a group of organic compounds that occur in animal tissues and provide a source of chemical energy in the form of high-energy phosphate bonds. Creatine phosphate is the most common and occurs in animal muscle and nerves. During activities, such as muscular contraction, the phosphagens break down to release their phosphate groups. This allows the generation of ATP from ADP. When ATP is available again, the phosphagens reform.

phosphate Any salt containing the phosphate ion, PO_4^{3-}. Phosphates can be made by the reaction of metals with phosphoric acid, for example:

$$3Mg + 2H_3PO_4 \rightarrow Mg_3(PO_4)_2 + 3H_2$$

Phosphates, particularly ammonium phosphate, are widely used as fertilizers. Phosphates are extensively used as BUFFER SOLUTIONS.

phosphide Any BINARY COMPOUND of PHOSPHORUS with a more electropositive element (*see* ELECTROPOSITIVITY), for example sodium phosphide, Na_3P.

phosphine (PH_3) A colourless gas; melting point –133°C; boiling point –88°C. Phosphine is similar to, but less reactive than, ammonia. It can form stable complexes (*see* COMPLEX ION) with TRANSITION METAL ions, which makes it highly toxic. Phosphine can be manufactured by the reaction between phosphorus and strong alkalis, for example:

$$8P + 15NaOH + 3H_2O \rightarrow 7PH_3 + 5Na_3PO_4$$

Phosphine is soluble in water, and will react with some acids to produce phosphinium salts, for example:

$$PH_3 + HCl \rightarrow PH_4Cl$$

phosphinic acid (H_3PO_3) A colourless solid; melting point 74°C; decomposes on further heating; relative density 1.7. Phosphinic acid may be made by hydrolysing (*see* HYDROLYSIS) phosphorus(III) chloride with water:

$$PCl_3 + 3H_2O \rightarrow H_3PO_3 + 3HCl$$

It is only dibasic, since one of the hydrogen atoms is covalently bonded to the phosphorus atom. Phosphinic acid decomposes on heating to give phosphine and phosphoric(V) acid:

$$4H_3PO_3 \rightarrow PH_3 + 3H_3PO_4$$

phosphodiester bond A COVALENT BOND linking a PHOSPHATE group to a sugar group, by means of an oxygen bridge, as occurs in the sugar-phosphate backbone of a NUCLEIC ACID.

phosphoglyceride A PHOSPHOLIPID based on GLYCEROL in which two of the three HYDROXYL GROUPS on glycerol contain FATTY ACIDS and the third has combined with a PHOSPHATE group.

phospholipid A LIPID comprising a PHOSPHATE group, with one or more FATTY ACIDS. Most phospholipids are based on the alcohol GLYCEROL and are termed PHOSPHOGLYCERIDES. SPHINGOLIPIDS are based on the alcohol SPHINGOSINE and contain only one fatty acid linked to an an AMINO GROUP. Phospholipids are found throughout living systems since they occur in cell membranes. The phosphate end is HYDROPHILIC (water-loving) and the other end is HYDROPHOBIC (water-repelling). Sphingomyelin is the only phospholipid in cell membranes that is not derived from glycerol. Another example of a phospholipid is LECITHIN.

phosphonium Any salt containing the ion PH_4^+ or the functional group $-PH_3^+$. The structure of phosphonium salts is similar to ammonium salts, but they are usually only stable in acidic solutions, for example phosphonium chloride, PH_4Cl, will dissociate into phosphine and hydrogen chloride.

phosphor Any material that, when struck by electrons, converts some of the KINETIC ENERGY of the electrons into visible light.

phosphor bronze An alloy of copper with up to 10 per cent tin and up to 1 per cent phosphorus. The alloy is hard-wearing and corrosion resistant, and is frequently used for metal parts that must be strong and will be exposed to water, for example in ships' propellers.

phosphorescence The production of visible light as a result of a chemical reaction, electron bombardment or other process in which electrons move from one ENERGY LEVEL to a lower energy level. Also, the emission of visible light some time after light of a shorter wavelength (e.g. ultraviolet) has been absorbed. *Compare* FLUORESCENCE.

phosphoric(V) acid (H_3PO_4) A white solid; melting point 42°C; decomposes on further heating; relative density 1.8. Phosphoric acid is formed by burning phosphorus in a mixture of air and steam:

$$4P + 6H_2O + 5O_2 \rightarrow 4H_3PO_4$$

Phosphoric acid is highly soluble in water. It is a weak tribasic acid (*see* BASICITY), and readily forms both ACIDIC SALTS and NORMAL SALTS.

phosphorus (P) The element with atomic number 15; relative atomic mass 31.0; melting point 44°C (white allotrope); boiling point 280°C (white allotrope), relative density 1.8 (white), 2.2 (red). It occurs in several ALLOTROPES, the most common of which are white phosphorus, which is toxic and highly reactive, and red phosphorus, which is less reactive and non toxic. White phosphorus ignites spontaneously on exposure to air.

Phosphorus is essential for life. It is required for the formation of NUCLEIC ACIDS and certain energy-carrying molecules (*see* ATP). Phosphorus is also an important constituent of bones and teeth. PHOSPHATES are used as fertilizers. *See also* PHOSPHORUS CYCLE.

phosphorus chloride Phosphorus(III) chloride (phosphorus trichloride), PCl_3, and phosphorus(V) chloride (phosphorus pentachloride), PCl_5.

Phosphorus(III) chloride is a colourless liquid; melting point −112°C; boiling point 76°C; relative density 1.6. It is produced by burning excess phosphorous in chlorine:

$$2P + 3Cl_2 \rightarrow 2PCl_3$$

Phosphorus(IV) chloride is a pale yellow solid; sublimes at 160°C; relative density 3.6. It is produced by heating phosphorus(III) chloride in chlorine:

$$PCl_3 + Cl_2 \rightarrow PCl_5$$

Both chlorides are hydrolysed (*see* HYDROLYSIS) by water:

$$PCl_3 + 3H_2O \rightarrow H_3PO_3 + 3HCl$$

$$PCl_5 + 4H_2O \rightarrow H_3PO_4 + 5HCl$$

Both of the above reactions are violent enough for the chloride to fume in moist air.

Phosphorus(III) chloride is an important material for the manufacture of organic phosphorus compounds. Phosphorus(IV) chloride is readily reduced to phosphorus(III) chloride and is used to introduce chlorine into organic molecules.

phosphorus cycle The recycling of phosphorous throughout the environment. A wide range of important biological chemicals contain phosphorus, including NUCLEOTIDES, ATP and proteins. The main source of phosphorous is from rocks, which release it into the ecosystem through erosion. Plants absorb dissolved PHOSPHATES, which are passed to other animals in the food chain. Phosphates are recycled through plant and animal waste and decomposing remains, including bones and shells, which can themselves be eroded to provide more dissolved phosphates or be deposited in rocks.

phosphorus oxide Phosphorus(III) oxide, P_4O_6, and phosphorus(V) oxide (phosphorus pentoxide), P_4O_{10}.

Phosphorus(III) oxide is a white waxy solid; melting point 24°C; boiling point 174°C. It can be formed by burning phosphorus in a limited supply of oxygen:

$$2P + 3O_2 \rightarrow P_2O_6$$

On heating it loses phosphorus and forms a polymeric form of phosphorus oxide. Phosphorus(III) oxide dissolves in water to give phosphonic acid:

$$P_4O_6 + 6H_2O \rightarrow 4H_3PO_3$$

Phosphorus(V) oxide is a white DELIQUESCENT powder; sublimes at 360°C; relative density 2.4. It can be formed by burning phosphorus in an excess of oxygen:

$$4P + 5O_2 \rightarrow P_4O_{10}$$

Phosphorus(V) oxide dissolves in water to produce phosphoric(V) acid:

$$P_4O_{10} + 6H_2O \rightarrow 4H_3PO_4$$

phosphorylase An ENZYME that catalyses the transfer of PHOSPHATE groups in PHOSPHORYLATION, for example in the conversion of ADP to ATP.

phosphorylation The transfer of a PHOSPHATE group by the enzyme phosphorylase, often from inorganic phosphate ions, to an organic compound. Phosphorylation can be photophosphorylation (using light energy; *see* PHOTOSYNTHESIS) or OXIDATIVE PHOSPHORYLATION (occurring during respiration in aerobic cells).

photocell *See* PHOTOELECTRIC CELL.

photochemistry The study of reactions that are affected by light or ultraviolet radiation. PHOTONS of light can excite certain molecules, or break them into FREE RADICALS, allowing a reaction to proceed which would otherwise have too high an ACTIVATION ENERGY. For example the reaction between hydrogen and chlorine,

$$H_2 + Cl_2 \rightarrow 2HCl$$

cannot proceed at normal temperatures unless ultraviolet light is used to DISSOCIATE the chlorine molecules into free atoms.

photoconductive (*adj.*) Describing a material, such as cadmium sulphide, that conducts electricity when light falls upon it. The conductivity of the material increases with the intensity of the light.

photoelectric cell, *photocell* One of several devices for detecting light and other forms of electromagnetic radiation. A photoemissive cell consists of a negative ELECTRODE (the photocathode) and a positive-collecting electrode (ANODE) in a vacuum. PHOTONS striking the photocathode liberate electrons, in a process known as photoelectricity (*see* PHOTOELECTRIC EFFECT). The electrons are attracted to the anode, and the resulting electric current is a measure of the light intensity.

In a photovoltaic cell, a potential difference is set up between two layers as a result of irradiation by light. In a photoconductive cell, the conductivity of a SEMICONDUCTOR increases on exposure to light.

photoelectric effect The emission of electrons from the surface of a metal on exposure to electromagnetic radiation. Electrons are emitted only if the wavelength is below some minimum which depends on the metal used. If light of too long a wavelength is used, the effect is not observed no matter how bright the light.

Einstein's explanation of the photoelectric effect (1901) suggested that light comes in quanta (*see* QUANTUM), or individual particles, called PHOTONS. Each electron gains the energy needed to escape by absorbing the energy of one photon. The energy of each photon is related to the frequency, f, of the light by

$$E = hf$$

where E is the photon energy and h is PLANCK'S CONSTANT. The minimum energy needed for an electron to escape from the surface of a metal is called the work function, and the photoelectric effect cannot take place for a given metal if the photon frequency is so low (i.e. the wavelength so long) that the photon energy is smaller than the work function.

The theory is further justified by experiments in which the energy of the ejected electrons is measured by collecting them on a plate held at a potential negative to the metal plate from which they are produced. The maximum POTENTIAL DIFFERENCE that can exist with photoelectrons still arriving is called the stopping potential and is proportional to the maximum energy with which photoelectrons are released. The photoelectric effect is the basis of many light-detecting devices.

See also PHOTOELECTRIC CELL.

photoelectric spectroscopy A technique for studying the ENERGY LEVELS of atoms and molecules using the PHOTOELECTRIC EFFECT. The sample being studied is illuminated with ULTRAVIOLET light and the energies of the ejected electrons are measured. The difference between the electron energy and the energy of the ultraviolet PHOTONS is a measure of the energy level from which the electron was ejected.

photographic film A light-sensitive surface that relies on a chemical reaction caused by light striking grains of a silver halide, which is made visible by chemical reactions when the film is developed.

photoionization The IONIZATION of a substance, usually a gas, by ELECTROMAGNETIC RADIATION. The PHOTONS involved must have an energy greater than or equal to the first IONIZATION ENERGY of the material involved, hence this effect is usually restricted to ultraviolet light and X-rays.

photolysis A chemical reaction or decomposition occurring on exposure to light or ultraviolet radiation. In PHOTOSYNTHESIS the photolysis of water constitutes the light-dependent stage, which generates oxygen, electrons and hydrogen ions with the simultaneous conversion of ADP to ATP, thus converting light energy to chemical energy.

photon The QUANTUM OF ELECTROMAGNETIC RADIATION, having an energy related to the frequency, f, of the light by

$$E = hf$$

where E is the photon energy and h is PLANCK'S CONSTANT. *See also* PHOTOELECTRIC EFFECT, WAVE-PARTICLE DUALITY.

photosynthesis The use of light energy by green plants to convert carbon dioxide, from the air, and water to CARBOHYDRATES and oxygen. Photosynthesis is the basis for most forms of life since it is the means by which basic food (sugar) is created, and all animals depend directly or indirectly on plants for food. Furthermore it releases oxygen for use by aerobic organisms. Photosynthesis occurs in structures called chloroplasts, mostly in the leaves, by means of various light-trapping pigments, most commonly CHLOROPHYLL.

There are two stages to photosynthesis – the light-dependent stage and the light-independent stage (also termed the dark stage or CALVIN CYCLE). The light-dependent stage involves the splitting of water by light (PHOTOLYSIS) to yield oxygen, hydrogen ions (protons) and electrons with the simultaneous conversion of ADP to ATP by a process known as photophosphorylation, thereby converting light energy to chemical energy. This can be non-cyclic photophosphorylation, in which electrons are lost from the chlorophyll to be passed into the light-independent stage and are replaced by electrons from a water molecule (with the production of ATP). Thus the same electrons are not recycled through the chlorophyll. Cyclic photophosphorylation also occurs, in which case electrons from the chlorophyll are returned to it via an electron carrier system. NADPH is also formed during the light-dependent stage by the reduction of NADP to NADPH and H^+. Electrons and protons from the light-dependent stage (NADPH + H^+) are used in the light-independent stage (along with the ATP generated) to convert carbon dioxide from the air into carbohydrates.

A number of factors affect photosynthesis – light intensity, carbon dioxide concentration, temperature (the light-independent stage is temperature-dependent), and chlorophyll and oxygen concentrations.

phthalic acid, *benzene-1,2-dicarboxylic acid* ($C_6H_4(COOH)_2$) A colourless, crystalline CARBOXYLIC ACID; melting point 207°C. It is usually used as phthalic anhydride in making PLASTICIZERS and RESINS. Phthalic anhydride is made from the oxidation of NAPHTHALENE.

physical chemistry The branch of chemistry that deals with the effect of chemical reactions on the physical properties of the materials involved, such as their melting points or electrical conductivity.

physisorption The ADSORPTION of a gas by a solid in which the molecules of the adsorbed gas are held on the surface of the adsorbing solid by VAN DER WAALS' FORCES.

phytomenandione *See* VITAMIN K.

pi-bond, *π-bond* A COVALENT BOND that has an electron distribution that is not symmetrical about the line joining the two atoms. Covalent DOUBLE BONDS and TRIPLE BONDS often consist of a SIGMA-BOND surrounded by one or two pi-bonds.

pico- (p) A prefix placed in front of a unit to denote that the size of that unit is to be multiplied by 10^{-12}. For instance a picofarad (pF) is one million millionth of a FARAD.

picric acid, *2,4,6-trinitrophenol* ($C_6H_2(NO_2)_3$ A bright yellow crystalline NITRO COMPOUND; melting point 122°C. It is highly explosive and has been used as such, although it has several disadvantages and has therefore been replaced as a high explosive. It is also used in the dyeing industry.

pi electrons, π *electrons* ELECTRONS associated with a PI-BOND.

piezoelectric The effect by which a POTENTIAL DIFFERENCE (p.d.) appears between the faces of certain materials when they are subject to stress. The reverse effect also occurs, with the material deforming in response to an applied p.d. Quartz is an example of a piezoelectric material.

pig iron The metal removed from the bottom of a BLAST FURNACE. It has too high a carbon content (3–5 per cent) to be used directly. Pig iron is refined and converted into CAST IRON and STEEL.

pigment Any substance that is used to give colour to surfaces, inks, plastics, etc. Pigments are usually insoluble, particulate and inorganic in nature. Examples include iron oxide to provide red/brown colour or cyanoferrates to give blue colours. Pigments are distinct from DYESTUFFS, which operate on a molecular level. Some organic dyestuffs can be used as pigments. Some natural pigments exist, such as HAEMOGLOBIN and CHLOROPHYLL, which give a characteristic colour to animal or plant tissue.

pipette A graduated glass tube for delivering a measured quantity of a liquid REAGENT. It consists of a glass stem with a bulb about half-way up the stem. Two lines are marked on either side of the bulb, the volume of liquid specified will have been delivered by the pipette when the liquid level has moved from one line to the other.

pitch A black, sticky substance derived from the DESTRUCTIVE DISTILLATION of COAL TAR or wood. It is a liquid when hot but solid when cold. Its uses include roofing, paving and waterproofing, for example, wooden ships.

pitchblende The principle ore of uranium, consisting mainly of uranium(IV) oxide, UO_2. Pitchblende is also a source of radium, thorium and polonium.

pK A measure of the strength of an ACID or BASE. For an acid, it is the extent to which the acid DISSOCIATES in aqueous solution, to release hydrogen ions. Likewise for a base, it is the extent to which the base dissociates to release HYDROXIDE ions.

$$\mathrm{pK} = -\log_{10} K$$

where K is the DISSOCIATION CONSTANT for the acid.

Planck's constant (h) A constant that determines the scale of the effects of WAVE-PARTICLE DUALITY. Since h is small, the quantum effects only become apparent at short distances or over short times. $h = 6.6 \times 10^{-34}$ J s. *See also* PHOTOELECTRIC EFFECT.

plane-polarized (*adj.*) Describing an ELECTROMAGNETIC WAVE in which the electric and magnetic fields each oscillate in a single plane, perpendicular to one another and to the direction of wave propagation. *See* POLARIZATION.

plaster of Paris *See* CALCIUM SULPHATE.

plastic 1. (*n.*) Any synthetic organic POLYMER that is liquid at some stage in its manufacture. Plastics are classified by their behaviour on heating. THERMOPLASTICS soften on heating and harden as they cool, for example POLYSTYRENE, POLYVINYL CHLORIDE. THERMOSETTING PLASTICS remain rigid in their final shape and do not soften on heating, for example RESINS, MELAMINE, POLYESTER, BAKELITE.

During manufacture plastics are moulded and shaped while they are in the heat-softening stage and then cooled for thermoplastics, or heated further for thermosetting plastics to yield the final product. The processing of plastics can produce a wide variety of materials from rigid and inflexible to soft and bendy. They can be extruded to make pipes and rods. Plastics can be strengthened by the addition of CARBON FIBRES for use in aircraft or engineering.

Due to the difficulties in disposal of plastics, a number of biodegradable plastics are now in use, including polyhydroxybutyrate (PHB), which is made from sugar and digested by micro-organisms in the soil.

2. (*adj.*) Describing a material or process in which the material is deformed by a force and does not return to its original shape once the force is removed.

plasticizers Organic liquids of high RELATIVE MOLECULAR MASS that are added to PLASTICS to increase their flexibility. Plasticizers have an important use in POLYVINYL CHLORIDE (PVC). PVC undergoes dramatic changes in the presence of a plasticizer, from the rigid substance used to make window frames (unplasticized uPVC) to the flexible plastic used in shoes.

plastoquinone A QUINONE that occurs in the chloroplasts of plants and is a carrier molecule in the electron system of the light-stage of PHOTOSYNTHESIS.

platinum (Pt) The element with atomic number 78; relative atomic mass 195.1; melting point

1,772°C; boiling point 3,800°C; relative density 21.3. Platinum is a very unreactive metal and is used to make INACTIVE ELECTRODES for ELECTROLYSIS. It also acts as a catalyst in a wide range of reactions.

platinum black Finely divided metallic PLATINUM. It is black in colour and is produced either as a precipitate or by the evaporation and condensation of metallic platinum on a metal surface in a vacuum chamber. Its large surface area maximizes the catalytic action of platinum in reactions between gases.

platinum metals The elements RUTHENIUM, RHODIUM PALLADIUM, OSMIUM, IRIDIUM and PLATINUM. They often occur together in alloys such as OSMIRIDIUM, and have very similar chemical properties, all being hard metals with low reactivity and catalytic action.

pleochroic (*adj.*) Describing a crystal whose colour appears different when it is viewed in different directions. The effect arises in ANISOTROPIC crystals as a result of the selective absorption of different directions of POLARIZATION in the light passing through the crystal.

plumbate Any compound containing the plumbate ion, PbO_3^{2-}. Plumbates are formed by the reaction between lead(IV) oxide and concentrated strong alkalis, for example:

$$PbO_2 + 2NaOH \rightarrow Na_2PbO_3 + H_2O$$

plumbic (*adj.*) An obsolete term describing compounds containing lead in its higher (+4) OXIDATION STATE, for example plumbic oxide, PbO_2.

plumbous (*adj.*) An obsolete term used to describe compounds containing the Pb^{2+} ion, particularly in those cases where there was a risk of confusion with PLUMBIC compounds, for example plumbous oxide, PbO.

plutonium (Pu) The element with atomic number 94; melting point 641°C; boiling point 3,232°C. It is chemically reactive and highly toxic. It has 13 ISOTOPES, all radioactive. The most important isotope (plutonium–239) has a HALF-LIFE of 24,000 years. Plutonium–239 is an important fuel source for nuclear power. It is produced after uranium–238 captures a neutron and then BETA DECAYS twice.

poison 1. A substance that damages the health of a living organism.

2. A material that, when present in very small quantities, can interfere with the action of a CATALYST. Many common poisons in this sense are also toxic to living creatures as they interfere with the functions of enzymes. Arsenic is an example.

polar (*adj.*) Describing a molecule or a substance, particularly a SOLVENT, that contains POLAR BONDS. Many IONIC solids are soluble in polar solvents, including water and ethanol.

polar bond A COVALENT BOND in which the electrons spend a higher proportion of their time closer to one atom in the bond than the other, thus one atom in effect carries a partial negative charge whilst the other has an equal positive charge. These are sometimes indicated on diagrams by δ+ and δ–. The degree of polarity in a bond depends on the difference in ELECTRONEGATIVITY between the two elements forming the bond. Typical examples are hydrogen chloride, HCl, and water H_2O, each of which have polar bonds with the hydrogen atom forming the positive end of the bond.

H – Cl
δ+ δ –

H δ+
/ \
O O
δ– δ–

The extreme version of a polar bond, with the electrons spending all their time attached to one atom, is an IONIC BOND. Polar bonds tend to lead to a greater attraction between molecules and materials with higher boiling points than would be expected from VAN DER WAALS' BONDING.

See also HYDROGEN BONDING.

polarimeter A device for analysing polarized light (*see* POLARIZATION), particularly for measuring the OPTICAL ACTIVITY of a solution. The sample to be analysed is placed between two pieces of polaroid. The first (the polarizer) polarizes the light and the second (the analyser) can be rotated until no light passes through the polarimeter. The degree of rotation required to prevent any light passing through the polarimeter is a measure of the optical activity of the sample.

polarizability A measure of the ease with which an ION can have its electron distribution distorted by a neighbouring ion of the opposite charge in an IONIC SOLID. A high value of polarizability indicates a tendency to form a polar COVALENT BOND rather than an IONIC BOND.

Larger ions are more polarizable than small ones, which means that ANIONS tend to have higher polarizabilities. TRANSITION METAL ions tend to have relatively high polarizabilities due to the more widely spread nature of the D-ORBITALS. Polarizability is measured on an arbitrary scale, with typical values ranging from 0.03 (Li^+) to 10 (S^{2-}).

polarization **1.** The direction of motion of the material through which a wave is travelling. In the case of an ELECTROMAGNETIC WAVE, the direction of polarization is the direction of the electric field. Electromagnetic waves may be plane-polarized, with the electric field in a single direction, or circularly polarized, in which case the electric field direction follows a circular path, which may be either left- or right-handed, as seen by an observer facing in the direction in which the wave is travelling. *See also* OPTICAL ACTIVITY, POLAROID.

2. A measure of the extent to which molecules have been polarized (*see* POLARIZE) by an electric field.

polarize (*vb.*) **1.** Of light, to transmit, reflect or scatter one direction of POLARIZATION in unpolarized light more strongly than another.

2. Of an electric field, to pull a molecule or atom away from symmetry so the centres of positive and negative charge no longer coincide, or to align POLAR MOLECULES.

polar molecule A molecule in which the electrons in a COVALENT BOND are not evenly distributed, so they have a higher probability of being found at one end of the molecule than the other. *See also* HYDROGEN BONDING, POLAR BOND.

polarography A technique of QUANTITATIVE ANALYSIS for the measurement of metal ions in solution. The solution under test is ELECTROLYSED using a cathode in the form of a steady flow of mercury, which produces a series of droplets. A graph is plotted of the current through the electrolyte as a function of the applied voltage. As the ELECTRODE POTENTIAL of each metal in solution is exceeded, there is a rise in current proportional to the concentration of the metal ion concerned. The steady flow of mercury to the cathode provides a self-cleaning electrode preventing complications due to the build up of material on the cathode.

Polaroid Trade name for a transparent plastic material that plane-polarizes light passing through it (*see* POLARIZATION). The plastic contains long-chain molecules along which electrons can travel, absorbing the energy from electromagnetic radiation. During manufacture, the plastic is stretched, aligning the molecules. Light that is polarized perpendicular to the direction of alignment will be transmitted through the Polaroid, whilst light polarized parallel to the molecules is absorbed. Polaroid filters are used in some sunglasses and are sometimes fitted to cameras to reduce the amount of reflected (rather than scattered) light and so reduce glare.

pollution Contamination of the environment by the by-products of human activity, mostly from industrial and agricultural processes. Air pollution is largely due to the burning of FOSSIL FUELS in the home, industry or the combustion engine of vehicles. Such pollutants include smoke (tiny particles of carbon), sulphur dioxide, carbon dioxide and carbon monoxide, nitrogen oxides and lead, particularly from car exhaust emissions. These contribute to the GREENHOUSE EFFECT and ACID RAIN. Other important air pollutants are CHLOROFLUOROCARBONS, which contribute to the thinning of the OZONE LAYER.

Water pollution can be the result of rain polluted by contaminants in the air, but is also due to the release of toxic chemicals, for example copper, zinc, lead, mercury and cyanide, into rivers and seas, killing fish and plant species. Oil pollution of water (usually accidental) is localized, but devastating to seabirds, shellfish and seaweed. Oil spillages can be degraded by some micro-organisms. Sewage is treated but still contains large amounts of PHOSPHATES from washing powders and detergents that remain a problem as a source of water pollution.

Land (terrestrial) pollution comes from the dumping of solid wastes, for example slag heaps from ore digging, metal refining and coal mining, which are unsightly and often cannot sustain any vegetation. Domestic rubbish is usually disposed of by burning; some plastics are now being used that can be degraded by micro-organisms to avoid the dangerous gases given off by burning. Other land pollutants are pesticides and noise.

Radioactive pollution can be from medical waste, televisions, watches, waste from the nuclear power industry, or from testing nuclear weapons. Disposal of nuclear waste has been at sea or by burial on land, but decay can take thousands of years and presents problems of safety, pollution and security.

See also EUTROPHICATION.

polonium (Po) The element with atomic number 84; relative atomic mass 210.0; melting point 254°C; boiling point 962°C. Polonium has no stable ISOTOPES, but occurs in nature from the decay of longer lived heavy radioactive elements. The longest lived isotope (polonium–209) has a HALF-LIFE of 100 years, which makes it far more radioactive than most other naturally occurring elements. It has been used as a heat source for radioactive electrical generators. Polonium was the first element to be discovered as a radioactive decay product, and was extracted from pitchblende, a uranium ore, by Pierre and Marie Curie in 1898.

polyacrylamide gel electrophoresis (PAGE) A type of ELECTROPHORESIS, widely used to determine the size and composition of protein mixtures. In this technique, proteins are placed on a gel matrix of polyacrylamide and an electric field is applied. The proteins migrate towards the positive pole of the electric field at a rate determined by their size as they pass through the pores of the gel. Various stains can be used to visualize the proteins. It is possible to purify some proteins by this technique, although it is more frequently used to analyse mixtures.

polyamide Any polymerized AMIDE, such as NYLON. The condensation polymer is produced by the action of an AMINO GROUP of one molecule and a CARBOXYLIC ACID group of another molecule. The polyamide chains are linked by HYDROGEN BONDS.

poly(chloroethene) *See* POLYVINYL CHLORIDE.

polychloroethylene *See* TETRACHLOROETHENE.

polychloroprene *See* NEOPRENE.

polycrystalline (*adj.*) Describing any solid substance that occurs in pieces with no regular shape, but on examination can be seen to be made of many small CRYSTALS. Within each crystal the molecules are arranged regularly, but the direction of alignment of this structure varies from one small crystal to the next, and the boundaries between one crystal and the next (GRAIN BOUNDARIES) are often irregular.

polyester A synthetic THERMOSETTING PLASTIC, formed by the condensation POLYMERIZATION of POLYHYDRIC alcohols with DIBASIC acids. Polyesters are used in the manufacture of synthetic fibres in textiles such as terylene and Dacron. They can be reinforced by the addition of glass fibres and then used in car bodies and boats.

poly(ethene), ***poly(ethylene)*** (Trade name ***Polythene***) An addition POLYMER of ETHENE (C_2H_4). It is a tough, white THERMOPLASTIC, repeatedly softened on heating. It is widely used, for example, in bottles, toys, packaging, pipes and electric cable.

Polythene may be manufactured by the high pressure POLYMERIZATION of ethene gas with trace amounts of oxygen. An unpaired electron from the oxygen molecule initiates the polymerization by attacking an ethene molecule. This produces a RADICAL that attacks another ethene molecule and so on until 1 to 10 thousand ethene molecules are joined together then the reaction is terminated. This method yields low-density polythene.

In another method, high-density polythene is made using catalysts at low pressure. This method was first used by Karl Ziegler (1898–1973) and the catalyst is suitably termed ZIEGLER–NATTA CATALYST. This yields polythene which is more rigid at low temperatures and softer at higher temperatures.

poly(ethylene) *See* POLY(ETHENE).

polyhydric Of ALCOHOLS, those containing more than one HYDROXYL GROUP. Dihydric alcohols contain two hydroxyl groups and trihydric contain three.

polymer A large molecule made up of two or more similar or identical repeated MONOMERS joined together by POLYMERIZATION to form a chain or branching matrix. There are many naturally occurring polymers, including PROTEINS, POLYSACCHARIDES, NUCLEIC ACIDS and also many synthetic polymers, including NYLON, POLY(ETHENE), POLYSTYRENE and POLYVINYL CHLORIDE (PVC).

Addition polymers are those in which the polymer chain is built up by the simple addition of identical monomers. Condensation polymers are those in which the chain can be built up of two or more different monomers with the loss of a small molecule, such as water, during the joining of the monomers. Biochemical polymers are usually condensation polymers, whereas many synthetic polymers, including polystyrene and PVC, are addition polymers.

See also ATACTIC POLYMER, ISOTACTIC POLYMER, RUBBER, SYNDIOTACTIC POLYMER.

polymerase An enzyme, for example DNA polymerase and RNA polymerase, that joins MONOMERS together to form POLYMERS.

polymerization The chemical process resulting in the joining together of MONOMERS to form POLYMERS. This can be addition polymerization in which a polymer chain is built up by the addition of identical monomers to each other, for example POLYSTYRENE, POLYVINYL CHLORIDE (PVC), PERSPEX. The repeating units in an addition polymer are identical to the initial reacting monomers.

The other type of polymerization is condensation polymerization, in which the joining of monomers results in the loss of a small molecule, usually water. Condensation polymerization can involve two or more different monomers (co-polymerization). In contrast to addition polymers, the repeating units in condensation polymers are not identical to the reacting monomers. Examples of the latter include NYLON and POLYESTER.

See also ZIEGLER–NATTA CATALYST.

poly(methyl 2-methylpropenoate) *See* PERSPEX.

poly(methylmethacrylate) *See* PERSPEX.

polymorphism The existence of an element or compound in more than one crystalline form. For example, sulphur forms rhombic crystals below 96°C and monoclinic crystals at higher temperatures, up to its melting point of 119°C. The distinction between polymorphism and ALLOTROPY is that polymorphism relates only to crystalline solids.

polynucleotide A long-chain molecule of NUCLEOTIDES.

polypeptide A PEPTIDE consisting of three or more AMINO ACIDS. Polypeptide chains can then fold or twist to form a PROTEIN.

poly(phenylethene) *See* POLYSTYRENE.

poly(propene) *See* POLYPROPYLENE.

polypropylene, ***poly(propene)*** A POLYMER of propene that is rigid, very strong and resistant to abrasion. It is used in moulded furniture and can be hinged for use in make-up compacts for example. Polypropylene can also be spun into fibres to make ropes or carpets.

polysaccharide A CARBOHYDRATE consisting of a variable number of MONOSACCHARIDES joined together in chains that can be branched or not and can fold for easy storage. Polysaccharides can be broken down into their constituent DISACCHARIDES or monosaccharides for use by an organism. In chains, they are insoluble. Examples include CELLULOSE, CHITIN, STARCH and GLYCOGEN.

polystyrene, ***poly(phenylethene)*** A THERMOPLASTIC that is , like POLY(ETHENE), an addition POLYMER but it is more rigid that poly(ethene), due to the presence of BENZENE RINGS. It is widely used in an expanded form produced by impregnating polystyrene with pentane and heating in steam where it is blown into a solid foam. This forms a light, white material which is used in packaging and for insulation.

polytetrafluoroethene (PTFE) (Trade name ***Teflon***) An addition POLYMER made from the MONOMER tetrafluoroethene (CF_2CF_2). It is a THERMOSETTING PLASTIC with a high melting point and is used for non-stick surfaces on kitchen pans.

Polythene The trade name for POLY(ETHENE).

polyurethane A major THERMOSETTING PLASTIC used as a liquid in paints or varnish and as a foam in upholstery. It is a condensation POLYMER consisting of the MONOMER –NHCOOH–.

polyvinyl chloride, ***PVC, poly(chloroethene)*** A THERMOPLASTIC that, like POLY(ETHENE), is an addition POLYMER. It is rigid and tough but can be made flexible by the addition of PLASTICIZERS. Unplasticized it is used in window frames (uPVC) and record discs. Plasticized PVC is widely used for shoes, luggage, flooring and toys.

p-orbital The second lowest energy ORBITAL for a given PRINCIPAL QUANTUM NUMBER. p-Orbitals exist only for principal quantum numbers of two or greater. There are three p-orbitals for a given principal quantum number, each formed of two lobes with each of the three orbitals at right angles to the other two.

porous (*adj.*) Describing any material that contains many small cracks or holes, able to absorb water, air or some other fluid.

porphyrin Any of a group of naturally occurring organic pigments. They have a characteristic structure consisting of four linked nitrogen-containing rings, often co-ordinated to a metal ion. The porphyrins include CHLOROPHYLL, which contains magnesium, the HAEM group of HAEMOGLOBIN, which contains iron, MYOGLOBIN and CYTOCHROME.

positron The positive counterpart of the ELECTRON. Although positrons are stable in isolation, when a positron meets an electron, they will annihilate one another, producing two or more gamma rays.

potash *See* POTASSIUM CARBONATE.

potash alum *See* ALUM.

potassium (K) The element with atomic number 19; relative atomic mass 39.1; melting point 64°C; boiling point 774°C; relative density 0.9. Potassium is a highly reactive ALKALI METAL, and its compounds give a characteristic violet colour when ionized in a flame. It is widespread in nature and slightly radioactive due to a small concentration (0.12 per cent) of the long-lived radioactive ISOTOPE potassium–40, which has a HALF-LIFE of 1.25×10^9 years.

Potassium is essential to all living organisms. In animals it aids in the transmission of nerve impulses, while in plants it is required for growth.

potassium bromide (KBr) A white solid; melting point 734°C; boiling point 1,435°C; relative density 2.7. Potassium bromide may be manufactured by the direct reaction between bromine and potassium:

$$2K + Br_2 \rightarrow 2KBr$$

It is widely used in the photographic industry for the manufacture of silver bromide, used in light-sensitive emulsions.

potassium carbonate, *potash* (K_2CO_3) A white solid; melting point 891°C; decomposes on further heating; relative density 2.4. Potassium carbonate is an important naturally occurring source of potassium. It is used in the manufacture of soaps and in textile processing.

potassium chlorate ($KClO_3$) A white powder; melting point 360°C; decomposes on further heating; relative density 2.3. Potassium chlorate is produced around the ANODE in the ELECTROLYSIS of concentrated potassium chloride:

$$KCl + 3H_2O \rightarrow KClO_3 + 3H_2$$

It decomposes on heating, releasing oxygen:

$$2KClO_3 \rightarrow 2KCl + 3O_2$$

This reaction has led to the use of potassium chlorate in some explosives. It is also a powerful OXIDIZING AGENT, a feature exploited in its use as a weedkiller and disinfectant.

potassium chloride (KCl) A white powder; sublimes at 1,700°C; relative density 2.0. Potassium chloride occurs naturally, and can be extracted from brine by FRACTIONAL CRYSTALLIZATION. It is used as a raw material for the manufacture of other potassium salts, particularly potassium chromate.

potassium chromate (K_2CrO_4) A yellow crystalline solid; melting point 971°C; decomposes on further heating; relative density 2.7. It can be prepared by heating CHROMITE with potassium carbonate and calcium carbonate. Potassium chromate is stable only in alkaline solutions; addition of acid produces POTASSIUM DICHROMATE, for example:

$$2K_2CrO_4 + 2HCl \rightarrow K_2Cr_2O_7 + H_2O + 2KCl$$

Potassium chromate is used in the processing of leather, as a pigment and as a MORDANT in dying.

potassium chromium sulphate, *chrome alum* ($K_2SO_4.Cr_2(SO_4)_3.12H_2O$) A crystalline DOUBLE SALT; melting point 89°C; loses water on further heating; relative density 1.8. Potassium chromium sulphate is deep red in colour and can be made by crystallizing a solution containing equal concentrations of potassium and chromium sulphates.

potassium cyanide (KCN) A white crystalline solid; melting point 634°C; decomposes on further heating; relative density 1.5. It can be prepared by passing hydrogen cyanide through a concentrated solution of potassium hydroxide,

$$HCN + KOH \rightarrow KCN + H_2O$$

Potassium cyanide is soluble in water and the aqueous solution hydrolyses with the gradual release of hydrogen cyanide,

$$KCN + H_2O \rightarrow HCN + KOH$$

Potassium cyanide is highly toxic, but is used industrially in the preparation of ELECTROLYTES for ELECTROPLATING and in the purification of gold and silver.

potassium dichromate ($K_2Cr_2O_7$) An orange solid; melting point 396°C; decomposes on further heating; relative density 2.7. Like all dichromates it is stable only in acidic solution. In the presence of alkali, chromates are produced, for example:

$$K_2Cr_2O_7 + 2NaOH \rightarrow K_2CrO_4 + Na_2CrO_4 + H_2O$$

Potassium dichromate is used as a laboratory OXIDIZING AGENT for certain analytical techniques. It is also used as a reagent for chromium ELECTROPLATING.

potassium hydrogencarbonate ($KHCO_3$) A white solid; decomposes on heating; relative density 2.2. Potassium hydrogencarbonate can be produced by passing carbon dioxide through potassium carbonate solution:

$$K_2CO_3 + CO_2 + H_2O \rightarrow 2KHCO_3$$

Potassium hydrogencarbonate reacts with acids to produce carbon dioxide, a reaction that is exploited in some cooking processes and in certain types of fire extinguisher, for example:

$$KHCO_3 + CH_3COOH \rightarrow CH_3COOK + H_2O + CO_2$$

Potassium hydrogencarbonate decomposes on heating:

$$KHCO_3 \rightarrow KOH + CO_2$$

potassium hydrogentartrate, *cream of tartar* (HOOC.CHOH.CHOH.COOK) A crystalline, weakly acidic salt, produced in the fermentation of wine and mixed with sodium hydrogencarbonate in baking powder to release carbon dioxide in baking,

$$NaHCO_3 + HOOC(CHOH)_2COOK \rightarrow NaOOC(CHOH)_2COOK + H_2O + CO_2$$

potassium hydroxide, *caustic potash* (KOH) A white solid; melting point 360°C; boiling point 1,320°C; relative density 2.0. Potassium hydroxide can be prepared by reacting potassium with water:

$$2K + 2H_2O \rightarrow 2KOH + H_2$$

Potassium hydroxide is used in many situations where a concentrated alkaline solution is required, but is highly damaging to body tissues, so must be handled with particular care – even small splashes in the eye can cause permanent damage.

potassium iodate (KIO_3) A white crystalline solid; melting point 561°C; decomposes on further heating; relative density 3.9. Potassium iodate can be prepared by heating iodine with concentrated potassium hydroxide solution,

$$6KOH + 6I \rightarrow KIO_3 + 3H_2O + 5KI$$

The potassium iodate can be separated from the potassium iodide by FRACTIONAL CRYSTALLIZATION. Potassium iodate is an OXIDIZING AGENT and is used as a preservative in some foods.

potassium iodide (KI) A white solid; melting point 686°C; boiling point 1,330°C; relative density 3.1. Potassium iodide can be made by the reaction between iodine and potassium hydroxide:

$$6KOH + 3I_2 \rightarrow KIO_3 + 5KI + 3H_2O$$

The potassium iodate so formed can be converted to potassium iodide by heating:

$$2KIO_3 \rightarrow 2KI + 3O_2$$

Potassium iodide is widely used in the manufacture of silver iodide, an important compound in light-sensitive emulsions.

potassium nitrate, *saltpetre* (KNO_3) A colourless solid; melting point 334°C; decomposes on further heating; relative density 2.1. Potassium nitrate occurs naturally, and is used as a fertilizer and as an OXIDIZING AGENT, especially in some explosives, such as gun powder.

potassium nitrite (KNO_2) A white crystalline solid; melting point 300°C; decomposes on further heating; relative density 1.9. It can be prepared by the reduction of potassium nitrate with lead,

$$KNO_3 + Pb \rightarrow KNO_2 + PbO$$

In the presence of strong acids it will dissociate to produce nitrous acid, for example:

$$H_2SO_4 + KNO_2 \rightarrow KHSO_4 + HNO_2$$

This leads to its use in DIAZOTIZATION reactions.

potassium sulphate (K_2SO_4) A colourless solid; melting point 1,062°C; decomposes on further heating; relative density 2.7. Potassium sulphate occurs in nature. It is manufactured by the reaction between potassium chloride and concentrated sulphuric acid:

$$2KCl + H_2SO_4 \rightarrow K_2SO_4 + 2HCl$$

It is used as a fertilizer and in the manufacture of cement.

potassium sulphite (K_2SO_3) A white crystalline solid; decomposes on heating; relative density 1.5. It can be prepared by passing sulphur dioxide through a concentrated solution of potassium hydroxide,

$$SO_2 + 2KOH \rightarrow K_2SO_3 + H_2O$$

It is a REDUCING AGENT and is widely used as an antioxidant in the food and brewing industries.

potential difference (p.d.) A more technical term for VOLTAGE, the p.d. between two points being the amount of energy converted from electrical energy to other forms when one COULOMB of charge flows between the two points.

Potential difference = energy transformed / charge flow.

potential energy The ENERGY possessed by an object or system as a result of its position or state. It is the amount of work done by the object or system moving from a state at which it is said to have no potential energy to a higher state.

powder coating A way of giving a plastic surface to a metal object, preventing corrosion, in a way similar to the application of paint. The object to be coated is given an ELECTROSTATIC charge and then sprayed with an oppositely charged powder. The opposite charges ensure that the powder sticks to the object. The object is then heated to melt the powder into a solid layer.

powder metallurgy The use of fine metal powders, which are often produced by the vaporization of a solid piece of metal, to manufacture objects by pressing and heating the metal powder in a mould (*see* SINTERING). The resultant objects can be made in a wide variety of shapes without complex machining and can be made porous to lubricating oils.

praseodymium (Pr) The element with atomic number 59; relative atomic mass 140.9; melting point 934°C; boiling point 3,512°C; relative density 6.8. Praseodymium is LANTHANIDE metal. Its IONIC compounds have a bright green colour and are used to colour glass and ceramics.

precipitate A solid separated from a solution, which may subsequently settle. In chemistry, a precipitate is a finely divided powder suspended in a liquid, which is formed when a reaction between two soluble salts in solution produce an insoluble salt.

precipitation The formation of a PRECIPITATE.

precursor Any compound produced in some intermediate step in a series of chemical reactions, and which is then involved in further reactions leading to the desired final product.

pressure The force acting on each square metre of area. The unit of pressure is the PASCAL (Pa), this being a pressure of one NEWTON per square metre. On a molecular level, the pressure of a fluid can be thought of in terms of the pressure exerted on a surface by the molecules of the fluid. The pressure is equal to the average change in momentum per molecular collision multiplied by the average number of collisions per second per square metre. Because there are a very large number of collisions, they are felt as a constant pressure, rather than a series of separate impacts. *See also* DALTON'S LAW OF PARTIAL PRESSURE, KINETIC THEORY.

pressure law A GAS LAW used to define the IDEAL GAS TEMPERATURE SCALE. It states that, for a fixed mass of gas held in a constant volume, the pressure is proportional to the ABSOLUTE TEMPERATURE, i.e. the pressure divided by the temperature is a constant. For a fixed mass of an ideal gas with a pressure p and absolute temperature T held at constant volume,

$$p/T = \text{constant}$$

See also BOYLE'S LAW, CHARLES' LAW, IDEAL GAS EQUATION.

primary cell Former name for a non-rechargeable electrochemical CELL.

primitive cell *See* UNIT CELL.

principal quantum number The QUANTUM NUMBER used to label an ORBITAL to give a broad indication of its energy. Thus in describing a 2s orbital, 2 is the principal quantum number.

probability amplitude A quantity used in QUANTUM MECHANICS to find the probability of a system being in a particular state or a particle in a particular position. Probability amplitudes can interfere, giving wave-like properties to the particles they describe. The probability of finding a particular state is the square of the probability amplitude. *See also* WAVE-PARTICLE DUALITY.

producer gas The gas formed by passing air over hot COKE. The coke is partially oxidized (*see* OXIDATION) by the oxygen, forming a mixture of nitrogen and carbon monoxide. This reaction heats the coke and is sometimes used in conjunction with the production of WATER GAS. *See also* SEMI-WATER GAS.

progesterone ($C_{21}H_{30}O_2$) A STEROID HORMONE in mammals that regulates the menstrual cycle in females and prepares the uterus for pregnancy. It is secreted by the ovary and when needed helps maintain early pregnancy. At high doses progesterone, or synthetic equivalents, can inhibit ovulation and this ability is used in oral contraceptives.

prolactin, *luteotrophic hormone, luteotrophin, LTH* A protein HORMONE secreted by the pituitary gland of vertebrates. It stimulates milk production and promotes secretion of PROGESTERONE. It has a relative molecular mass of about 33,000.

promethium (Pm) The element with atomic number 61; relative atomic mass 145.0; melting point 1,042°C; boiling point 3,000°C; relative density 7.3. Promethium has only radioactive ISOTOPES, all of which have HALF-LIVES of 20 years or less. Promethium–147 occurs in nature in minute amounts from the spontaneous fission of uranium.

promoter In chemistry, any substance that, when present in very small quantities, can increase the effectiveness of a CATALYST.

proof spirit A mixture of ETHANOL and water containing 49 per cent ethanol by mass. This is the standard used in measuring the level of ethanol in certain alcoholic drinks, including gin and whiskey, which are usually sold at 70° proof; that is, at 70 per cent of the proof level of alcohol content.

propanal, *propionaldehyde* (C_2H_5CHO) A colourless liquid ALDEHYDE; melting point –81°C; boiling point 49°C.

propane (C_3H_8) A colourless gas with an odour, of the ALKANE series of HYDROCARBONS; boiling point –45°C; melting point –190°C. It is a component of NATURAL GAS and can also be made from reduction of propene. Propane is used as a fuel and a refrigerant.

propanedioic acid, *malonic acid* ($C_3H_4O_4$) A colourless, crystalline DICARBOXYLIC ACID; melting point 132°C. It is obtained during the processing of sugar beet. It decomposes above its melting point to give ethanoic acid. It is used in organic syntheses.

propanoic acid, *propionic acid* ($C_3H_6O_2$) A colourless liquid CARBOXYLIC ACID; boiling point 141°C; melting point 24°C. It is used in the manufacture of ESTERS and POLYMERS and to make calcium propanate, a food additive.

propanol, *propyl alcohol* (C_3H_7OH) The third member in the series of ALCOHOLS, a colourless liquid which is a mixture of the ISOMERS propan-1-ol ($CH_3CH_2CH_2OH$) and propan-2-ol ($CH_3CHOHCH_3$). The boiling points are 97°C and 82°C respectively. Propanol is used in perfumery.

propanone, *acetone* (CH_3COCH_3) A colourless, flammable liquid; boiling point 56°C. It is miscible with water and widely used as nail varnish remover. Propanone is a KETONE with a characteristic odour. It is manufactured by the CUMENE PROCESS.

propan-1, 2, 3-triol *See* GLYCEROL.

propene, *propylene* (C_3H_6,) A colourless organic gas; boiling point –48°C; melting point –185°C. It is the second member of the ALKENE series of HYDROCARBONS. Propene is made from petroleum by CRACKING. It is widely used in the manufacture of resins and plastics, such as POLYPROPYLENE.

propenoic acid, *acrylic acid* (CH_2=CHCOOH) A colourless liquid made from the oxidation of the ALDEHYDE propanal; boiling point 141°C; melting point 13°C. Propenoic acid can be polymerized (*see* POLYMERIZATION) to produce important synthetic POLYMERS used as artificial fibres (ACRYLICS), acrylic artist's paint and adhesives. The derivative methyl propenoate (methyl acrylate) is polymerized to produce glass-like RESINS for use as lenses, dentures and transparent parts. Another derivative is polymeric methyl methyl-acrylate, better known under its trade name of PERSPEX.

propenonitrile, *acrylonitrile* (CH_2=CHCN) An organic liquid; boiling point 78°C. It readily POLYMERIZES and is used in the manufacture of synthetic RUBBER.

propionaldehyde *See* PROPANAL.

propionic acid *See* PROPANOIC ACID.

propyl alcohol *See* PROPANOL.

propylene *See* PROPENE.

prostaglandin Any one of a group of complex FATTY ACIDS synthesized continuously in mammals by most nucleated cells. They act in a similar way to HORMONES in being chemical messengers between cells, but usually only act locally.

Prostaglandins are made all over the body and can be released directly into the blood. They were first discovered in semen, which is an especially rich source. Their effects include stimulating the contraction of smooth muscle (useful in inducing labour in childbirth or abortions), regulation of production of stomach acid, modifying other hormonal activity, assisting in blood clotting and being responsible for inflammation following injury, infection or allergy. In excess they may be involved in causing inflammatory disorders such as arthritis. Pain-relieving drugs such as aspirin act by inhibiting prostaglandins. They are of great potential importance in the alteration of blood pressure and broncodilation and constriction. Many synthetic prostaglandins are available. *See also* ARACHIDONIC ACID.

prosthetic group A non-protein group that firmly attaches to a PROTEIN or ENZYME to create a functional complex, in contrast to a COENZYME. Examples include HAEMOGLOBIN, which contains iron as its prosthetic group, and GLYCOPROTEINS, which contain CARBOHYDRATES as the prosthetic groups. Metal ions such as Zn^{2+}, K^+ and Na^+ are often prosthetic groups for enzymes, providing a charge needed in an active site.

protactinium (Pa) The element with atomic number 91. Only radioactive ISOTOPES are known, the longest lived of which (protactinium–231) has a HALF-LIFE of 34,000 years. It exists in minute amounts in uranium ores, produced by the decay of uranium–235.

protamine Any of a group of basic PROTEINS of low molecular mass that are associated with NUCLEIC ACIDS. They consist of a single POLYPEPTIDE chain that contains high levels of ARGININE but no sulphur-containing AMINO ACIDS. Protamines are associated with the DNA of sperm cells.

protease *See* PROTEOLYTIC ENZYME.

protein Any one of a group of complex organic compounds that have a large relative molecular mass and consist of AMINO ACIDS linked together. Proteins always contain carbon, hydrogen, oxygen, nitrogen, usually sulphur and sometimes phosphorous. They are essential to all living organisms.

The amino acids in a protein link together to form PEPTIDES or POLYPEPTIDES, and this sequence of amino acids is the primary structure of the protein. There is an amino end (NH_2) and carboxyl end (COOH) to any protein molecule. The ultimate shape of the protein molecule depends on the types of bonds within it. HYDROGEN BONDS can form, which often results in a polypeptide chain coiling into an ALPHA HELIX or BETA PLEATED SHEETS. The shape of the polypeptide chain is called the secondary structure. Other types of bonding can occur, including IONIC BONDS and HYDROPHOBIC interactions between non-polar R groups (*see* AMINO ACIDS), which cause folding of the protein to shield these groups from water. Together these cause folding and twisting of the constituent polypeptide chains of a protein into a three-dimensional structure termed the tertiary structure. A large complex protein molecule has many polypeptide chains combined and incorporates non-protein groups (*see* PROSTHETIC GROUP, COFACTOR) vital to its function into its structure which is then referred to as the quarternary structure. The shape of a protein is crucial to its functioning, for example in providing enzyme-binding sites.

There are a seemingly limitless number of proteins and unlike CARBOHYDRATES they vary from species to species. Their functions are numerous. FIBROUS PROTEINS, such as COLLAGEN, provide a structural role, whilst GLOBULAR PROTEINS, which include ENZYMES, provide a metabolic role. CONJUGATED PROTEINS incorporate non-protein groups into their structure which play a vital role in their functioning, for example the HAEM group in HAEMOGLOBIN. Proteins can combine with carbohydrates to give GLYCOPROTEINS, with LIPIDS to give LIPOPROTEINS or with NUCLEIC ACIDS to give NUCLEOPROTEINS. Proteins can be denatured, for example by heat, or by strong acid or alkali, which breaks up the three-dimensional structure and prevents it from functioning.

Protein is an essential requirement of the human diet providing energy; it is not usually stored in the body so needs to be included regularly in the diet (60 g per day).

proteinase *See* PROTEOLYTIC ENZYME.

protein synthesis The manufacture of all the PROTEINS needed by an organism. This is ultimately controlled by the DNA contained within the cell nucleus. The genetic information contained within the DNA is transferred to messenger RNA (mRNA) in the cell nucleus. Messenger RNA carries this information to the cell cytoplasm, where it determines which AMINO ACIDS are to be linked in which order. This in turn determines the type of protein made. Animals obtain the ESSENTIAL AMINO ACIDS from their diet and synthesize the others. Plants synthesize their own amino acids from NITRATES in the soil and CARBOHYDRATE products from, for example, the KREBS CYCLE. The amino acids combine with another specific RNA, called transfer RNA (tRNA), by which they are carried to the cell cytoplasm. Here they meet the mRNA that attaches at one end to a cell structure called a ribosome and attracts a specific amino acid carried by the tRNA. In this way amino acids are linked in a specific order to form POLYPEPTIDES. The polypeptides are then assembled into proteins.

proteolytic enzyme, ***protease, proteinase, peptidase*** A general term for any ENZYME that catalyses the breakdown of PROTEINS to PEPTIDES and then into their constituent AMINO ACIDS. Several enzymes act sequentially to effect the complete breakdown of a protein. The term endopeptidase is often used to describe enzymes that break internal links of the protein chain, whilst exopeptidases attack both ends of the chain. Examples of proteolytic enzymes include trypsin and pepsin.

proton The positively charged ELEMENTARY PARTICLE found in the NUCLEUS of an ATOM. The number of protons in a nucleus is called the ATOMIC NUMBER and fixes the number of ELECTRONS needed to produce a neutral atom, which in turn determines the chemical properties of the element.

The mass of a proton is 1.6×10^{-27} kg. It has a charge of 1.6×10^{-19} C, equal but opposite to the charge on an electron.

protonic acid A term used to distinguish ACIDS that actually produce hydrogen ions in aqueous solution, as opposed to LEWIS ACIDS or Lowry–Brønsted acids, which do not do this but are termed acids because they respectively accept electrons or give up protons in specific reactions.

Prussian blue A blue pigment, the DOUBLE SALT potassium iron hexacyanoferrate, $KFe[Fe(CN)_6]$.

pseudohalogen A compound that behaves in many ways like a HALOGEN but contains more than one element. The most important example is CYANOGEN, $(CN)_2$. Like the halogens, pseudohalogens form covalent molecules with fairly low melting points and react with hydrogen to form MONOBASIC acids, for example:

$$(CN)_2 + H_2 \rightarrow 2HCN$$

PTFE *See* POLYTETRAFLUOROETHENE.

puddling The process of producing WROUGHT IRON in a REVERBERATORY FURNACE using a mixture of PIG IRON and haematite ore, which comprises mostly iron(III) oxide. Oxygen from the haematite oxidizes the carbon in the pig iron (*see* OXIDATION) and produces almost pure iron.

pump A device for moving or changing the pressure of a fluid. A pump normally consists of a chamber together with two valves through which fluid can flow in one direction only, or some other mechanism for ensuring that the fluid can flow in one direction only. Fluid enters the chamber through one of the valves. The volume of the chamber is then reduced and fluid expelled through the other valve.

purine A type of organic base occurring in NUCLEIC ACIDS and NUCLEOTIDES that consists of a double ring, one with six sides and one with five. Adenine and guanine are the most common.

PVC *See* POLYVINYL CHLORIDE.

pyranose A sugar with a six-sided ring structure containing five carbon atoms and one oxygen atom. Glucose is an example of a sugar that forms a pyranose ring structure. Glucose can exist as a straight chain or carbon atom 1 may combine with the oxygen on carbon atom 5 to form a pyranose ring. These can exist in α and β forms and are more stable than five-sided FURANOSE rings.

pyridine (C_5H_5N) An AROMATIC HETEROCYCLIC compound found in COAL TAR, melting point –42°C, boiling point 115°C. It is a colourless liquid with a strong characteristic smell. It is used as a solvent in the plastics industry and in the manufacture of other organic chemicals, including nicotinic acid.

pyridoxine, ***vitamin*** B_6 ($C_8H_{11}NO_3$) A VITAMIN of the VITAMIN B COMPLEX. Pyridoxine is a COENZYME in AMINO ACID metabolism and lack of it causes nervous disorders, dermatitis and anaemia. It is found in cereals and yeast.

pyrimidine A type of organic base occurring in NUCLEIC ACIDS and NUCLEOTIDES that consists of two single, six-sided rings. The most common are cytosine, thymine and uracil.

pyrites A mineral SULPHIDE of some metals, having a cubic crystal form. Iron pyrites is iron(II) sulphide, FeS, sometimes called fools' gold as its metallic lustre has led to it being mistaken for NATIVE gold.

pyrolysis The DECOMPOSITION of a compound (usually organic) under the action of heat.

pyrometer Any device for measuring high temperatures, especially one that operates by analysing visible light or infrared radiation given off by a hot object.

pyrometry The measurement of high temperatures.

pyrones Organic compounds containing a six-membered ring system with an oxygen HETEROATOM and a CARBONYL GROUP attached to the ring (in one of two positions). This ring system occurs in nature, for example in ANTHOCYANINS.

pyrophoric (*adj.*) A substance, particularly a metal produced in a porous state by a chemical reaction, that will ignite spontaneously in air.

pyrophoric alloy An alloy of cerium (typically 70 per cent) and iron that will produce sparks when struck or rubbed with a piece of iron, used as a source of ignition in the 'flint' of a cigarette lighter.

pyroxenes A group of silicate minerals found in basic igneous rocks with a complex crystal chemistry. They consist of long chains of silicon and oxygen atoms linked by a variety of other elements. An example is diopside, $CaMgSi_2O_6$

pyrrole (C_4H_5N) A colourless oil that forms part of the structure of PORPHYRINS; boiling point 130°C. Many pyrrole derivatives occur naturally, for example HAEM and CHLOROPHYLL.

pyruvate Any salt or ester of pyruvic acid.

pyruvic acid, *2-oxopropanoic acid* ($CH_3COCOOH$) A colourless, pleasant-smelling organic acid; melting point 13°C. It is an important product in the metabolism of carbohydrates and proteins.

qualitative analysis The branch of chemical ANALYSIS that aims to discover which elements or compounds are present in a sample, without measuring the amount of the materials present. A wide range of techniques can be used, from simple chemical tests for certain ions, to sophisticated SPECTROSCOPY and MASS SPECTROSCOPY techniques.

quantitative analysis The branch of chemical ANALYSIS that uses techniques enabling the quantity of a certain element or compound present in a sample to be measured. The chief techniques of quantitative analysis are GRAVIMETRIC ANALYSIS, in which a component, such as a PRECIPITATE, is separated from the sample and weighed, and VOLUMETRIC ANALYSIS, in which the volume of material that can react with the sample in a particular way is measured, for instance by TITRATION.

quantum (*pl.* ***quanta***) A particle, or the group of waves associated with such a particle, which has a fixed (quantized) value of some quantity. Thus, for example, the PHOTON is a quantum of ELECTROMAGNETIC RADIATION with a fixed amount of energy. The term quantum also refers to the minimum amount by which certain properties, such as the energy, of a system can change.

quantum mechanics A system of mechanics developed from QUANTUM THEORY. Quantum mechanics explains the properties of atoms, molecules and ELECTROMAGNETIC RADIATION.

quantum number A number that represents the value of some quantity such as CHARGE, which is conserved in certain types of interactions and which is only found in whole number multiples of some basic quantity.

quantum theory The theory that energy is absorbed or released in discrete, indivisible units called quanta. According to quantum theory, there is no real distinction between effects traditionally described in terms of waves, such as light, and objects that are more usually thought of as particles, such as ELECTRONS. This double nature of waves and particles is referred to as WAVE-PARTICLE DUALITY. *See also* BAND THEORY, ENERGY LEVEL, HYDROGEN SPECTRUM, PAULI EXCLUSION PRINCIPLE, PHOTOELECTRIC EFFECT, SPIN.

quark The fundametal particle from which PROTONS and NEUTRONS are composed.

quartz The most abundant mineral in the Earth's crust, containing mostly silicon dioxide. In its purest form, quartz is colourless, but small amounts of impurities produce coloured gemstones, such as amethyst. Quartz is PIEZOELECTRIC, and mechanically resonant quartz crystals are used as oscillators in clocks and watches.

quenching A process in which a material is heated and then cooled suddenly, usually using oil or water. The result is that the large numbers of DISLOCATIONS produced by thermal vibrations at high temperatures become 'frozen' in place and tangled with one another, so are unable to move through the metal. The result is a material that is very hard, but brittle. *See also* ANNEALING.

quicklime *See* CALCIUM OXIDE.

quinone Specifically, CYCLOHEXADIENE-1,4-DIONE. Generally, any similar compound that contains C=O groups in an unsaturated ring.

quinhydrone electrode An electrode used for measuring pH values based on a HALF-CELL in which a platinum electrode is immersed in a solution containing equal concentrations of quinone and hydroquinone. The equilibrium of these materials is affected by the reaction

$$C_6H_4(OH)_2 \Leftrightarrow C_6H_4O_2 + 2H^+ + 2e^-$$

enabling the pH to be measured. The method has largely been replaced by the use of glass electrodes.

R

racemic mixture A 1:1 mixture of the two ENANTIOMERS of a compound exhibiting OPTICAL ISOMERISM. A racemic mixture does not rotate PLANE-POLARIZED light since one enantiomer rotates it in one direction and the other equally in the opposite direction. A racemic compound has the prefix (+/–) denoting the presence of equal amounts of the (+) and (–) enantiomers.

racemization The process by which an optically active (*see* OPTICAL ACTIVITY) compound is converted into a RACEMIC MIXTURE.

radiation In general, the emission of rays, waves or particles from a source, with intensity falling off according to an INVERSE SQUARE LAW. In particular, the term radiation refers to IONIZING RADIATION and ELECTROMAGNETIC RADIATION. *See also* BACKGROUND RADIATION, RADIOACTIVITY.

radical A group of atoms, such as CH_3 or NH_2 that frequently occur together and pass unchanged through many chemical reactions, being held together by strong COVALENT BONDS. *See also* FREE RADICAL.

radioactive series, *decay series* A series of NUCLIDES, all but the last radioactive, where the decay of the first nuclide gives rise to the second, which decays into the third and so on. In general, the first member of such a series has a much longer HALF-LIFE than the others, so that a sample of this nuclide will create the other nuclides in the series with an equilibrium between the rates of production and decay. Thus the amounts of each of the members of the decay series, apart from the final product, will be proportional to their half-lives. *See also* RADIOACTIVITY.

radioactivity The spontaneous decay of unstable atomic nuclei with the emission of IONIZING RADIATION, either ALPHA PARTICLES, BETA PARTICLES or GAMMA RADIATION. Radioactivity occurs spontaneously in many naturally-occurring radioisotopes without any external influence, and may also be induced in certain unstable nuclei by bombarding with NEUTRONS or other particles.

Radioactivity can be harmful to living tissue because of the damage done to living cells by the ionizing radiation. In particular, the formation of FREE RADICALS in the vicinity of the DNA in a cell can cause cancer. However, radioactivity can also be used to kill cancerous cells: a cell is most vulnerable to genetic damage when it is dividing, and since cancer cells divide more rapidly than healthy cells, they are more easily killed. To ensure that healthy tissue does not receive a dose which may create mutations leading to further cancers, the radiation source is either implanted in the patient or is in the form of a beam aimed at the patient from several directions, overlapping to form a large dose at the location of the tumour.

See also ACTIVITY, ALPHA DECAY, BACKGROUND RADIATION, BETA DECAY, DECAY CONSTANT, HALF-LIFE, RADIOACTIVE SERIES.

radiocarbon dating An important example of the use of the decrease in activity of a radioactive ISOTOPE to find the age of some objects. All living organisms extract carbon from their surroundings. Some of this carbon will be the unstable isotope carbon–14, which is produced in the atmosphere by the interaction of cosmic radiation with nitrogen nuclei. When the organism dies, it stops taking in carbon from its surroundings, and since carbon–14 has a HALF-LIFE of about 5,700 years, measurement of the proportion of carbon–14 compared to stable carbon–12 enables the age to be determined. One technique is to measure the level of radioactivity, but since this is very small, greater precision can be achieved and smaller samples are needed if a MASS SPECTROMETER is used to detect the nuclei.

Radiocarbon dating assumes that the proportion of carbon–14 in the atmosphere has remained constant. This is not quite true, as has been shown by the comparison of radiocarbon dates with those obtained by other methods, such as examining the growth rings of trees – a technique called dendrochronology.

radiochemistry The study and use of RADIOACTIVITY in chemistry. It includes the use of radioactive ISOTOPES to label certain atoms in a molecule – so the path of a reaction can be followed more clearly – and radiolysis, the study of the effects of IONIZING RADIATION on chemical reactions.

radioisotope Any radioactive ISOTOPE.

radium (Ra) The element with atomic number 88; relative atomic mass 226.0; melting point 700°C; boiling point 1,140°C; relative density 5.1. Only radioactive ISOTOPES are known, with HALF-LIVES up to 1,600 years (radium–226). Radium was one of the first radioactive elements to be discovered and extracted from pitchblende, a uranium ore in which it is produced by the ALPHA DECAY of uranium, via thorium. Radium was widely used in luminous paints until the cancer risks of IONIZING RADIATION became known.

radon (Rn) The element with atomic number 86; relative atomic mass 222.0; melting point –72°C; boiling point –62°C. Radon is a naturally occurring radioactive element produced by the ALPHA DECAY of radium. It is an inert gas and thus can easily be separated chemically from the elements that produce it. The longest lived ISOTOPE (radon–222) has a HALF-LIFE of just over 3 days. Concern has been expressed about the level of radon in some mines and in homes situated over some rock formations that may release significant amounts of radon into the atmosphere.

Raney nickel A black spongy PYROPHORIC form of NICKEL, widely used as a catalyst for gas or liquid phase reactions. It is manufactured by the action of sodium hydroxide on a nickel aluminium alloy. The aluminium is dissolved, leaving a spongy mass of nickel.

Raoult's Law In any mixture of liquids the SATURATED VAPOUR PRESSURE above the liquid is equal to the sum of the vapour pressure of each liquid multiplied by its MOLE FRACTION. Thus for a solution of a volatile liquid in a non-volatile solvent, the major contribution to the vapour pressure will be proportional to the concentration of the volatile liquid. The truth of this law can be seen by applying KINETIC THEORY to the escape of molecules from the surface of the mixture, but in practice Raoult's law is only approximately true due to interactions between the molecules in the mixture.

rare earth element *See* LANTHANIDE.

rare gas *See* NOBLE GAS.

rate constant The constant of proportionality in the LAW OF MASS ACTION. For all reactions, the rate constant may be affected by temperature and CATALYSTS. For aqueous reactions it is virtually independent of pressure, but in many reactions involving gases it is pressure dependent. *See* LE CHATELIER'S PRINCIPLE.

rate-determining step In a CHEMICAL REACTION that proceeds via a number of steps, the slowest step, which effectively limits the rate of the overall reaction. In a reaction A→B→C, the reaction B→C may be limited by the rate at which B is produced by the first stage in the reaction, which is then called the rate-determining step. Alternatively, the reaction A→B may go to completion almost immediately, but B→C proceed only slowly, in which case this is the rate-limiting step in the production of A from C.

When considering techniques to increase the rate of a reaction, for example in an industrial process, it is most important to increase the rate of the rate-determining step.

See also RATE OF REACTION.

rate of reaction The rate at which a CHEMICAL REACTION takes place, usually expressed in moles of product produced per second from one decimetre cubed of reacting material. The rate of reaction may be altered by the presence of a CATALYST, and will also depend on the concentrations of reagents and products present and on the physical conditions.

See also ACTIVATION ENERGY, ARRHENIUS' EQUATION, KINETICS, LAW OF MASS ACTION, LE CHATELIER'S PRINCIPLE, RATE CONSTANT, RATE-DETERMINING STEP.

ratio of specific heats The HEAT CAPACITY of a substance at constant pressure divided by the heat capacity at constant volume. For air, this ratio, usually given the symbol γ, is about 1.4. For a material with a MOLAR HEAT CAPACITY at constant pressure of C_P and a molar heat capacity at constant volume of C_V

$$\gamma = C_P / C_V$$

rayon The general name for a group of textiles and artificial fibres made from CELLULOSE. The cellulose used is usually obtained from wood fibre, made into a solution and filaments regenerated from this. Viscose and acetate (*see* ETHANOATE) are types of rayon.

reaction *See* CHEMICAL REACTION.

reaction profile A diagrammatic way of describing a chemical process that proceeds via some intermediate state, usually of higher energy, i.e. a process having an ACTIVATION ENERGY. The energy of the reagents is shown as a line at the left of the diagram, with the energy of the products shown as a line to the right, lower than the reagents line if the reaction is EXOTHERMIC. A 'hill' between them represents the energy of the intermediate state, which must be reached before the reaction can proceed. The effect of any CATALYST is to reduce the height of the barrier. An increase in temperature increases the number of reagent molecules with sufficient energy to cross the barrier and therefore increases the rate of the reaction. *See also* ACTIVATION PROCESS.

reactivity In chemistry, the ease with which a particular element takes part in chemical reactions and the rate at which such reactions proceed. For example sodium is said to be a reactive metal – it reacts violently with cold water and will spontaneously burn in air. Zinc is a less reactive metal – it reacts with water only at high temperatures and will only burn if ignited in a pure oxygen atmosphere.

In the case of these reactions, the reactivity can be related to the relative ease with which sodium forms IONS compared to zinc. This lower IONIZATION ENERGY makes sodium more reactive. In turn, this stems from the fact that sodium has only a single VALENCE ELECTRON in a 3s ORBITAL, where it is almost completely screened from the nuclear charge. In zinc there are two 4s electrons, each incompletely screened from the nuclear charge by the 3d electrons.

See also REACTIVITY SERIES.

reactivity series A list of elements, usually metals, in order of REACTIVITY. The details of a reactivity series may differ depending on which reaction is being considered, but the broad features are the same. Generally speaking, the group 1 elements (ALKALI METALS, formerly group I) are the most reactive of the metals, followed by the group 2 elements (ALKALINE EARTHS, formerly group II) followed by aluminium, zinc, iron and copper. Within groups 1 and 2 the reactivity increases with increasing ATOMIC NUMBER, thus potassium is more reactive than sodium and calcium is more reactive than magnesium. These trends are due to the decreasing IONIZATION ENERGY caused by the increasingly effective screening of the nuclear charge from the valence electrons.

Amongst the HALOGENS, the lighter elements are the most reactive, with chlorine being more reactive than bromine for example. This is a consequence of the fact that these elements generally gain electrons in reactions, unlike metals, which tend to lose them. The less complete screening in the lighter elements increases the ELECTRON AFFINITY of these elements, making them more reactive.

See also ELECTROCHEMICAL SERIES.

reactor Any vessel in which some kind of reaction takes place.

reagent Any substance that takes part in a CHEMICAL REACTION.

real gas Any gas that shows significant departures from the behaviour predicted by the IDEAL GAS EQUATION. The behaviour of such gases can be modelled fairly well by VAN DER WAALS' EQUATION, which takes account of the volume taken up by the gas molecules themselves and of the attractive forces between the molecules (VAN DER WAALS' FORCE).

If the density of the gas is high enough and the temperature low enough, van der Waals' equation predicts a region where volume would increase with increasing pressure. This is unrealistic, and this region is in fact the area where the gas would exist as a SATURATED VAPOUR along with some liquid. The temperature at which this is first seen is called the critical temperature – above this temperature there is no distinction between the liquid and gas states.

See also IDEAL GAS.

recrystallization A process for obtaining a pure sample of a crystalline material. The material is produced in solution and then crystallized. The crystals are then dissolved again and recrystallized, possibly several times.

rectification An old term for the process of purifying a liquid, especially ETHANOL, by DISTILLATION.

red lead *See* DILEAD(II) LEAD(IV) OXIDE.

redox couple A pair of sets of reagents, one of which represents an oxidized form and the other a reduced form of the same set of elements (*see* OXIDATION, REDUCTION). For example:

$$Fe^{3+} + e^- \rightarrow Fe^{2+}$$

is a redox couple. By convention, the more oxidized form (the OXIDIZING AGENT) is written on the left of the equation, so the more positive the REDOX POTENTIAL the more likely it is to proceed from left to right.

redox half-cell A device for measuring or comparing REDOX POTENTIALS. In its simplest form, a pair of HALF-CELLS comprise two vessels each containing an INACTIVE ELECTRODE (such as a platinum electrode). Each cell contains the reagents for a different redox process. The two cells are linked by a conducting bridge, typically a piece of POROUS material soaked in aqueous potassium chloride solution. The bridge provides an electrical connection without allowing any extra chemical reaction. The POTENTIAL DIFFERENCE between the two electrodes is the difference in redox potentials for the two REDOX COUPLES. The redox potential for hydrogen,

$$2H^+ + 2e^- \rightarrow H_2$$

is assigned the value zero.

redox potential An ELECTRODE POTENTIAL deduced for a reaction that does not necessarily involve a metal entering an ionic solution. The redox potential is the POTENTIAL DIFFERENCE, measured relative to a standard HYDROGEN ELECTRODE, needed to bring about a specified OXIDATION (at an ANODE), or REDUCTION (at a CATHODE). Such potentials can be deduced from known STANDARD ELECTRODE POTENTIALS in a REDOX HALF-CELL.

The more positive the redox potential for a REDOX COUPLE, the more likely it is to proceed in the direction of reduction, whilst the lower (less positive or more negative) redox potentials indicate a tendency to oxidation. When a reaction involves a redox process, it will proceed in the direction that involves the redox couple with the more positive redox couple being reduced. Thus in the reaction

$$2Na + Cl_2 \rightarrow Na^+ + Cl^-$$

the redox potential for

$$Na^+ + e^- \rightarrow Na$$

is –2.71 V, whilst for

$$Cl_2 + 2e^- \rightarrow 2Cl^-$$

it is +1.36 V. Thus this reaction proceeds in the direction that oxidizes sodium and reduces chlorine.

See also NERNST EQUATION, REDOX REACTION, STANDARD REDOX POTENTIAL.

redox reaction A reaction in which one material is reduced whilst another is oxidized. Since REDUCTION and OXIDATION are reverse processes they always occur together and any reaction can be regarded as a redox reaction. Thus to consider a process in these terms is really a particular way of analysing it, particularly the analysis of a reaction in terms of REDOX POTENTIALS.

reducing agent A material that brings about a REDUCTION, being itself oxidized (*see* OXIDATION). Hydrogen is an important reducing agent.

reducing sugar A SUGAR that can act as a REDUCING AGENT in solution, as indicated by a positive BENEDICT'S TEST or FEHLING'S TEST. This depends on the presence of a free ALDEHYDE or KETONE group. Most MONOSACCHARIDES are reducing sugars, as are most DISACCHARIDES except SUCROSE. *See also* CARBOHYDRATE.

reduction The opposite process to OXIDATION. It is the reduction of the OXIDATION STATE of an element, typically by the addition of hydrogen, the removal of oxygen or the addition of electrons to form a negative (or less positive) ion. For example, the reaction between copper oxide and hydrogen involves the reduction of the copper:

$$CuO + H_2 \rightarrow Cu + H_2O$$

The copper has oxygen removed and is converted from ionic Cu^{2+} to neutral copper. This represents a change in oxidation state from +2 to 0. Chlorine is reduced when it is burnt with hydrogen to form hydrogen chloride, which in aqueous solution involves the formation of the Cl^- ion. *See also* CATHODIC REDUCTION, REDOX REACTION, REDUCING AGENT.

refine (*vb.*) To increase the purity of a material by some chemical process. For example, copper can be refined by ELECTROLYSIS. In this process, the impure copper is made the ANODE of an electrochemical CELL containing copper(II) sulphate as an ELECTROLYTE. Pure copper is deposited on the CATHODE whilst impurities form a sludge that sinks to the bottom of the electrolyte.

Many organic compounds can be refined by FRACTIONAL DISTILLATION, which takes advantage of differences in boiling point. *See also* FRACTIONAL CRYSTALLIZATION, RECRYSTALLIZATION.

reflux A chemical reaction in which the reagents are heated and the vapour collected, condensed and returned to the reaction vessel.

reforming The process by which straight-chain ALKANES are converted into branched-chain alkanes. This is achieved by CRACKING or by a catalytic reaction. *See* CATALYTIC REFORMING.

refractory (*adj.*) Describing a material that can be heated to a high temperature without losing its mechanical strength. Refractory materials are used for lining furnaces and kilns.

Regnault's method A method for measuring the density of a gas by weighing a glass vessel after all air has been pumped out of it, and again when the gas to be measured has been allowed to enter at a known pressure and temperature.

relative atomic mass, *atomic weight* The mass of an ATOM measured in ATOMIC MASS UNITS; that is, on a scale where a single atom of carbon–12 has a mass of 12 exactly. One MOLE of atoms will have a mass in grams that is equal to the relative atomic mass of the atom in atomic mass units.

In the case of an element that occurs with several ISOTOPES, the relative atomic mass is normally given as an average of the isotopes, weighted by their natural abundances. For example, chlorine occurs with two isotopes with relative atomic masses of approximately 35 and 37. However, the lighter isotope is roughly three times more common than the heavier one, so any compound made from a naturally occurring chlorine sample will have this ratio of isotopes. Thus the relative atomic mass of chlorine is 35.5.

relative density (r.d.), ***specific gravity*** The DENSITY of a substance at a specified temperature, divided by the maximum density of water (its density at 4°C). Since the density of water is close to 1 g cm^{-3}, the relative density is close to the density in grams per centimetre cubed. If its relative density is less than 1, a substance will float on water; if its density is greater than 1, it will sink. The relative density of a gas is usually quoted relative to dry air – both at the same temperature and pressure.

relative molecular mass (rmm), ***molecular weight*** The mass of a MOLECULE measured in ATOMIC MASS UNITS. The relative molecular mass of a molecule is equal to the sum of the RELATIVE ATOMIC MASSES of the atoms from which the molecule is composed. *See also* DUMAS' METHOD.

rennin An ENZYME that coagulates milk in the stomach and is therefore important to young animals in their digestion of milk. Rennin is secreted as the inactive prorennin, which is activated in the stomach to form rennin. Rennin acts on the soluble milk protein caseinogen, which it converts into the insoluble CASEIN.

resin ($(C_5H_8)_n$) A material with a high RELATIVE MOLECULAR MASS that softens at high temperatures. Natural resins are exuded from trees such as pines and firs and harden in air. Synthetic resins are produced by POLYMERIZATION and used in adhesives, plastics and varnishes. Examples include EPOXY RESINS and PHENOL-FORMALDEHYDE RESINS. Soft resins are used in ointments.

resolution The separation of a RACEMIC MIXTURE into its two ENANTIOMERS.

resonance hybrid A molecule or set of COVALENT BONDS within a molecule that cannot be described in terms of simple single and double covalent bonds, but which is a quantum mechanical mixture of a number of possibilities, with the resonance hybrid having a lower energy.

The BENZENE molecule can be regarded as a resonance hybrid between two alternative structures in which single and double bonds alternate about the ring, and the carbon dioxide molecule is a resonance hybrid involving double bonds and CO-ORDINATE BONDS.

respiration The biochemical processes occurring within cells to break down food molecules to release energy. There are three stages to respiration: GLYCOLYSIS, the KREBS CYCLE and the ELECTRON TRANSPORT SYSTEM.

Respiration is more correctly called 'internal cellular respiration' and contrasts with 'external respiration', which is the exchange of oxygen and carbon dioxide during breathing. Most food is converted into the sugar GLUCOSE, which is then converted into carbon dioxide and water with the release of energy. FATS and PROTEINS can also be used as respiratory substrates (without being first converted to CARBOHYDRATES) but only during starvation (or dieting).

Respiration is usually aerobic (requiring oxygen) and occurs in living cells. Some cells can function for short periods without oxygen, but most die. Some organisms, such as certain bacteria, yeast and parasites, are anaerobes and can use GLUCOSE to make energy without the use of oxygen. This is called anaerobic respiration and is less efficient than aerobic respiration, but because it produces alcohol and carbon dioxide it is of great use in the baking and brewing industry (*see* FERMENTATION).

respiratory chain *See* ELECTRON TRANSPORT SYSTEM.

restriction endonuclease, *restriction enzyme* An ENZYME, derived from a bacterium, that cuts a chain of DNA between specific NUCLEOTIDE base sequences. Many such enzymes exist that are specific for different nucleotide sequences. They are all nuclease enzymes. Any fragment of DNA produced in this way can be joined to other DNA by the use of another enzyme called DNA LIGASE. Hence manipulation of DNA is possible, which is a valuable technique in genetic engineering.

restriction enzyme *See* RESTRICTION ENDONUCLEASE.

retinal, *retinene* ($C_{20}H_{28}O$) An ALDEHYDE derivative of VITAMIN A that forms part of the visual pigment RHODOPSIN. It is an intermediate in the conversion of β-CAROTENE to vitamin A in the mammalian liver.

retinene *See* RETINAL.

retinol *See* VITAMIN A.

retort A vessel used for DISTILLATION, traditionally in the form of a glass bulb with a long stem.

reverberatory furnace A device for heating solid materials, particularly metal ores, in which a mixture of gas and air is burned above a hearth containing the material to be heated. Gas and air are fed in at one end and leave through a chimney, or flue, at the other. The furnace has a low roof, lined with a REFRACTORY material to direct as much heat as possible onto the contents of the hearth.

reverse osmosis The passage of a solvent through a SEMIPERMEABLE MEMBRANE from a region of high SOLUTE concentration to one lower – that is, in the reverse direction to OSMOSIS. Reverse osmosis occurs on the application of a pressure greater than the OSMOTIC PRESSURE.

reversible reaction Any CHEMICAL REACTION that can proceed in either direction. For example the reaction:

$$3H_2 + 2N_2 \Leftrightarrow 2NH_3$$

is said to be reversible as nitrogen and hydrogen can combine to form ammonia, or ammonia can dissociate to form nitrogen and hydrogen. All reactions are to some extent reversible, but a reaction is considered reversible if the equilibrium state, where the reaction proceeds in both directions at the same rate, is reached with comparable concentrations of the reagents and products. The position of the equilibrium will vary depending on the conditions of the reaction, such as temperature and pressure. *See also* CHEMICAL EQUILIBRIUM, EQUILIBRIUM CONSTANT, FREE ENERGY, LE CHATELIER'S PRINCIPLE.

RF value In CHROMATOGRAPHY, the ratio between the distance moved by the leading edge of the SOLVENT (the solvent front) and a specified material. For a particular solvent and stationary phase at a constant temperature, the RF value is constant, so an unknown material can be identified by measuring its RF value.

rhenium (Re) The element with atomic number 75; relative atomic mass 186.2; melting point 3,180°C; boiling point 5,620°C; relative density 22.0. It is a TRANSITION METAL, whose hardness and high melting point have led to its use in some alloys.

rheology The study of the deformation and flow of matter, especially plastic solids and viscous liquids.

rheopexy The property of some THIXOTROPIC SOLS, such as calcium sulphate in water (plaster of Paris), that set far more rapidly if gently stirred or tapped.

rhodium (Rh) The element with atomic number 45; relative atomic mass 102.9; melting point 1,966°C; boiling point 3,727°C; relative density 12.4. Rhodium is chemically fairly unreactive. Rhodium salts have a characteristic red colour.

rhodopsin A light-sensitive pigment found in the eye. It consists of the protein opsin and a form of RETINAL.

riboflavin, *vitamin B_{12}* ($C_{17}H_{20}N_4O_6$) A VITAMIN of the VITAMIN B COMPLEX. Riboflavin is a precursor of FAD, which is particularly important in the ELECTRON TRANSPORT SYSTEM. It is found in liver, milk and egg whites. Deficiency causes mouth sores and swollen tongue.

ribonuclease *See* RNASE.

ribonucleic acid *See* RNA.

ribose ($C_5H_{10}O_5$) A PENTOSE sugar that is a component of RNA.

RNA, *ribonucleic acid* A NUCLEIC ACID associated mainly with the synthesis of PROTEINS from DNA. RNA is found in the nucleus and cytoplasm of cells. It is usually a single-stranded chain of NUCLEOTIDES synthesized from DNA by the formation of BASE PAIRS. The organic bases in RNA are adenine, guanine, cytosine and uracil (which replaces the thymine of DNA) and a PENTOSE sugar that is always RIBOSE. There are three main forms of RNA, all concerned with PROTEIN SYNTHESIS. These are ribosomal RNA (rRNA), transfer RNA (tRNA) and messenger RNA (mRNA). In some viruses, RNA can make up the hereditary material, instead of DNA.

RNA polymerase *See* POLYMERASE.

RNAse (ribonuclease) One of many enzymes that hydrolyse RNA by breaking down the sugar–phosphate bonds. *See also* RESTRICTION ENDONUCLEASE.

roasting The process of heating an ore in air. This can be used to extract some impurities, such as sulphur and sulphur dioxide.

Rochelle salt, *sodium potassium tartrate tetrahydrate* ($KOOC.CHOH.CHOH.COONa.4H_2O$) A white crystalline solid; melting point 74°C; decomposes on further heating. It is traditionally used for its PIEZOELECTRIC properties, though its high solubility is a disadvantage in this context, so quartz or synthetic materials are more often used.

röntgen, *roentgen* Former unit for DOSE of IONIZING RADIATION. One röntgen is approximately 8.7×10^{-7} GRAY.

root mean square A form of average in which a quantity is squared, the mean value found and then the square root taken.

rubber A natural or synthetic POLYMER that has elasticity at room temperature. Natural rubber comes mostly from the tree *Hevea brasilienis* where it exudes from grooves cut in the trunk. It is then coagulated with methanoic or ethanoic acid to form a solid rubber. It is treated before use by a process called VULCANIZATION. This yields a harder, stronger end-product that is less affected by external temperatures than untreated rubber. Most of the natural supply of rubber comes from Malaysia. There is a large variety of synthetic rubbers that are cheaper than natural rubber and can be adapted to special purposes. For example SBR, styrene-butadiene rubber, is used for car tyres, shoe soles and can be blended with natural rubber.

rubidium (Rb) The element with atomic number 37; relative atomic mass 85.5; melting point 38°C; boiling point 688°C; relative density 1.5. Rubidium is an ALKALI METAL, highly reactive and combining violently with water or oxygen in the atmosphere.

ruby A transparent red gemstone form of aluminium oxide, the red colour being caused by chromium impurities. *See also* CORUNDUM.

rust (*n., vb.*) The powdery red material formed when iron CORRODES. The chemical nature of rust is complex, but it is essentially a mixture of HYDRATED iron(III) oxide, Fe_2O_3 with iron(III) oxyhydroxide, FeO(OH).

ruthenium (Ru) The element with atomic number 44; relative atomic mass 101.1; melting point 2,310°C; boiling point 3,900°C; relative density 12.4. Ruthenium is a TRANSITION METAL. It often occurs in association with platinum. It is used in some platinum alloys and as a catalyst.

Rutherford–Bohr atom A model of the ATOM, incorporating the idea of an atomic NUCLEUS and the idea that ELECTRONS only occupy certain ENERGY LEVELS with quantized values of angular momentum. *See also* BOHR THEORY, HYDROGEN SPECTRUM.

rutherfordium The name proposed for the element with ATOMIC NUMBER 104, also called unnilquadium. It has only radioactive ISOTOPES, with very short HALF-LIVES. It is synthesized by bombarding heavy nuclei with fast moving ions. Its discovery was first claimed by a Russian group in 1964, but its existence was not independently confirmed until 1967. Little is known about its physical or chemical properties.

rutile A mineral form of TITANIUM DIOXIDE, TiO_2.

Rydberg constant The constant R in the RYDBERG EQUATION, equal to $1.10 \times 10^7\,m^{-1}$.

Rydberg equation An equation that gives the wavelengths λ, of the lines in the EMISSION SPECTRUM of the hydrogen atom.

$$1/\lambda = R\,(1/n^2 - 1/m^2)$$

where R is the RYDBERG CONSTANT and n and m are positive integers, with m greater than n.

S

saccharide *See* SUGAR.

saccharin, *ortho-sulpho benzimide* ($C_7H_5NO_3S$) An artificial sweetener derived from AMINES. It is a white solid 500 times sweeter than sugar but with a bitter aftertaste and potentially carcinogenic (*see* CARCINOGEN). It has largely been replaced by other sweetening agents.

sacrificial cathode A piece of fairly reactive metal, such as zinc, attached to a steel structure exposed to moisture. The sacrificial cathode CORRODES in preference to the steel. *See* SACRIFICIAL CORROSION.

sacrificial corrosion A CORROSION process in which two pieces of metal are in electrical contact. Surrounding moisture sets up an electrochemical CELL between the two metals and the one with the more negative ELECTRODE POTENTIAL (that is, the more reactive metal), called a SACRIFICIAL CATHODE, is attacked by corrosion far more rapidly than the other metal.

Zinc is used as a sacrificial cathode to prevent iron from corroding. In steel ships for example, a piece of zinc is attached to the outside of the hull and replaced as it is eroded by corrosion. *See also* GALVANIZING.

salicylic acid, *2-hydroxybenzoic acid* (HOC_6H_4COOH) An organic acid that occurs as white, crystalline needles when pure; melting point 159°C. At 200°C it decomposes to phenol and carbon dioxide. Salicylic acid is the active ingredient in aspirin and is also used in dyes. It occurs naturally in willow tree bark and oil of wintergreen.

salt Any compound produced by a reaction in which some or all of the hydrogen in an ACID is replaced by a metal or other positive ion. An example is the action of sulphuric(VI) acid on magnesium. Hydrogen is displaced from the acid to produce magnesium sulphate, a salt:

$$Mg + H_2SO_4 \rightarrow MgSO_4 + H_2$$

Salts are also formed when an acid reacts with a base, for example:

$$2HCl + Ca(OH)_2 \rightarrow CaCl_2 + 2H_2O$$

In general,

$$\text{acid} + \text{base} \rightarrow \text{salt} + \text{water}$$

See also ACIDIC HYDROGEN, ACIDIC SALT, BASIC SALT, NORMAL SALT.

salt bridge A connection between two HALF-CELLS, comprising a piece of porous material, such as filter paper, or a gel-filled glass U-tube, with a solution of a salt, usually potassium chloride, forming the conducting path between the two half-cells.

salting out The process of removing an organic substance from solution in water or ethanol by the addition of a concentrated sodium chloride solution. This increases the POLAR nature of the solvent and reduces the solubility of non-polar compounds.

saltpetre *See* POTASSIUM NITRATE.

samarium (Sm) The element with atomic number 62; relative atomic mass 150.3; melting point 1,073°C; boiling point 1,791°C; relative density 7.5. Samarium is one of the more widely occurring of the LANTHANIDE metals. It is used for its catalytic properties and in some FERROMAGNETIC alloys.

sand Small particles of rock, usually comprised mainly of QUARTZ, formed by the weathering of quartz-rich rocks. Sand particles are defined as having diameter in the range 0.06–2.0 mm.

Sandmeyer reaction *See* DIAZONIUM SALTS.

sandstone A SEDIMENTARY rock formed by the deposition of layers of sand, fused into rock by the pressure from further layers. Sandstone is porous and permeable and used primarily as a building material.

sandwich compound A compound in which a TRANSITION METAL atom or ion is trapped between two rings of carbon atoms, for example ferrocene, $Fe(C_5H_5)_2$. Such compounds are bound by the interaction between the electrons in the P-ORBITALS and the D-ORBITALS of the iron atoms.

saponification A process used in the manufacture of SOAP. It consists of the HYDROLYSIS of an ESTER by treatment with a strong ALKALI. The ester is split to yield the ALCOHOL from which the ester was derived and a salt of the constituent FATTY ACID.

sapphire Any of a number of gemstone forms of CORUNDUM, particularly the blue form, where the colour arises from cobalt impurities.

saturated (*adj.*) Describing a SOLUTION or VAPOUR that can hold no more dissolved or evaporated substance of a specified type. A saturated solution is one in which the dissolved substance is in perfect EQUILIBRIUM with the undissolved substance. This is an example of a dynamic equilibrium: the rate at which solute molecules come out of solution is exactly balanced by the rate at which they dissolve. A solution with more than the equilibrium amount of solute is said to be supersaturated. *See also* SOLUBILITY PRODUCT.

saturated compound Any organic compound in which the carbon atoms are linked by single COVALENT BONDS only, such as ethane and other ALKANES. They therefore can only react further by SUBSTITUTION REACTIONS. *Compare* UNSATURATED COMPOUNDS.

saturated vapour The state of a VAPOUR whose PARTIAL PRESSURE is equal to its SATURATED VAPOUR PRESSURE, the maximum density that the vapour can have at that temperature. If a saturated vapour is cooled, the liquid will condense.

saturated vapour pressure The PRESSURE in the VAPOUR above a liquid at which the molecules leave and re-enter the liquid at the same rate. The saturated vapour pressure increases with temperature. Once the saturated vapour pressure reaches the pressure of the atmosphere above the liquid, bubbles can form and the liquid will be at its boiling point. *See also* CLAPEYRON CLAUSIUS EQUATION, RAOULT'S LAW, SUPERSATURATED VAPOUR.

s-block element Any element in GROUPS 1 or 2 of the PERIODIC TABLE, with the outer electrons in an S-ORBITAL.

SBR (styrene-butadiene rubber) An artificial RUBBER that is a copolymer (*see* COPLYMERIZATION) made from BUTADIENE combined with PHENYLETHENE (styrene). It is rubbery in texture, similar to natural rubber, and is the most important synthetic rubber.

scandium (Sc) The element with atomic number 21; relative atomic mass 45.0; melting point 1,540°C; boiling point 2,850°C; relative density 3.0. Scandium has some use as a catalyst and in the manufacture of REFRACTORY ceramics.

scanning electron microscope A type of ELECTRON MICROSCOPE developed in the 1960s. A fine beam of electrons passes over the surface of the specimen; some electrons are absorbed and others reflected. Secondary electrons may also be emitted by the specimen and these and the reflected electrons are amplified to form an image showing the three-dimensional exterior of the specimen on a screen. The resolving power is not as good as in the TRANSMISSION ELECTRON MICROSCOPE (about 10 nm) and the overall magnification is 10–200,000 times. In the scanning tunnelling microscope a small tungsten probe is passed over the surface of the specimen and electrons jump (or tunnel) between the specimen and probe. This allows images of 100 million times to be obtained, so that individual atoms can be resolved.

scheelite A mineral form of calcium tungstate, $CaWO_4$, mined as an ore of tungsten.

Schiff's bases Weakly basic compounds prepared from the reaction of AROMATIC AMINES with ALIPHATIC or aromatic ALDEHYDES and KETONES.

Schiff's reagent A solution that distinguishes ALDEHYDES and KETONES. ALIPHATIC aldehydes and aldose sugars (*see* MONOSACCHARIDES) cause the colourless solution to change to magenta. In AROMATIC aldehydes and aliphatic ketones the colour develops more slowly. Aromatic ketones do not react.

Schrödinger's equation An equation that provides information about the wave nature of particles and describes the behaviour of a quantity called the WAVEFUNCTION. The wavefunction represents a PROBABILITY AMPLITUDE, the square of which is a measure of the probability of finding the particle at a given point.

scrubber A vessel in which water is passed through a mixture of gasses to remove a soluble component from the gas mixture. In the production of COAL GAS, for example, water is used in a scrubber to remove ammonia, which is often present as an impurity.

seam A layer of a particular MINERAL, particularly coal, that is thick enough to be exploited by mining.

second The SI UNIT of time. One second is defined as being equal to the time taken for 9,192,631,770 oscillations of the electromagnetic radiation produced in a transition between two specified ENERGY LEVELS of an atom of caesium–133.

secondary cell Former term for a rechargeable CELL.

second law of thermodynamics Any change will bring about an increase in the total ENTROPY of the system if the change is irreversible, or produces no entropy change if it is reversible. Systems that appear to produce order out of chaos, such as living organisms, do not actually result in a net decrease in entropy, as they give out sufficient heat to increase the entropy of their surroundings by an amount that more than compensates for the decrease in entropy of the living organism itself. *See also* CLAUSIUS STATEMENT OF THE SECOND LAW OF THERMODYNAMICS, KELVIN STATEMENT OF THE SECOND LAW OF THERMODYNAMICS.

sedimentary (*adj.*) Describing a rock that has formed from the action of pressure forcing together small particles of other rocks that had been broken down by the forces of erosion.

sedimentation The process of separating small particles from a SUSPENSION by allowing them to settle under their own weight, or with assistance from a CENTRIFUGE.

seed crystal A small crystal added to a supersaturated solution to start the process of CRYSTALLIZATION. Once the seed crystal is added, ions leave the solution under the influence of the electric field surrounding the ions already in the lattice of the seed crystal.

selenide Any BINARY COMPOUND containing selenium and a more electropositive element (*see* ELECTROPOSITIVITY). Hydrogen selenide is covalent, whilst the selenides of the ALKALI METALS have ionic structures. Selenides of the TRANSITION METALS tend to be NON-STOICHIOMETRIC COMPOUNDS.

selenium (Se) The element with atomic number 34; relative atomic mass 79.0; melting point 217°C; boiling point 685°C; relative density 4.8. Selenium is a METALLOID, and its ability to conduct electricity on exposure to light has led to its use in some photoconductive cells.

semicarbazones Organic compounds containing the group $=C:N.NH.CONH_2$. They are formed by reacting ALDEHYDES or KETONES with a semicarbazide ($H_2N.NH.CO.NH_2$). Semicarbazones are used to isolate and identify aldehydes and ketones.

semiconductor A material intermediate between a CONDUCTOR and an INSULATOR, which conducts electricity but not very well. By far the most important semiconductor is silicon, the material from which many electronic devices, particularly integrated circuits, are made.

Pure semiconductors do not conduct at all at very low temperatures, but thermal vibrations of the lattice, or the addition of certain impurities, can make the material conduct. The addition of impurities, a process called doping, must be carefully controlled if the semiconductor is to have predictable properties. In the case of silicon, the lattice is held together by each atom forming four COVALENT BONDS with its neighbours. In such a structure there are no free CHARGE CARRIERS. The addition of an impurity with five VALENCE ELECTRONS, called a donor impurity, can release a FREE ELECTRON that can carry charge through the material. Such a material is called an n-type semiconductor. An impurity with only three valence electrons, called an acceptor impurity, will create a gap in the electron structure of the lattice, which may be filled by an electron from a neighbouring bond, In effect, the shortage of an electron, called a hole, moves through the lattice like a positive charge carrier. A semiconductor in which current is carried mostly by holes is called a p-type semiconductor.

The action of thermal vibrations or light will release electrons from the lattice structure, meaning that semiconductors conduct better at high temperatures, or on exposure to light.

See also BAND THEORY.

semi-metal *See* METALLOID.

semipermeable membrane A material through which one type of molecule can pass but not another. Typically, such membranes are considered in processes such as OSMOSIS, where the SOLVENT molecules can pass through the membrane, but the SOLUTE cannot.

One explanation of this effect is that the solute molecules are too large too pass through the membrane, which acts as a 'molecular sieve'. However, some membranes still work with IONIC solutions where the solute ions are smaller than the solvent molecules, suggesting that this cannot be the explanation for every case.

semi-water gas, *Mond gas* A mixture of WATER GAS and PRODUCER GAS formed by mixing air and steam and passing them over hot COKE. If the balance of the gases is right, a reasonable fuel gas can be produced without having to supply any extra energy to maintain the temperature of the coke.

sequestration The process of forming a complex around an ion in solution, particularly in order to reduce the chemical activity of the ion. An example is the sequestration of calcium ions in WATER SOFTENING. Sequestration is also a way of supplying ions in a protected form, for example sequestered iron is used to deliver iron to plants in an alkaline soil, where the iron would otherwise be from a precipitate within the soil. The process of CHELATION is another example of sequestration. *See also* COMPLEX ION.

serine ($C_3H_7NO_3$) An AMINO ACID found in many proteins.

serotonin, *5-hydroxytryptamine* A substance, derived from the amino acid TRYPTOPHAN, that is secreted by the brain and acts as a NEUROTRANSMITTER. It causes a general lack of activity (opposes NORADRENALINE), influences mood and also induces the constriction of blood vessels. The hallucinogenic drug LSD acts as an antagonist of serotonin.

shale A SEDIMENTARY rock formed by the deposition of layers of mud, fused into rock by pressure from further layers.

shell A series of atomic ORBITALS of roughly similar energies. Completing the filling of one shell and starting to fill the next results in a sudden change in chemical properties, atomic size, IONIZATION ENERGY, etc. This represents the start of a new PERIOD in the PERIODIC TABLE. The shells are labelled K, L, M, N, O, P. The K-shell is closest to the nucleus and can hold two electrons, the next shell is called the L-shell and holds a further eight electrons. The *n*th shell out from the nucleus can hold $2n^2$ electrons.

Elements with a full outermost shell of electrons are particularly stable; these are the NOBLE GASES. Elements that have a few electrons beyond a full shell will tend to lose them to form positive ions (for example sodium, which has 11 electrons, one beyond a full shell). Those with a few electrons too few, will tend to gain electrons to complete a shell, forming negative ions (for example fluorine with nine electrons, one short of a full shell). Those that are not close to a full shell tend to form COVALENT BONDS (carbon for example, with six electrons, four short of a full shell, will form four covalent bonds) or else exhibit more complex behaviour, for example the TRANSITION METALS show multiple VALENCY.

The shell model does not take full account of the energy differences within a single shell and so does not fully account for the behaviour of the elements of higher ATOMIC NUMBERS, where a full consideration of the orbitals (sometimes called sub-shells) must be used. However, the shell model provides a useful description of the X-ray spectra of all elements (*see* SPECTRUM), which depend on the motion of electrons in the innermost shells.

shell model A model of the structure of the atomic NUCLEUS that treats individual PROTONS and NEUTRONS as existing in ENERGY LEVELS, similar in pattern to those occupied by electrons in an atom. The shell model works well for small nuclei, but the greater complexity of the interactions in the nucleus mean that the model breaks down for larger nuclei, for which the LIQUID DROP MODEL is more often used.

sherardizing A process of providing a protective coating of zinc on an iron object by coating it in zinc powder and then heating. The zinc forms a layer of zinc-iron alloy coated with pure zinc. *See also* GALVANIZING.

short period Any of the first three PERIODS in the PERIODIC TABLE, which do not contain any TRANSITION METALS. The short periods are periods 1, 2 and 3, and involve elements with electrons only in S-ORBITALS and P-ORBITALS.

side reaction A reaction between a set of reagents that proceeds alongside the main reaction between those regents and which produces different products. Many organic reactions have side reactions, which mean that some additional process must take place before the product can be obtained in a sufficiently pure form.

Siemens–Martin process The process of producing steel from a mixture of steel scrap and haematite (an ore containing iron oxide) in an OPEN HEARTH FURNACE. The iron oxide oxidizes some of the carbon in the steel.

sigma-bond, *σ-bond* A COVALENT BOND that has an electron distribution symmetrical about the line joining the bound atoms. Single covalent bonds are sigma bonds. *See also* PI-BOND.

silane 1. (SiH_4) A colourless gas; melting point –184°C; boiling point –112°C. Silane can be produced by the reaction of magnesium silicide with acids:

$$Mg_2Si + 4HCl \rightarrow SiH_4 + 4HCl$$

Silane is a powerful REDUCING AGENT, and ignites spontaneously on contact with air:

$$SiH_4 + 2O_2 \rightarrow SiO_2 + 2H_2O$$

It is used in the semiconductor industry to deposit silicon in the manufacture of integrated circuits.

2. A generic term for any compound containing silicon and hydrogen, with the general formula Si_nH_{2n+2}, for example disilane, Si_2H_6. They are all highly unstable.

silica A common mineral, composed mostly of SILICON DIOXIDE, SiO_2. Sand is composed mostly of fine grains of silica.

silica gel A water-absorbent material made by drying a GEL of sodium silicate in water. The dried gel will absorb moisture from the air, and can be regenerated by heating. Cobalt chloride is often included with the gel as an INDICATOR, turning from blue to pink as the gel hydrates.

silicate Any compound of SILICA, SO_2, and one or more metal ions. Natural silicates form the major components of most rocks.

silicide Any BINARY COMPOUND containing silicon and a more electropositive element (*see* ELECTROPOSITIVITY), for example magnesium silicide, Mg_2Si.

silicon (Si) The element with atomic number 14; relative atomic mass 28.1; melting point 1,410°C; boiling point 2,355°C; relative density 2.3. Silicon is widespread in the Earth's crust and core. It is a METALLOID and a SEMICONDUCTOR. The most important use of silicon is in the electronics industry, where relatively small amounts of highly pure silicon are used for the manufacture of integrated circuits (silicon chips). The production of large single crystal silicon wafers has been at the forefront of many recent developments in materials technology.

silicon carbide (SiC) A black solid; melting point 2,700°C; decomposes on further heating; relative density 3.2. Silicon carbide is made by heating silicon oxide with carbon:

$$SiO_2 + 2C \rightarrow SiC + CO_2$$

Silicon carbide is a very hard material and is widely used in ABRASIVES.

silicon dioxide (SiO_2) A colourless glassy solid; melting point 1,610°C; boiling point 2,230°C; relative density 2.3. Various impure forms of silicon dioxide are extremely common in the Earth's crust. Silicon dioxide is widely used in the manufacture of glass.

silicones Organic POLYMERS derived from silicon. They are used in greases, sealing compounds, resins, synthetic rubber.

silicon oxide Any one of the BINARY COMPOUNDS of silicon and oxygen, in particular silicon(II) oxide, SiO, and silicon(IV) oxide, SiO_2. Silicon(II) oxide is believed to exist only as gaseous molecules; on condensing it disproportionates to silicon(IV) oxide and silicon. Silicon monoxide can be formed by the partial REDUCTION of an excess of silicon dioxide with carbon,

$$SiO_2 + C \rightarrow SiO + CO$$

See also SILICON DIOXIDE.

siloxane Any of a group of organic compounds containing the FUNCTIONAL GROUP –O–Si–O–

silver (Ag) The element with atomic number 47; relative atomic mass 107.9; melting point 962°C; boiling point 2,212°C; relative density 10.5. Silver is easily worked and tarnishes fairly slowly, which has lead to its widespread use in the manufacture of jewellery and other ornamental work. It is also a very good conductor of heat and electricity, having the lowest resistivity of any metal.

Silver HALIDES are decomposed by the action of light, and photographic developers can be used to amplify this change, enabling small amounts of light to control large changes. This chemistry is the basis of the photographic process.

silver bromide (AgBr) A pale yellow solid; melting point 432°C; decomposes on further heating; relative density 6.5. Silver bromide may be precipitated from other silver salts, for example:

$$AgNO_3 + KBr \rightarrow AgBr + KNO_3$$

Silver bromide is sensitive to light, and a suspension will slowly decompose to give a black precipitate of finely divided silver. It is less light sensitive than silver iodide, but more sensitive than silver chloride.

silver chloride (AgCl) A white solid; melting point 455°C; boiling point 1,350°C; relative density 5.6. Silver chloride can be precipitated from other silver salts, for example:

$$AgNO_3 + KCl \rightarrow AgCl + KNO_3$$

It decomposes slowly on exposure to light, a suspension of silver chloride slowly turning black due to the production of a fine precipitate of metallic silver. Silver chloride will dissolve readily in dilute ammonia due to the formation of the silver diamine complex, $[Ag(NH_3)_2]^+$.

silver iodide (AgI) A yellow solid; melting point 556°C; boiling point 1,506°C. Silver iodide can be precipitated from other silver salts, for example:

$$AgNO_3 + KI \rightarrow AgI + KNO_3$$

It is the most light sensitive of the silver halides and rapidly decomposes on exposure to light, a fact that has led to its widespread use in photographic emulsions.

silver nitrate ($AgNO_3$) A colourless solid; melting point 212°C; decomposes on further heating; relative density 4.3. Silver nitrate is important as it is the only soluble salt of silver. It is widely used in the photographic industry to prepare silver iodide. It is also used in QUALITATIVE ANALYSIS to test for halides, forming a white or yellow precipitate, soluble in dilute ammonia in the case of a chloride, soluble in concentrated ammonia for a bromide, and insoluble in ammonia for an iodide.

silver oxide Either of the BINARY COMPOUNDS of silver and oxygen, silver(I) oxide, Ag_2O, and silver(II) oxide, AgO. Silver(I) oxide is a brown solid; decomposes on heating; relative density 7.1. It can be formed as a precipitate in the reaction of sodium hydroxide with silver(I) nitrate,

$$2AgNO_3 + 2NaOH \rightarrow Ag_2O + 2NaNO_3 + H_2O$$

Silver(II) oxide is a black powder; decomposes on heating; relative density 7.4. It is formed by reacting silver(I) oxide with ozone,

$$Ag_2O + O_3 \rightarrow 2AgO + O_2$$

It is paramagnetic (*see* PARAMAGNETISM), and is believed to be more correctly described as being a mixed oxide, silver(I) silver(III) oxide.

single bond A COVALENT BOND formed by the overlap of a single pair of ORBITALS, a SIGMA-BOND with no PI-BOND.

singlet A line in the absorption or emission SPECTRUM of an atom, ion or molecule that has no FINE STRUCTURE. *Compare* MULTIPLET.

sintering The process of forming a solid object by heating a powder, especially a metal powder, under pressure in a mould.

sitosterol, ***24-(24-R)ethylcholesterol*** ($C_{29}H_{50}O$) A major plant STEROL; melting point 137°C. It is difficult to separate from other related sterols.

SI unit The unit of measurement in the internationally agreed METRIC SYSTEM (*Système International*). In any physical equation, if all quantities are substituted into the equation in SI units, the result will also be in SI units. All units in the SI system are expressed in terms of seven base units and two supplementary units. The base units are metre (length), second (time), kilogram (mass), ampere (current), kelvin (temperature), mole (amount of substance) and candela (luminous intensity). The supplementary units are the radian (angle) and steradian (solid angle).

Any quantity that cannot be expressed directly in terms of one of these units can be expressed in terms of a derived unit, such as the metre per second for velocity. Eighteen of the derived units are given special names: the newton (force), pascal (pressure), joule (energy), watt (power), coulomb (charge), farad (capacitance), ohm (resistance), siemens (conductance), volt (potential difference), hertz (frequency), tesla (magnetic field strength), weber (magnetic flux), henry (magnetic inductance), lumen (luminous flux), lux (illuminance), becquerel (activity), gray (radiation dose) and sievert (radiation dose equivalent).

slag Non-metallic impurities released in smelting metals. These impurities are often removed by the addition of a suitable material, such as calcium carbonate, which reacts to form a molten layer on top of the metal.

The term slag particularly refers to impurities extracted from iron ore and produced as a waste product in a BLAST FURNACE. The waste material is mostly in the form of SILICATES and SILICA, which combine with the calcium in the calcium carbonate to form calcium silicate.

slaked lime *See* CALCIUM HYDROXIDE.

slip plane In a metallic CRYSTAL, a surface along which the layers of atoms can move relatively easily. Thus a metal crystal is more DUCTILE when being pulled along a slip plane than at an angle to one. The POLYCRYSTALLINE nature of many metal samples makes the presence of slip planes less obvious than the CLEAVAGE PLANES in non-metallic crystals.

smelt (*vb.*) To extract a metal from its ore by heating in combination with a suitable REDUCING AGENT.

SNG *See* SYNTHETIC NATURAL GAS.

soap A cleansing agent consisting of a mixture of sodium salts of various FATTY ACIDS, usually sodium stearate, sodium oleate and sodium palmitate. It is made by the action of a strong ALKALI (for example sodium hydroxide or potassium hydroxide) on animal or vegetable fats, by the process of SAPONIFICATION.

soda A term now largely obsolete, used in the common names of many sodium compounds, such as caustic soda, which is sodium hydroxide, and washing soda (sodium carbonate).

soda lime A mixture of calcium hydroxide and sodium hydroxide, formed by the addition of sodium hydroxide solution to calcium oxide, from which crystals of soda lime are crystallized. It is widely used in industry for the absorption of carbon dioxide and water vapour.

soda water *See* CARBONIC ACID.

sodium (Na) The element with atomic number 11; relative atomic mass 23.0; melting point 98°C; boiling point 892°C, relative density 1.0. Sodium is very common in the Earth's crust, occurring mostly as dissolved sodium chloride in sea water, from which it is extracted as sodium hydroxide by ELECTROLYSIS in a MERCURY-CATHODE CELL. Sodium is an ALKALI METAL, reacting with water and burning in air.

Sodium compounds produce a characteristic bright orange colour in a flame test, and this is also the colour of the gas discharge in sodium vapour, which is widely used as an electric light source more efficient than filament lamps. The vast majority of sodium compounds are soluble.

sodium acetate *See* SODIUM ETHANOATE.

sodium amide ($NaNH_2$) A white powder; melting point 210°C; boiling point 400°C. It is formed by heating sodium in a stream of ammonia,

$$2Na + 2NH_3 \rightarrow 2NaNH_2 + H_2$$

Sodium amide decomposes in water,

$$NaNH_2 + H_2O \rightarrow NaOH + NH_3$$

sodium benzenecarboxylate *See* SODIUM BENZOATE.

sodium benzoate, *sodium benzenecarboxylate* (C_6H_5COONa) A white powder, soluble in water and sparingly soluble in ethanol. It used to be used as an antiseptic and is now used in the DYE industry and as a food preservative.

sodium bicarbonate *See* SODIUM HYDROGENCARBONATE.

sodium bisulphate *See* SODIUM HYDROGENSULPHATE.

sodium bisulphite *See* SODIUM HYDROGENSULPHITE.

sodium carbonate (Na_2CO_3) A white solid; melting point 851°C; decomposes on further heating. Sodium carbonate is most frequently found as the decahydrate, $Na_2CO_3.10H_2O$, commonly known as washing soda, an efflorescent (*see* EFFLORESCENCE) colourless crystalline solid; relative density 1.4. Sodium carbonate is produced in the SOLVAY PROCESS. Sodium carbonate is widely used in water treatment as it removes dissolved calcium and magnesium salts from hard water. *See also* WATER SOFTENING.

sodium chlorate(V) ($NaClO_3$) A white solid; melting point 250°C; decomposes on further heating; relative density 2.5. Sodium chlorate is manufactured by the reaction between chlorine and concentrated sodium hydroxide:

$$6NaOH + 3Cl_2 \rightarrow NaClO_3 + 5NaCl + 3H_2O$$

It decomposes on heating:

$$2NaClO_3 \rightarrow 2NaCl + 3O_2$$

Sodium chlorate is an powerful OXIDIZING AGENT. It is used in matches and as a weedkiller.

sodium chloride, *common salt* (NaCl) A white solid; melting point 801°C; boiling point 1,413°C; relative density 2.2. Sodium chloride is the chief dissolved component of BRINE, from which it is extracted by evaporation. It is an essential component of the ELECTROLYTE structure of biological organisms, and is used commercially as the raw material for the manufacture of other sodium salts.

sodium cyanide (NaCN) A white solid; melting point 564°C; boiling point 1,496°C. Sodium cyanide is made by the reaction between hydrogen cyanide and sodium hydroxide:

$$HCN + NaOH \rightarrow NaCN + H_2O$$

Like most cyanides, sodium cyanide is highly toxic.

sodium dichromate ($Na_2Cr_2O_7$) An orange-red crystalline solid that decomposes on heating.

It is usually found as the DIHYDRATE, $Na_2Cr_2O_7.2H_2O$; relative density 2.5. It is widely used as an OXIDIZING AGENT and as a MORDANT in dying. It is used in QUANTITATIVE ANALYSIS, though being slightly HYGROSCOPIC it is less suitable than the more expensive POTASSIUM DICHROMATE.

sodium ethanoate, *sodium acetate* (CH_3COONa) A colourless, crystalline compound that occurs either as an anhydrate or a trihydrate. Both forms are soluble in water and sparingly soluble in ethanol. Sodium ethanoate is prepared by reacting ethanoic acid with sodium carbonate or sodium hydroxide. It is used in BUFFERS for pH control, in dye manufacture, soaps, drugs and in photography.

sodium fluoride (NaF) A white solid; melting point 993°C; boiling point 1,695°C; relative density 2.6. It can be manufactured by the reaction of sodium hydroxide with hydrogen fluoride:

$$NaOH + HF \rightarrow NaF + H_2O$$

It is used in the manufacture of ceramic glazes, and is introduced into drinking water at low concentrations, where it helps prevent tooth decay.

sodium hexaflouraluminate (Na_3AlF_6) A white crystalline solid; melting point 1,000°C; decomposes on further heating; relative density 2.9. It occurs naturally as the mineral CRYOLITE, and is an important in the extraction of aluminium.

sodium hydride (NaH) A white crystalline solid; decomposes on heating; relative density 0.92. It can be formed by heating hydrogen with sodium,

$$2Na + H_2 \rightarrow 2NaH$$

It is a powerful REDUCING AGENT, reacting violently with water,

$$NaH + H_2O \rightarrow NaOH + H_2$$

Its reducing power makes it an important reagent in many organic reactions. ELECTROLYSIS of sodium hydride dissolved in a molten potassium and lithium chloride mixture produces hydrogen at the ANODE, suggesting the presence of the H^- ion.

sodium hydrogencarbonate, *sodium bicarbonate* ($NaHCO_3$) A white solid; decomposes on heating; relative density 2.2. Sodium hydrogencarbonate can be prepared by passing carbon dioxide through sodium carbonate solution:

$$Na_2CO_3 + CO_2 + H_2O \rightarrow 2NaHCO_3$$

Sodium carbonate reacts with acids to liberate carbon dioxide, and is widely used as a remedy for excess stomach acid:

$$NaHCO_3 + HCl \rightarrow NaCl + H_2O + CO_2$$

It decomposes on heating, again releasing carbon dioxide:

$$NaHCO_3 \rightarrow NaOH + CO_2$$

This reaction is exploited in 'dry powder' fire extinguishers, which use sodium hydrogencarbonate propelled by carbon dioxide.

sodium hydrogensulphate, *sodium bisulphate* ($NaHSO_4$) A white crystalline solid; melting point 315°C; decomposes on further heating; relative density 2.4 (ANHYDROUS). It can be crystallized from a mixture containing equal numbers of moles of sodium hydroxide and sulphuric acid,

$$NaOH + H_2SO_4 \rightarrow NaHSO_4 + H_2O$$

Solutions are ACIDIC due to the formation of separate ions Na^+, SO_4^{2-} and H^+. On heating the solid form, it decomposes first to $Na_2S_2O_7$ (sodium pyrosulphate) then to sulphur trioxide, SO_3:

$$2NaHSO_4 \rightarrow Na_2S_2O_7 + H_2O$$

$$Na_2S_2O_7 \rightarrow Na_2O + 2SO_3$$

Sodium hydrogensulphate is used in the paper and textile industries.

sodium hydrogensulphite, *sodium bisulphite* ($NaHSO_3$) A white crystalline solid; decomposes on heating; relative density 1.48. It can be crystallized from a solution of sodium carbonate in which sulphur dioxide has been dissolved,

$$Na_2CO_3 + 2SO_2 + H_2O \rightarrow 2NaHSO_3 + CO_2$$

It decomposes on heating to give sodium sulphate, sulphur dioxide and sulphur:

$$4NaHSO_3 \rightarrow 2Na_2SO_4 + SO_2 + S + 2H_2O$$

Sodium hydrogensulphite is widely used as a sterilizing agent in the brewing industry and as a bleach.

sodium hydroxide, *caustic soda* (NaOH) A white soapy solid; melting point 318°C; boiling point 1,390°C; relative density 2.1. Sodium hydroxide is produced as a by-product of the ELECTROLYSIS of BRINE in MERCURY-CATHODE CELLS. It is a strong alkali, DELIQUESCENT and widely used in the manufacture of soap. It is extremely corrosive to living tissue, and even small splashes in the eye can cause permanent damage.

sodium iodide (NaI) A white solid; melting point 661°C; boiling point 1,304°C; relative density 3.7. It is formed by the reaction of sodium hydroxide with hydrogen iodide,

$$NaOH + HI \rightarrow NaI + H_2O$$

It is soluble in water and an aqueous solution of sodium iodide will dissolve iodine to give a brown solution containing the I_3^- ion. Sodium iodide is used in the photographic industry for the production of silver iodide emulsions and is an important source of iodine in the human diet.

sodium methanoate (HCOONa) A white solid; melting point 253°C; decomposes on further heating; relative density 1.9. It can be prepared by reacting methanoic acid with sodium hydroxide,

$$HCOOH + NaOH \rightarrow HCOONa + H_2O$$

On heating, it decomposes with the release of carbon monoxide,

$$HCOONa \rightarrow NaOH + CO$$

sodium nitrate ($NaNO_3$) A white solid; melting point 306°C; decomposes on further heating; relative density 2.3. Sodium nitrate can be prepared by the reaction between sodium carbonate and nitric acid:

$$Na_2CO_3 + 2HNO_3 \rightarrow 2NaNO_3 + H_2O + CO_2.$$

Sodium nitrate decomposes on heating:

$$4NaNO_3 \rightarrow 2Na_2O + 2N_2O_4 + O_2$$

It is used as a fertilizer.

sodium nitrite ($NaNO_2$) A yellow crystalline solid; melting point 271°C; decomposes on further heating; relative density 2.2. It can be formed by heating sodium nitrate:

$$2NaNO_3 \rightarrow 2NaNO_2 + O_2$$

Sodium nitrite is used as a reagent in DIAZOTIZATION reactions.

sodium peroxide (Na_2O_2) A creamy white solid; decomposes on heating; relative density 2.8. Sodium peroxide is formed by burning sodium in excess oxygen:

$$2Na + O_2 \rightarrow Na_2O_2$$

It decomposes in water to give sodium hydroxide and hydrogen peroxide:

$$Na_2O_2 + 2H_2O \rightarrow 2NaOH + H_2O_2$$

It is a powerful OXIDIZING AGENT, and releases oxygen on heating:

$$2Na_2O_2 \rightarrow 2Na_2O + O_2$$

sodium potassium tartrate tetrahydrate *See* ROCHELLE SALT.

sodium pump, ***cation pump*** A mechanism in the cell membrane of animal cells in which sodium ions (Na^+) are pumped out of the cell and potassium ions (K^+) are pumped into the cell by ACTIVE TRANSPORT.

sodium sulphate (Na_2SO_4) A white powder; melting point 888°C; decomposes on further heating; relative density 2.7. Sodium sulphate is formed in the reaction between sodium chloride and concentrated sulphuric acid:

$$NaCl + H_2SO_4 \rightarrow Na_2SO_4 + 2HCl$$

It is used in the manufacture of glass.

sodium sulphide (Na_2S) An orange solid; melting point 1,180°C; decomposes on further heating, relative density 1.9. Sodium sulphide can be formed by the REDUCTION of sodium sulphate with carbon:

$$Na_2SO_4 + 2C \rightarrow Na_2S + 2CO_2$$

It is hydrolysed by water (*see* HYDROLYSIS):

$$Na_2S + 2H_2O \rightarrow 2NaOH + H_2S$$

It is used in the paper industry to soften woodpulp.

sodium sulphite (Na_2SO_3) A white solid; decomposes on heating; relative density 2.6. Sodium sulphite is made by the reaction between sulphur dioxide and sodium carbonate:

$$Na_2CO_3 + SO_2 \rightarrow Na_2SO_3 + CO_2$$

It is a mild REDUCING AGENT and is used as a food preservative. It liberates sulphur dioxide on heating, or on reaction with strong acids:

$$Na_2SO_3 \rightarrow Na_2O + SO_2$$

$$Na_2SO_3 + 2HCl \rightarrow 2NaCl + H_2O + SO_2$$

sodium thiosulphate, *hypo* ($Na_2S_2O_3$) A white solid; decomposes on heating; relative density 1.7. It is more frequently encountered as clear crystals of the PENTAHYDRATE, $Na_2S_2O_3.5H_2O$; melting point 42°C. Sodium thiosulphate is manufactured by the reaction between sulphur dioxide, sulphur and sodium hydroxide:

$$SO_2 + S + 2NaOH \rightarrow Na_2S_2O_3 + H_2O$$

Its chief use is in photographic processing, where its ability to dissolve unexposed silver halides from a photographic emulsion leads to its use as a photographic fixer.

soft water Water that is not HARD WATER; in other words, which does not contain dissolved calcium and magnesium SALTS.

sol A COLLOID in which solid particles are suspended in a liquid. *See also* HYDROSOL, LYOPHILIC, LYOPHOBIC.

solder A metal alloy, usually of tin and lead (soft solder), with a low melting point. Molten solder is used to join together metals with higher melting points, particularly for the purpose of forming permanent electrical connections. Higher melting point solders, which often contain silver (hard solder or silver solder), are used in applications where a stronger joint is needed or where a joint must retain its strength at high temperatures.

solid The state of matter in which a substance retains its shape. The molecules are closely packed, so a solid is not easily compressed, and rigidly held together. Solids may be either CRYSTALLINE, in which case the molecules are arranged in a regular lattice, or AMORPHOUS, in which there is no regular arrangement of the atoms. A crystalline solid has a definite melting point at which it becomes liquid, whereas an amorphous solid becomes increasingly pliable over a range of temperatures until it assumes liquid properties.

Crystalline solids can occur as CRYSTALS, with the lattice ordering being maintained over long distances to produce pieces of material with symmetrical shapes reflecting the ordered nature of the lattice. This is particularly the case with ionic materials.

See also BAND THEORY, COVALENT CRYSTAL, IONIC SOLID, POLYCRYSTALLINE.

solid solution A solid MIXTURE in which the atoms, ions or molecules of the constituents are entirely intermixed, rather than appearing as small crystals of each type of substance. Certain ALLOYS, such as those formed between gold and silver, are solid solutions. When such a solution is heated, it does not have a single melting point, but melts over a range of temperatures. *See* LIQUIDUS, SOLIDUS.

solidus In a SOLID SOLUTION, the line on a graph of temperature against composition below which the material is entirely solid.

solubility A measure of the amount of substance that can DISSOLVE in a given amount of SOLVENT. Usually measured in gram per decimetre cubed, or mole per decimetre cubed. The solubility of ionic salts in water generally increases with temperature, but gases become less soluble at high temperatures. *See also* HENRY'S LAW.

solubility product For a material A_xB_y that DISSOLVES to form a saturated IONIC solution, the solubility product K_s is

$$K_s = [A^{m+}]^x[B^{n-}]^y$$

where the square brackets, [ ] denotes CONCENTRATION. This value is constant for a SATURATED solution of a given salt at a specified temperature. If the actual value of this concentration product is less than the solubility product, the solution will not be saturated. If the value of this concentration product exceeds the solubility product, the solid will start to be formed as a PRECIPITATE. This explains the COMMON ION EFFECT: to precipitate solid A_xB_y, the concentration of B ions that is required can be reduced by increasing the concentration of A ions and vice versa.

soluble (*adj.*) Describing a material that will dissolve in a particular SOLVENT (usually water) to form a SOLUTION. Many, but by no means all, IONIC salts will dissolve in water, forming HYDRATED ions. The breaking up of the crystal lattice requires an input of energy that is obtained from the HYDRATION process. Almost all salts of ALKALI METALS are soluble, as are many salts of the TRANSITION METALS. The ALKALINE EARTH metals form salts with far lower solubilities. Organic solids are generally insoluble in POLAR solvents such as water. *See also* HEAT OF SOLUTION.

solute A solid that is dissolved to form a SOLUTION. *See* SOLUBLE.

solution A liquid that comprises a SOLVENT and a dissolved solid or gas. *See also* DISSOLVE, HEAT OF SOLUTION, SATURATED, SOLUBILITY, SOLUBILITY PRODUCT, SOLUBLE.

solvation The process of forming a SOLUTION, particularly the breaking down of an IONIC SOLID as it DISSOLVES in water.

Solvay process The reaction between sodium chloride solution and calcium carbonate to produce sodium carbonate:

$$2NaCl + CaCO_3 \rightarrow Na_2CO_3 + CaCl_2$$

This cannot be carried out directly since calcium carbonate is insoluble. Instead, sodium chloride solution (brine) is saturated with ammonia in a scrubber and trickled down a column called a SOLVAY TOWER, whilst the calcium carbonate is heated to form carbon dioxide:

$$CaCO_3 \rightarrow CaO + CO_2$$

A solution containing ammonium chloride is formed, whilst sodium hydrogencarbonate, which is less soluble, is formed as a PRECIPITATE. The rate of formation of the precipitate is increased by cooling the Solvay tower and by a COMMON ION EFFECT with the sodium ions in the brine. The precipitate is extracted by filtering, and heated to form sodium carbonate:

$$2NaHCO_3 \rightarrow Na_2CO_3 + H_2O + CO_2$$

The ammonium chloride is reacted with calcium oxide from the heated calcium carbonate, regenerating the ammonia and producing calcium chloride as a waste product:

$$CaO + 2NH_4Cl \rightarrow CaCl_2 + 2NH_3 + H_2O$$

Solvay tower A vessel used in the SOLVAY PROCESS. It comprises a tall column fitted with metal plates to enable a liquid flowing down the tower to react with a gas being passed up the tower.

solvent A liquid in which a substance will DISSOLVE to form a SOLUTION. The term solvent particularly refers to volatile organic liquids, such as ALKANES, in which a wide range of organic materials will dissolve.

solvent extraction A method of extracting a SOLUTE from a SOLUTION containing a mixture of materials by mixing the solution with a second solvent in which the material being extracted is more SOLUBLE. Organic molecules can be extracted from water by using an IMMISCIBLE organic solvent such as benzene. The PARTITION CONSTANT is used as a measure of how much more readily the material will dissolve in the new solvent than the original one.

somatotrophin *See* GROWTH HORMONE.

s-orbital The lowest energy ORBITAL for a given PRINCIPAL QUANTUM NUMBER. There is only a single s-orbital for each principal quantum number.

sorption pump A device for creating or improving a vacuum by the absorption of gas molecules into a solid, such as charcoal or a ZEOLITE.

spatula A laboratory tool shaped like a small spoon, or having a flat surface, used to handle small amounts of powdered material

specific gravity *See* RELATIVE DENSITY.

specific heat capacity The HEAT CAPACITY per unit mass of a substance. Specific heat capacity is measured in $J\,kg^{-1}K^{-1}$.

$$\text{Energy flow} = \text{mass} \times \text{specific heat capacity} \times \text{temperature change}$$

The specific heat capacity of water is unusually large at 4200 $J\,kg^{-1}K^{-1}$. This is due to the HYDROGEN BONDS in water, which absorb energy as the water is heated. The consequence of this is that water heats up and cools down more slowly than most other substances.

For solids and liquids it makes little difference whether the heat capacity is measured under conditions of constant pressure or constant volume, but gases expand substantially when they are heated under a constant pressure. The work then done in pushing back the atmosphere makes the specific heat capacity at constant pressure greater.

Specific heat capacities can be measured by finding the amount of heat needed to change the temperature of a certain quantity of the material by a measured amount. If the temperature of the container in which the specific heat capacity is being measured also changes during the experiment, it is important to take account of the energy involved in heating the container rather than the material under test. This is done by performing the measurement in a CALORIMETER.

See also MOLAR HEAT CAPACITY, RATIO OF SPECIFIC HEATS.

specific volume The reciprocal of DENSITY, the volume occupied by unit mass of a material, usually at a specified temperature and pressure.

spectral line A narrow range of wavelengths present in an EMISSION SPECTRUM or absent from an ABSORPTION SPECTRUM. The wavelength of a spectral line corresponds to the energy of a transition between two ENERGY LEVELS in the atom or ion which produced the line. *See also* FINE STRUCTURE, SPECTROSCOPY, SPECTRUM.

spectrochemical series A list of LIGANDS arranged in order according to the size of the effect they produce in LIGAND-FIELD THEORY. Experimentally it is found that the change in the ENERGY LEVELS of an ion produced by a particular ligand depends very little on the type of ion involved and far more on the nature of the ligand. Of the common ligands, the strongest effects are produced by CN^-, NO^{2-} and NH_3, whilst H_2O, OH^- and halogen ions produce successively weaker effects. The spectrochemical series can be used to make predictions about the effect of complex formation on the colour of ionic materials.

spectrometer An instrument for forming and recording a SPECTRUM. *See* SPECTROSCOPY.

spectroscope An instrument for forming a SPECTRUM and viewing it directly. *See* SPECTROSCOPY.

spectroscopy The study of the ELECTROMAGNETIC RADIATION produced by a sample, usually in the INFRARED, visible and ULTRAVIOLET regions of the ELECTROMAGNETIC SPECTRUM. Spectroscopy is a powerful tool for chemical analysis. INFRARED SPECTROSCOPY gives information about the chemical bonds in organic molecules, whilst visible and ultraviolet spectroscopy provides information about which elements are present and their IONIZATION states. Visible spectroscopy is also used in astronomy to provide information about the surface temperatures of stars and, via the DOPPLER EFFECT, about their motion.

A SPECTROMETER is a device used to produce a record of a SPECTRUM, whilst a SPECTROSCOPE enables a visible spectrum to be viewed directly. In either instrument, light from the sample is used to illuminate a slit. Light diffracting from this slit (the source slit) is focused into a parallel beam by a converging lens. The lens and slit assembly is called a collimator. The light is separated into its different wavelengths by a diffraction grating.

See also ATOMIC ABSORPTION SPECTROSCOPY, ATOMIC EMISSION SPECTROSCOPY, MICROWAVE SPECTROSCOPY.

spectrum (*pl. **spectra***) The arrangement of ELECTROMAGNETIC RADIATION into its constituent energies in order of wavelength or frequency. White light separated into its component wavelengths gives a characteristic spectrum of coloured bands.

When a sample is heated, or bombarded with ions or electrons, or absorbs photons of electromagnetic radiation, it emits radiation of wavelengths characteristic to the sample. This type of spectrum is called an emission spectrum. If radiation of a continuous range of wavelengths is passed through a sample, the sample absorbs certain characteristic wavelengths. When the transmitted radiation is viewed by a SPECTROSCOPE, the absorbed wavelengths show up as dark bands or lines. This is called an absorption spectrum. A line spectrum is one in which only certain wavelengths appear, while a continuous spectrum is one in which all the wavelengths in a certain range appear.

The term 'spectrum' can also apply to any distribution of entities or properties arranged in order of increasing (or decreasing value). For example, a mass spectrum is an arrangement of molecules, ions or isotopes by mass (*see* MASS SPECTROSCOPY).

See also ELECTROMAGNETIC SPECTRUM, HYDROGEN SPECTRUM, RYDBERG EQUATION, SPECTRAL LINE.

sphalerite An alternative name for ZINC BLENDE, a mineral form of zinc sulphide, commonly mined as a source of zinc.

sphingolipid A PHOSPHOLIPID based on the alcohol SPHINGOSINE.

sphingosine A complex ALCOHOL containing a long hydrocarbon chain and including an AMINO GROUP. The amino group can link to a FATTY ACID to form SPHINGOLIPIDS.

spin The inherent angular momentum that a particle, atom, nucleus, etc. possesses rather as if it were spinning. The spin of a particle is always a whole number times $h/4\pi$, where h is PLANCK'S CONSTANT. Particles with half-integral spin (odd number times this basic unit) are called fermions, of which the electron is an example. Particles with no spin or integral spin (an even number times $h/4\pi$), such as photons, are called bosons.

spinel Any of a group of minerals of the form AB_2O_4 where A is a DIVALENT metal and B a TRIVALENT metal. They are generally found in IGNEOUS rocks. Spinel is also the name given to the mineral $MgAl_2O_4$.

spontaneous combustion COMBUSTION that takes place without an external triggering factor, such as a flame. The heat needed to start the combustion process is generated internally by the reagents, usually by a slow OXIDATION process.

squalene ($C_{30}H_{50}$) A TERPENE that is the natural precursor of CHOLESTEROL and other STEROLS in animals, plants and fungi. It is the major hydrocarbon of human skin surface lipid. Squalene is used in cosmetics and for facilitating absorption of drugs applied to the skin.

square-planar (*adj.*) Describing a CO-ORDINATION COMPOUND with four LIGANDS, each at the corners of a square, that co-ordinate to a metal ion at the centre of the square.

stabilizer 1. A negative CATALYST used to reduce the rate of an undesirable reaction, for example in the decay of foodstuffs.

2. A material used to prevent the COAGULATION of a COLLOID, for example in paints.

stalactite A downward-growing calcium carbonate column formed by the seepage and evaporation of water containing dissolved calcium compounds in limestone caves. *See also* STALAGMITE.

stalagmite An upward-growing calcium carbonate column formed by the dripping and evaporation of water containing dissolved calcium compounds in limestone caves. *See also* STALACTITE.

standard atmosphere *See* ATMOSPHERE.

standard electrode A HALF-CELL with a known ELECTRODE POTENTIAL used together with a second half-cell and a voltmeter to measure an unknown electrode potential. *See also* CALOMEL HALF-CELL, HYDROGEN ELECTRODE.

standard electrode potential The POTENTIAL DIFFERENCE in an electrochemical CELL with a concentration of one MOLE of metal ions per decimetre cubed, and a temperature of 25°C, with an ANODE made of the same metal, measured relative to a HYDROGEN ELECTRODE. *See also* ELECTROCHEMICAL SERIES.

standard enthalpy of formation *See* STANDARD HEAT OF FORMATION.

standard heat of formation The HEAT OF FORMATION measured with all reagents and products at atmospheric pressure and a temperature of 25°C.

standard redox potential A REDOX POTENTIAL measured at one MOLE per decimetre cubed concentration of all reagents and at a temperature of 25°C. *See also* NERNST EQUATION.

standard state The STATE OF MATTER in which a substance is in equilibrium under specified conditions of pressure and temperature (usually one atmosphere and 25°C). Quantities such as the HEAT OF FORMATION of a substance usually relate to the substance being formed in its standard state from elements in their standard states.

standard temperature and pressure (s.t.p.) Conditions where the pressure is one ATMOSPHERE (1.01×10^5 Pa) and the temperature is 0°C.

stannane *See* TIN HYDRIDE.

stannate Any compound containing the stannate ion, SnO_4^{2-}. Stannates are formed by the reaction between tin oxide and alkalis, for example:

$$SnO_2 + 2NaOH \rightarrow Na_2SnO_4 + H_2O$$

stannic (*adj.*) An obsolete term describing a compound containing TIN in its +4 oxidation state, for example stannic chloride $SnCl_4$. Such compounds are often predominantly covalent in nature, though the Sn^{4+} ion does exist in aqueous solution.

stannous (*adj.*) An obsolete term describing a compound containing TIN in its +2 oxidation state, for example stannous chloride, $SnCl_2$. Such compounds contain the Sn^{2+} ion.

starch $(C_6H_{10}O_5)_x$ A complex POLYSACCHARIDE consisting of various proportions of two GLUCOSE polymers, AMYLOSE and AMYLOPECTIN. Amylose is composed of long straight chains containing 200–1000 molecules of glucose, linked by GLYCOSIDIC BONDS. Amylopectin consists of shorter chains of 20 glucose molecules in a branched structure, also cross-linked by glycosidic bonds. Starch is insoluble in cold water, but if heated the granules disrupt to form a gelatinous solution. If it is completely hydrolysed (*see* HYDROLYSIS), starch yields only glucose, but it can be broken down to MALTOSE, DEXTRIN or amylose and amylopectin under other conditions.

Starch is the main food reserve of green plants and a major energy source for animals. There are many uses for starch in industry, for example to stiffen paper or textiles, as a thickening agent in foodstuffs and as glucose syrups. The test to identify starch in solution is the addition of iodine in potassium iodide, which integrates into the polymer resulting in the formation of an intense blue colour.

See also CARBOHYDRATE.

states of matter The three physical forms in which a substance can usually exist: SOLID,

LIQUID and GAS. The different properties of these three states are explained by the KINETIC THEORY of matter.

state symbol *See* CHEMICAL EQUATION.

stationary phase *See* CHROMATOGRAPHY.

statistical mechanics A development of KINETIC THEORY that applies the laws of mechanics to systems containing large numbers of particles in order to calculate quantities such as TEMPERATURE and PRESSURE from the average values of the ENERGY and momentum of these particles. *See also* MAXWELL–BOLTZMANN DISTRIBUTION.

steam distillation The DISTILLATION of a mixture of liquids that are immiscible with water by heating them in a current of steam. The more volatile components of the mixture vaporize faster than the less volatile liquids. The technique is most effective at removing volatile components to purify the less volatile component.

steam reforming The process of heating METHANE from NATURAL GAS with steam to provide a source of hydrogen,

$$CH_4 + H_2O \rightarrow CO + 3H_2$$

The reaction is carried out at about 900°C in the presence of a nickel catalyst. The mixture of gases produced is called synthesis gas.

stearic acid, *octadecanoic acid* ($C_{18}H_{36}O_2$) A solid saturated FATTY ACID; melting point 71.5–72°C. It is one of the most common fatty acids and is found in most animal and vegetable fats. It is used to make candles and soap.

steel An alloy in which IRON is the predominant component. *See also* BESSEMER CONVERTER, ELECTRIC ARC FURNACE, OPEN HEARTH FURNACE, OXYGEN FURNACE.

stereochemistry The branch of chemistry that concerns itself with the shapes of MOLECULES, particularly organic molecules.

stereoisomerism The existence of two or more ISOMERS that differ only in the spatial orientation of the atoms in the molecule. *See* GEOMETRIC ISOMERISM, OPTICAL ISOMERISM.

stereospecific (*adj.*) Describing a reaction that produces a particular spatial arrangement of atoms in the molecule produced. Many reactions in biochemistry are stereospecific. *See also* STEREOISOMERISM.

steric hindrance A reduction in the rate at which a particular FUNCTIONAL GROUP reacts when it is surrounded by other functional groups, which physically interfere with the approach of a reacting molecule.

steroid hormone Any one of a group of LIPID hormones derived from a compound called cyclopentanoperhydrophenanthrene. They have a complex structure consisting of four carbon rings. A large subgroup of steroids is the steroid alcohols, or STEROLS, of which CHOLESTEROL is a member. The sex hormones OESTROGEN, PROGESTERONE and TESTOSTERONE are another important group of steroid hormones, as are the CORTICOSTEROIDS, which are produced by the adrenal gland and include CORTISONE and ALDOSTERONE. Many of these hormones, or their synthetic variants, have therapeutic uses. *See also* ANABOLIC STEROIDS.

sterol A large subgroup of the STEROID HORMONES, which are themselves steroid-based ALCOHOLS. Sterols have a side-chain of 8–10 carbon atoms and may contain one or more DOUBLE BONDS. They are found in all animal and plant cells and in some bacteria and fungi, either as free sterols or as ESTERS of FATTY ACIDS. Animal sterols are called zoosterols, of which CHOLESTEROL is an example. Plant sterols are called phytosterols, of which SITOSTEROL is the major example.

stibnite An ore of antimony, containing mostly antimony sulphide.

stoichiometric (*adj.*) Describing a compound in which the proportions of the elements reacting to form the compound are always the same. Such compounds obey the LAW OF CONSTANT PROPORTIONS.

Some solids are effectively mixtures of more than one compound. These are sometimes stoichiometric, such as the mixed iron oxide Fe_3O_4, which is effectively a mixture of iron (II) oxide and iron (III) oxide, but many minerals occur as non-stoichiometric crystals.

stoichiometry The proportions of various ELEMENTS that are involved in the formation of a COMPOUND.

streptomycin ($C_{21}H_{39}N_7O_{12}$) An ANTIBIOTIC produced by the bacteria *Streptomyces griseus* that is active against many bacteria. It is used in the form of its hydrochloride or sulphate, which are white solids. Bacteria may become resistant to the drug during treatment and it has some toxicity on the auditory nerves.

strong acid An ACID that is almost completely IONIZED when dissolved in water. An example is sulphuric acid, H_2SO_4, which forms H^+ ions (which associate with water molecules to form the oxonium ion, H_3O^+) and sulphate ions SO_4^{2-}. A strong acid is not the same as a concentrated acid – indeed at high concentrations the level of IONIZATION of many acids falls.

strontium (Sr) The element with atomic number 38; relative atomic mass 87.6°C; boiling point 1,300°C; relative density 2.5. Strontium is an ALKALINE EARTH metal. Strontium ISOTOPES are a common constituent of radioactive fallout. Since strontium is chemically similar to, but more reactive than calcium, there are concerns about radioactive strontium entering the FOOD CHAIN and being incorporated into the bone structure of those exposed to the fallout.

strontium carbonate ($SrCO_3$) A white solid; decomposes on heating; relative density 3.7. It can be made by passing carbon dioxide through strontium hydroxide solution:

$$Sr(OH)_2 + CO_2 \rightarrow SrCO_3 + H_2O$$

On heating, it decomposes to strontium oxide:

$$SrCO_3 \rightarrow SrO + CO_2$$

Strontium carbonate is used as a red colouring in pyrotechnic flares and fireworks, and as a PHOSPHOR in cathode ray tubes.

strontium chloride ($SrCl_2$) A white solid; melting point 872°C; boiling point 1,250°C; relative density 3.1. It can be formed by burning strontium in chlorine,

$$Sr + Cl_2 \rightarrow SrCl_2$$

Strontium chloride is used as a source of red colour in flares and fireworks.

strontium oxide (SrO) A white solid; melting point 2,430°C; boiling point 3,000°C. It can be manufactured by heating strontium carbonate, or by burning strontium in oxygen:

$$2Sr + O_2 \rightarrow 2SrO$$

It is used as a raw material in the manufacture of other strontium salts.

strontium sulphate ($SrSO_4$) A white solid; melting point 1,605°C; decomposes on further heating; relative density 4.0. Strontium sulphate is insoluble in water and can be produced as a precipitate in the reaction between aqueous strontium chloride and sulphuric acid,

$$SrCl_2 + H_2SO_4 \rightarrow SrSO_4 + 2HCl$$

It is used as a pigment in ceramic glazes and as a colouring agent in fireworks.

structural formula A chemical formula that gives some indication of the way in which the atoms are arranged in the molecule. The two ISOMERS of dichloroethane for example, both have the MOLECULAR FORMULA $C_2H_4Cl_2$, but they can be written as different structural formulae, CH_2ClCH_2Cl or CH_3CHCl_2. *See also* EMPIRICAL FORMULA.

structural isomerism *See* ISOMER.

strychnine ($C_{21}H_{22}N_2O_2$) A colourless ALKALOID obtained from certain plants. It is poisonous, causing stimulation of the nervous system and convulsions in high doses. It is used to kill vermin.

styrene *See* PHENYLETHENE.

styrene-butadiene rubber *See* SBR.

subatomic particle Any particle, in particular the PROTON, NEUTRON and ELECTRON, from which atoms are made.

sublimation The direct change from solid to gas without passing through a liquid phase when a substance is heated. Solid carbon dioxide (dry ice) and iodine are two examples of materials that sublime rather than melt at ATMOSPHERIC PRESSURE.

sublime(*vb.*) To change directly from a solid to a gas without passing through a liquid phase.

subsidiary quantum number The QUANTUM NUMBER used to label an ORBITAL giving a measure of the angular momentum of an electron in the orbital, and describing the shape of the orbital. The subsidiary quantum number is usually expressed as a letter in the sequence s, p, d, f, from an early mistaken categorization of lines in SPECTRA as sharp, principal, diffuse and fine.

substitution reaction A reaction in which one atom or group of atoms in a molecule is replaced by another. An example is the chlorination of methane:

$$CH_4 + Cl_2 \rightarrow CH_3Cl + HCl$$

where a hydrogen atom is replaced by a chlorine atom. *See also* ELECTROPHILIC SUBSTITUTION, NUCLEOPHILIC SUBSTITUTION.

substrate A substance that is acted upon by an ENZYME.

succinic acid *See* BUTANEDIOIC ACID.

suckback The state of an apparatus that has been heated, to produce gas in a chemical reaction, and then allowed to cool, drawing water up the DELIVERY TUBE and producing violent boiling.

sucrase An enzyme that breaks down SUCROSE into GLUCOSE and FRUCTOSE.

sucrose ($C_{12}H_{22}O_{10}$) A DISACCHARIDE sugar made up of the MONOSACCHARIDE units GLUCOSE and FRUCTOSE. Sucrose is found in the pith of sugar cane and in sugar beet. Sucrose is what is commonly referred to as 'sugar'.

sugar, *saccharide* Any one of a group of CARBOHYDRATES with relatively low RELATIVE MOLECULAR MASS and a typically sweet taste. The term sugar commonly refers to SUCROSE. *See also* DISACCHARIDE, MONOSACCHARIDE, POLYSACCHARIDE.

sugar of lead *See* LEAD(II) ETHANOATE.

sulphane Any of the polymeric (*see* POLYMER) forms of hydrogen sulphide, having the general formula H_2S_n.

sulphate A salt containing the sulphate ion, SO_4^{2-}, together with a CATION, either a metal or the ammonium ion, NH_4^+. Sulphates form stable crystals that contain WATER OF CRYSTALLIZATION.

sulphate(IV) *See* SULPHITE.

sulphide Any BINARY COMPOUND containing sulphur and a more electropositive element (*see* ELECTROPOSITIVITY). Non-metallic sulphides are covalent (*see* COVALENT BOND) and generally unstable. Metallic sulphides are mostly IONIC, though the TRANSITION METAL sulphides have some covalent character and are insoluble.

sulphite, *sulphate(IV)* A compound containing the sulphite ion, SO_3^{2-}, together with a cation, either a metal or the ammonium ion, NH_4^+. Sulphites are easily oxidized to SULPHATES.

sulphonamides Organic compounds containing the group $-SO_2NH_2$. They are AMIDES of sulphonic acids and many have antibacterial properties. Sulphonamides constitute a group of drugs called the sulpha drugs, which are used to treat a variety of infections, particularly of the gut and urinary system. They act by preventing bacterial reproduction rather than being bacteriocidal.

sulphonation In organic chemistry, the introduction of a sulphonic group, HSO_3, to ARENES. For example, the reaction of concentrated sulphuric acid with benzene at room temperature gives benzenesulphonic acid. The reaction is an ELECTROPHILIC SUBSTITUTION in which sulphur trioxide, SO_3, is generated from the sulphuric acid, which serves as the ELECTROPHILE:

H H
$+ SO_3 \rightarrow$ SO_3^-
O O
S
$\pm H^+$
$\rightarrow$ OH

benzenesulphonic acid

Sulphonation reactions are used in the manufacture of detergents and dyes.

sulphonic acids AROMATIC organic compounds containing the group $-SO_2OH$.

sulphonic group The HSO_3 group.

sulphoxides Organic compounds containing the group =S=O linked to two other groups, for example DIMETHYLSULPHOXIDE.

sulphur (S) The element with atomic number 16; relative atomic mass 32.1; melting point 112°C; boiling point 445°C; relative density 2.1. Sulphur exists mainly as a yellow non-metallic solid with a distinctive odour, but a red ALLOTROPE also exists that is stable at higher temperatures.

Sulphur burns in air and is soluble in carbon disulphide, but is otherwise fairly unreactive. It forms sulphides with the more reactive metals, but occurs more often in association with oxygen as sulphates. Large amounts of sulphur are used industrially in the manufacture of sulphuric acid.

sulphur cycle The natural cycling of sulphur between the biological (living) and the geological (non-living) components of the environment. Most sulphur in the non-living environment is found in rocks with some in the atmosphere as sulphur dioxide, SO_2, from volcanic activity and the burning of FOSSIL FUELS. Sulphates, SO_4^{2-}, are formed from the weathering of rocks and oxidation of sulphur or sulphur dioxide. Certain sulphur-oxidizing bacteria such as *Thiobacillus* are able to convert sulphur, or its compounds, to sulphates. Sulphates can then be taken up by plants and used to make certain sulphur-containing proteins. Thus sulphur can pass along the food chain to animals. Dead animals and their faeces are decomposed by the action of anaerobic (non-oxygen requiring) sulphate-reducing bacteria such as *Desulfovibrio,* which are able to reduce sulphur to generate hydrogen sulphide. This can then be converted back to sulphates by the sulphur-oxidizing bacteria. Another group of anaerobic bacteria, the green sulphur and the purple sulphur bacteria can convert hydrogen sulphide to sulphur. Sulphur can then again become incorporated into rocks.

sulphur dichloride dioxide (SO_2Cl_2) A colourless liquid; melting point –51°C; boiling point 69°C; relative density 1.7. It is formed by the reaction of sulphur dioxide with chlorine in the presence of ultraviolet light or iron(III) chloride as a catalyst. Its structure is TETRAHEDRAL, consisting of central sulphur atom forming DOUBLE BONDS with two oxygen atoms and single bonds with two chlorine atoms. It decomposes in water:

$$SO_2Cl_2 + 2H_2O \rightarrow H_2SO_4 + 2HCl$$

Sulphur dichloride dioxide is used as a chlorinating agent in the manufacture of chlorinated hydrocarbons.

sulphur dichloride oxide ($SOCl_2$) A colourless liquid; melting point –100°C; boiling point 76°C. It is formed by the action of phosphorus(V) chloride on sulphur dioxide:

$$SO_2 + PCl_5 \rightarrow SOCl_2 + POCl_3$$

Sulphur dichloride oxide hydrolyses rapidly:

$$SOCl_2 + H_2O \rightarrow SO_2 + 2HCl$$

sulphur dioxide (SO_2) A colourless gas with a distinct odour; melting point –73°C; boiling point –10°C. Sulphur dioxide is made by roasting iron sulphide in air:

$$FeS + O_2 \rightarrow Fe + SO_2$$

It dissolves in water to give sulphurous acid:

$$SO_2 + H_2O \rightarrow H_2SO_3$$

Large amounts of sulphur dioxide are used in the manufacture of sulphuric acid.

sulphuric acid (H_2SO_4) An important MINERAL ACID; melting point 10°C; boiling point 338°C; relative density 1.8. Sulphuric acid is manufactured by the CONTACT PROCESS or the LEAD CHAMBER process. Both processes involve oxidizing sulphur dioxide, SO_2, usually obtained by burning sulphur in oxygen, to sulphur trioxide, SO_3, which is then dissolved in water to form sulphuric acid.

Sulphuric acid is a very important chemical FEEDSTOCK, particularly in the fertilizer and paint industries. It is also important as an ELECTROLYTE in LEAD-ACID CELLS. In its concentrated form it has a great affinity for water, which makes it useful as a drying agent, though the heat given off can cause boiling, which means that great care must be taken in handling the concentrated acid. In particular, it should always be diluted by adding acid to water, never water to acid.

sulphurous acid (H_2SO_3) A weak DIBASIC ACID, known only in aqueous solution and from its salts, called sulphites. Sulphurous acid is formed when sulphur dioxide dissolves in water:

$$SO_2 + H_2O \rightarrow H_2SO_3$$

sulphur trioxide (SO_3) A colourless solid; melting point 17°C; boiling point 45°C; relative density 2.0. Sulphur dioxide is obtained by oxidizing sulphur dioxide with oxygen in the presence of a vanadium(V) oxide catalyst:

$$2SO_2 + O_2 \rightarrow 2SO_3$$

Sulphur trioxide dissolves readily in water to give sulphuric acid:

$$SO_3 + H_2O \rightarrow H_2SO_4$$

superconductivity The disappearance of electrical resistance, exhibited by some materials, called SUPERCONDUCTORS, at low temperatures.

superconductor Any material that exhibits SUPERCONDUCTIVITY. Many metals become superconductors at low enough temperatures, but the temperatures below which superconductivity occurs are often very low, only a few KELVIN. One of the best metallic superconductors is Niobium-tin alloy, which retains its superconductivity up to 22 K in a zero magnetic field.

supercooled (*adj.*) Describing a material that has cooled below its melting point, but remained liquid. This effect is usually demonstrated with sodium thiosulphate crystals, $Na_2S_2O_3.5H_2O$, which can be melted and then cooled in a clean test tube to well below their melting point. Adding a single SEED CRYSTAL of sodium thiosulphate will then trigger a rapid CRYSTALLIZATION and the release of large amounts of heat (LATENT HEAT of fusion).

superfluidity The disappearance of all VISCOSITY in liquid helium–4 below 2.2 K. Superfluid helium–4 also displays other unexpected properties, such as the ability to escape from an open vessel by forming a thin film and climbing up the inside of the vessel and down the outside, to collect in droplets at the base of the vessel.

superheated The state of a liquid heated above its boiling point without boiling. When a superheated liquid does boil, it will do so rather violently – an effect called BUMPING.

superoxide An OXIDE formed by the ALKALI METALS, except lithium, containing the ion O_2^-. An example is potassium superoxide, KO_2, which is formed by burning potassium in excess oxygen. All the superoxides are extremely powerful OXIDIZING AGENTS.

superphosphate A PHOSPHATE fertilizer formed by the addition of sulphuric or phosphoric acid to insoluble calcium phosphate, $Ca_3(PO_4)_2$, to produce soluble calcium hydrogenphosphate, $Ca(H_2PO_4)_2$;

$$Ca_3(PO_4)_2 + 2H_2SO_4 \rightarrow Ca(H_2PO_4)_2 + 2CaSO_4$$

(single superphosphate), or

$$Ca_3(PO_4)_2 + 4H_3PO_4 \rightarrow 3Ca(H_2PO_4)_2$$

(triple superphosphate). The soluble calcium hydrogenphosphate is used as a source of phosphorus for plants. In each case, the amount of acid added is carefully controlled to prevent excess acidity in the finished product, and the fertilizer also contains impurities that result from other materials present in the original phosphate rocks.

supersaturated vapour The state of a VAPOUR with a PARTIAL PRESSURE higher than its SATURATED VAPOUR PRESSURE. This is an unstable state, and droplets of liquid will form on any small particles or any irregularities in the surface of the container.

surface tension The force that appears at the surface of a liquid and tends to pull the liquid into spherical droplets. It results from the imbalance of INTERMOLECULAR FORCES acting on a molecule near the surface of a liquid. Such molecules are attracted by all the molecules below it in the liquid, whereas a molecule inside the liquid is attracted by molecules from all sides. The result is physically the same as if the surface were made of an elastic sheet with a constant force across any imaginary line on that surface, proportional to the length of this line. The surface tension – which is temperature dependent, falling to zero at the CRITICAL TEMPERATURE of the liquid – is equal to the force per unit length along such a line.

The effect of surface tension allows small objects, such as some insects, to rest on a water surface. Another result is that the pressure inside a bubble is higher than that outside, by an extent which depends on the surface tension of the liquid and the radius of the bubble. It is highest when the bubble is smallest; for similar reasons, a rubber balloon requires the most effort to blow it up when it is small.

surfactant A material, such as a DETERGENT, added to a liquid, usually water, to reduce its SURFACE TENSION, enabling it to wet a solid surface effectively, rather than running off the surface in droplets.

suspension A mixture in which small particles of solid or liquid are suspended in a liquid or gas. The term usually refers to particles of a solid in a liquid, for example a precipitate, which is a suspension of solid particles formed by a chemical reaction.

syndiotactic polymer, *syntactic polymer* A POLYMER in which the substituted carbons are arranged alternately above and below the plane of the carbon chain, if it is considered that the carbon atoms all lie in the same plane. One form of POLYVINYL CHLORIDE has this arrangement. *Compare* ATACTIC POLYMER, ISOTACTIC POLYMER.

syntactic polymer *See* SYNDIOTACTIC POLYMER.

synthesis A chemical reaction in which a more complex compound, such as an organic molecule, is manufactured from simpler substances.

synthesis gas The mixture of hydrogen and carbon monoxide produced in the STEAM REFORMING of methane.

synthetic natural gas (SNG) METHANE produced from other materials and used as a fuel. Methane can be produced from coal by either the FISCHER-TROPSCH PROCESS or directly by heating coal with hydrogen in the presence of a nickel catalyst. Synthetic natural gas is also produced from crude oil by the process of CRACKING.

T

2,4,5-T, ***2,4,5-trichlorophenoxyacetic acid*** ($C_8H_5Cl_3O_3$) A selective weedkiller that kills broad-leafed plants; melting point 155°C. It is made from 2,4,5-trichlorophenol and sodium chloroacetate and is used as a spray, sometimes combined with 2,4-D. Dioxin is a highly toxic impurity formed in the production of 2,4,5-T.

tactosol A COLLOID containing non-spherical particles that can be aligned by an applied electric or magnetic field.

talc A mineral form of magnesium silicate, $Mg_3Si_4O_{10}(OH)_2$. It is soft and slightly greasy to the touch and is easily ground into a fine powder that is used in cosmetics and as a filler in paints.

tannic acid A yellow-white organic compound with an unpleasant taste; melting point 210°C. It occurs in certain plants. Tannic acid is used as a MORDANT in the textile industry.

tannins A large group of complex organic compounds found in plants, in the leaves, unripe fruit or tree bark. They have an unpleasant taste, and their function in plants is uncertain. Tannins are used in the production of leather and as MORDANTS in the textile industry.

tantalum (Ta) The element with atomic number 73; relative atomic mass 180.9; melting point 2,996°C; boiling point 5,427°C; relative density 16.7. Tantalum is a TRANSITION METAL, used in some chemical process for its catalytic properties. It is also used in the manufacture of some small electronic components.

tar Any of the various dark sticky substances obtained by the DESTRUCTIVE DISTILLATION of organic matter, such as coal, wood or peat, or by the refining of petroleum. Tar is used to cover roads and to stop timber from rotting.

tarnish (*vb.*) Of a metal, particularly silver, to lose the characteristic metallic LUSTRE, due to the formation of compounds, usually OXIDES, by reaction with the atmosphere.

tartaric acid, ***2,3-dihydroxybutendioic acid*** ($C_4H_6O_6$; $(CHOH)_2(COOH)_2$) A CARBOXYLIC ACID that occurs naturally in grapes and other fruit. It is optically active (*see* OPTICAL ACTIVITY) and exists in more than one form. It can be prepared from tartar deposits in wine vats. Its main use is in baking powder and effervescent drinks, but it also has use in textiles and dyeing.

tartrazine, ***E102*** A DYE used for giving yellow colour to food. It is associated with hyperactivity in children and skin and respiratory problems in those with an allergy to it.

tautomerism A form of isomerism (*see* ISOMER) in which a material is able to turn from one of its isomers into the other, with an equilibrium being reached between the isomers. Such isomers are called tautomers.

TCA cycle *See* KREBS CYCLE.

technetium (Tc) The element with atomic number 43, melting point 2,171°C; boiling point 4,876°C. Technetium is a TRANSITION METAL, unusual in having no stable ISOTOPES despite a fairly low atomic number. It is extracted from the nuclear waste of uranium fission and some isotopes are used as radioactive tracers in medical diagnosis.

Teflon The trade name for POLYTETRAFLUOROETHENE.

telluride Any compound containing TELLURIUM and a more electropositive element (*see* ELECTROPOSITIVITY). Hydrogen telluride is covalent, but the tellurides of the ALKALI METALS are IONIC, containing the Te^{2-} ion.

tellurium (Te) The element with atomic number 52; relative atomic mass 127.6; melting point 451°C; boiling point 1,390°C; relative density 6.2. Tellurium is a METALLOID, used for its semiconducting properties in some electronic devices.

temperature A measure of the hotness or coldness of an object. If two objects are placed in contact with one another, HEAT will flow from the one at the higher temperature to the one at the lower temperature. If they are at the same temperature, there will be no heat flow. The SI UNIT of temperature is the KELVIN.

tempering Any heat treatment designed to increase the toughness of a metal alloy, especially steel that has previously been hardened by QUENCHING and would otherwise be unacceptably brittle. Tempering involves heating a material to a specified temperature for a certain time and then cooling it at a controlled rate. *See also* ANNEALING.

temporary hardness HARDNESS in water that can be removed by boiling. It is caused by the presence of calcium (and sometimes magnesium) hydrogencarbonate in the water, which decomposes on heating to give calcium carbonate (which is insoluble):

$$Ca(HCO_3)_2 \rightarrow CaCO_3 + CO_2 + H_2O$$

See HARD WATER.

terbium (Tb) The element with atomic number 65; relative atomic mass 158.9; melting point 1,365°C; boiling point 3,230°C; relative density 8.2. Terbium is a LANTHANIDE metal with some uses in electronics as a PHOSPHOR in cathode ray tubes and in some semiconducting devices.

terephthalic acid *See* 1,4-BENZENEDICARBOXYLIC ACID.

ternary compound Any compound made up of three different elements. *Compare* BINARY COMPOUND.

terpene An important group of unsaturated HYDROCARBONS consisting of ISOPRENE units, CH_2:$C(CH_3)CH$:CH_2. The term is used rather loosely to include other compounds derived from terpene hydrocarbons. Monoterpenes consist of two isoprene units, $C_{10}H_{16}$; sesquiterpenes have three units, $C_{15}H_{24}$; and diterpenes have four units, $C_{20}H_{32}$. Terpenes may be open-chain compounds or contain rings. They are very reactive.

Terpenes occur in many plants. Most have a pleasant odour, leading to their extensive use in perfumery. Other terpenes are constituents of menthol, camphor and eucalyptic oils.

tervalent *See* TRIVALENT.

Terylene The trade name for a synthetic POLYESTER fibre. It is made by the POLYMERIZATION of GLYCOL and 1,4-BENZENEDICARBOXYLIC ACID (terephthalic acid). Terylene is the most widely produced synthetic fibre and textiles made from these fibres are hard-wearing and wash well.

testosterone, ***17β-hydroxy-4-androsten-3-one*** ($C_{19}H_{28}O_2$) A STEROID HORMONE of male vertebrates produced in the testes. It is responsible for the development of secondary sexual characteristics, for example hair growth, deepening of voice, sexual behaviour, muscle development and also for stimulating sperm production. Synthetic testosterone has been used to aid muscular development in athletes, although its use has now been banned.

tetrachloroethene, ***tetrachloroethylene, polychloroethylene*** (CCl_2:CCl_2) A colourless, nonflammable liquid; melting point –22°C; boiling point 121°C. It is used as a dry-cleaning solvent.

tetrachloroethylene *See* TETRACHLOROETHENE.

tetrachloromethane *See* CARBON TETRACHLORIDE.

tetracycline A group of ANTIBIOTICS produced by bacteria of the genus *Streptomyces.* They are all based on a naphthacene skeleton (*see* NAPHTHACENE RING SYSTEM). The parent compound is tetracycline and there are various derivatives of this. The tetracyclines are effective against a wide variety of bacterial infections.

tetraethyl lead *See* LEAD TETRAETHYL.

tetragonal (*adj.*) Describing a CRYSTAL structure where the UNIT CELL has all its faces at right angles to one another, with two of the faces being square and the other four rectangular, so there are two different lengths characterizing the shape and size of the unit cell.

tetrahedral (*adj.*) Having the structure of a tetrahedron; that is, a figure with four triangular faces, each side having the same length. In chemistry, the word is applied to molecules or RADICALS which would fill a tetrahedron, with one central atom covalently bonded to an atom at each corner of the tetrahedron. The angle between the bonds is 109.5°. This structure is very common in organic chemistry as the carbon atom forms four single bonds with this structure. Thus methane, CH_4, has a tetrahedral structure. Other molecules, such as chloromethane, $CClH_3$, have similar structures, though the POLAR nature of the carbon–chlorine bond means that the bond angles are no longer equal.

tetrahedron A three-dimensional shape with four faces, each of which is a triangle.

tetrahydrofuran (C_4H_8O) A colourless liquid; boiling point 66°C. It is prepared from POLYSACCHARIDES in oat husks and widely used as a solvent for RESINS and PLASTICS.

tetravalent (*adj.*) Having a VALENCY of 4.

thallium (Ti) The element with atomic number 81; relative atomic mass 204.4; melting point 303°C; boiling point 1,460°C; relative density 11.9. Thallium's highly toxic nature has been exploited in some pesticides.

thermal analysis A technique of QUALITATIVE ANALYSIS based on the detection of gases given off, or changes in mass of a solid sample, as it is slowly heated.

Thermit process A reaction used to extract magnesium and chromium from their ores and sometimes also used to produce high temperatures for welding. The oxide of the metal concerned is mixed with finely powdered aluminium and the mixture ignited by setting fire to a strip of magnesium. The aluminium is oxidized and reduces the less reactive metal, for example:

$$Cr_2O_3 + 2Al \rightarrow 2Cr + Al_2O_3$$

thermochemistry The branch of physical chemistry that concerns itself with energy changes in chemical reactions, such as the calculation of HEATS OF FORMATION.

thermodynamic equilibrium The state in KINETIC THEORY where individual molecules are exchanging quantities such as energy and momentum, or reacting chemically, but the total amount of any chemical present, or the total energy, is unchanging. Thus the system can be meaningfully described by quantities such as temperature or the chemical concentration of its constituents.

thermodynamics The study of thermal ENERGY changes and ENTROPY.

Thermodynamics is based on four laws. The ZEROTH LAW defines the concept of temperature, by stating that two objects are at the same temperature if there is no net heat flow between them when they are in thermal contact. The FIRST LAW encapsulates the LAW OF CONSERVATION OF ENERGY including INTERNAL ENERGY and the recognition that heat is a form of energy. The SECOND LAW defines the concept of entropy, a measure of the degree of disorder in a system, and states that the entropy of a closed system can never decrease. The consequence of the third law, the NERNST HEAT THEOREM, is that it is imposssible to reach ABSOLUTE ZERO in a finite number of steps.

See also CARNOT ENGINE.

thermometer Any device for measuring TEMPERATURE.

thermoplastic, *thermosoftening plastic* Any PLASTIC that can be repeatedly softened on heating and hardened on cooling. In contrast to THERMOSETTING PLASTIC, these thermoplastics do not undergo cross-linking on heating and can therefore be resoftened. Examples include POLY(ETHENE), POLYSTYRENE, POLYVINYL CHLORIDE (PVC).

thermoset *See* THERMOSETTING PLASTIC.

thermosetting plastic, *thermoset* Any PLASTIC that can be moulded to shape during manufacture but which sets permanently rigid on further heating. This is due to extensive cross-linking that occurs on heating and cannot be reversed by reheating. Examples include PHENOL-FORMALDEHYDE RESINS, EPOXY RESINS, BAKELITE, POLYESTERS, POLYURETHANE and SILICONES.

thermosoftening plastic *See* THERMOPLASTIC.

thiamine, *vitamin B_1* A VITAMIN of the VITAMIN B COMPLEX that is a precursor of a coenzyme involved in carbohydrate metabolism. Thiamine is found in seeds, grain, yeast and eggs. Deficiency causes beri-beri, a disease causing inflammation of nerve endings resulting particularly in difficulty in walking.

thin-layer chromatography (TLC) A CHROMATOGRAPHY technique widely used for analysing the components in liquid mixtures. The stationary phase is a thin layer of an absorbent solid, such as aluminium oxide, supported on a vertical glass plate.

thioalcohol *See* THIOL.

thiocyanate Any salt of THIOCYANIC ACID, containing the ion SCN^-, or an organic compound containing –SCN as a FUNCTIONAL GROUP. The salts can be formed by the direct reaction of CYANIDES with sulphur, for example

$$KCN + S \rightarrow KSCN$$

thiocyanic acid (HSCN) An unstable gas that POLYMERIZES on heating. Thiocyanic acid can be formed by the reaction of potassium hydrogensulphate on potassium thiocynate,

$$KHSO_4 + KSCN \rightarrow K_2SO_4 + HSCN$$

thiol, *mercaptan, thioalcohol* Any one of a group of organic compounds containing the thiol group –SH (also called the mercapto group or sulphydryl group). Thiols have strong, unpleasant odours. They are sulphur analogues of ALCOHOLS in which the oxygen atom has been replaced by a sulphur atom. An example of a thiol is ethane thiol, C_2H_5SH. Thiols are easily oxidized to disulphides.

thiophene (CH_4N_2S) A colourless liquid with a odour similar to that of benzene; melting point –38°C; boiling point 84°C. It occurs in commercial benzene (about 0.5 per cent). The ring system is called the thienyl ring.

thiosulphate Any compound containing the thiosulphate ion, $S_2O_3^{2-}$. Sodium thiosulphate is important in the photographic process as a 'fixer', since it dissolves silver HALIDES, the light sensitive component of photographic films.

thiosulphuric acid ($H_2S_2O_3$) An unstable DIBASIC ACID, known only in aqueous solution. It is the parent acid of the THIOSULPHATES. Thiosulphuric acid can be formed by the reaction between hydrogen sulphide and sulphuric acid:

$$H_2S + H_2SO_4 \rightarrow H_2S_2O_3 + H_2O$$

thiourea (CH_4N_2S) A colourless crystalline compound; melting point 172°C. It is used as a fixer in photography. Many of its properties are similar to those of UREA.

thixotropic (*adj.*) Describing the property of some GELS that appear to have a lower VISCOSITY for fast flow rates than for slow ones. This property is exploited in non-drip paint, which will flow more easily at high speeds, as when being applied by a brush, than at low speeds, as in the formation of a drip.

thorium (Th) The element with atomic number 90; relative atomic mass 232.0; melting point 1,750°C; boiling point 4,790°C; relative density 11.7. Thorium is radioactive and produced in nature form the ALPHA DECAY of long-lived ISOTOPES of uranium. Along with uranium and plutonium, it is one of the three fissile elements, but this has not been exploited commercially.

threonine ($C_4H_9NO_3$) An ESSENTIAL AMINO ACID, produced by the HYDROLYSIS of PROTEINS.

thulium (Tm) The element with atomic number 69; relative atomic mass 168.9; melting point 1,545°C; boiling point 1,950°C; relative density 9.3. Thulium is the least abundant of the LANTHANIDE metals and as such has found few commercial applications.

thymine, *5-methyl-2,6-dioxytetrahydropyrimidine* ($C_5H_6N_2O_2$) An organic base called a PYRIMIDINE that occurs in DNA but not in RNA.

thyroxine ($C_{15}H_{11}I_4NO_4$) The main HORMONE of the thyroid gland, concerned with regulation of the body's metabolic rate. Thyroxine is derived from the amino acid TYROSINE and contains four molecules of iodine. If iodine is limited in the diet, another hormone called TRIIODOTHYRONINE (with three molecules of iodine) is made instead. These two hormones work together to stimulate the growth and metabolic rate of cells.

tin (Sn) The element with atomic number 50; relative atomic mass 118.7; melting point 232°C; boiling point 2,270°C; relative density 7.3. Tin is an abundant metal, occurring mostly as tin(IV) oxide, from which it is extracted by reducing the oxide with carbon:

$$SnO_2 + C \rightarrow Sn + CO_2$$

Tin was traditionally alloyed with copper to form bronze, which is much harder than either tin or copper, and more recently it has been used as a corrosion resistant coating to steel in tin cans which are actually made of steel dipped in molten tin.

Chemically, tin is a moderately reactive metal. It reacts with dilute acids to give hydrogen and tin(II) or tin(IV) salts, for example:

$$Sn + 2HCl \rightarrow SnCl_2 + H_2$$

tin chloride Tin(II) chloride, $SnCl_2$, or tin(IV) chloride, $SnCl_4$. Tin(II) chloride is a white solid; melting point 246°C; boiling point 652°C; relative density 2.2. It can be formed by reacting tin with hydrochloric acid:

$$Sn + 2HCl \rightarrow SnCl_2 + H_2$$

Tin(IV) chloride is a colourless liquid; melting point –33°C; boiling point 114°C; relative density 2.2. It can be produced by burning tin in chlorine:

$$Sn + 2Cl_2 \rightarrow SnCl_4$$

Tin (IV) chloride is rapidly hydrolysed (*see* HYDROLYSIS) by water:

$$SnCl_4 + 4H_2O \rightarrow Sn(OH)_4 + 4HCl$$

tin hydride, *stannane* (SnH_4) A colourless gas; melting point –150°C; boiling point –54°C. It can be formed by the action of lithium tetrahydroaluminate on tin chloride,

$$SnCl_4 + LiAlH_4 \rightarrow SnH_4 + SnCl + AlCl_3$$

It is a powerful REDUCING AGENT used in some organic reactions.

tin(IV) oxide (SnO_2) A white solid; sublimes at 1,850°C; relative density 7.0. Tin(IV) oxide is insoluble in water and occurs naturally. It is AMPHOTERIC, reacting with acids to form tin salts, and with bases to form STANNATES, for example:

$$SnO_2 + 4HNO_3 \rightarrow Sn(NO_3)_4 + 2H_2O$$

$$SnO_2 + 2NaOH \rightarrow Na_2SnO_3 + H_2O$$

tin sulphide Tin(II) sulphide, SnS, and tin(IV) sulphide, SnS_2. Tin(II) sulphide is a dark grey solid; melting point 882°C; boiling point 1,230°C; relative density 5.2. It can be made by heating tin and sulphur together:

$$Sn + S \rightarrow SnS$$

At high temperatures tin(II) sulphide slowly disproportionates to tin(IV) sulphide:

$$2SnS \rightarrow SnS_2 + Sn$$

Tin(IV) sulphide is a golden yellow solid; decomposes on heating; relative density 4.5. It is made by passing hydrogen sulphide through a solution of a soluble tin (IV) salt, for example:

$$Sn(NO_3)_4 + 2H_2S \rightarrow SnS_2 + 4HNO_3$$

titanium (Ti) The element with atomic number 22; relative atomic mass 47.9; melting point 1,660°C; boiling point 3,280°C; relative density 4.5. Titanium's high strength and low density make it an important component of some alloys used in the aerospace industry, but its high cost and difficult engineering properties have prevented its more widespread use.

titanium dioxide *See* TITANIUM(IV) OXIDE.

titanium(IV) oxide, *titanium dioxide* (TiO_2) A white solid; decomposes on heating. Titanium dioxide occurs in nature and is widely used as a white pigment in paint and paper.

titration A technique used in QUANTITATIVE ANALYSIS in which a measured quantity of one reagent is added to another until the chemical reaction between them is complete. The volume of the first reagent is usually measured with a PIPETTE, whilst the second regent is added through a BURETTE until the reaction reaches the end point, which is usually determined by a colour change. If the reaction does not naturally produce a colour change, an INDICATOR is used. *See also* END-POINT, EQUIVALENCE POINT.

TNT, *trinitrotoluene* ($CH_3C_6H_2(NO_2)_3$) A yellow solid; melting point 81°C. It is prepared from methylbenzene (toluene) by using sulphuric and nitric acids. TNT is a powerful explosive, used as a filling for shells and bombs.

tocopherol *See* VITAMIN E.

Tollen's reagent A solution used to distinguish ALDEHYDES from KETONES. The reagent is prepared by mixing sodium hydroxide and silver nitrate to give a brown PRECIPITATE of silver oxide. This silver oxide is then dissolved in aqueous ammonia and when warmed with aldehydes a deposit of silver is left on the test-tube like a mirror. Ketones do not do this.

toluene The common name for METHYLBENZENE.

topaz A mineral or gemstone form of aluminium silicate or fluorosilicate, $Al_2SiO_4(OH)_2$ or $Al_2SiO_4F_2$. The presence of impurities gives topaz a range of colours including pale pink, green and brown.

top-pan balance *See* BALANCE.

trace element An element that is required in small quantities for certain metabolic functions (*see* METABOLISM) in a plant or animal. Most are metals and iron, copper, zinc and magnesium are amongst the most important.

transactinide element Any element with an ATOMIC NUMBER of 104 and above. Only those with atomic numbers 104, 105 and possibly 106 have been synthesized.

transamination The conversion of one AMINO ACID to another to replace deficient non-essential amino acids. Transamination occurs in the vertebrate liver. The ENZYMES needed for this are transaminases and they transfer an AMINE group from an amino acid to a keto acid to form a new amino acid. Often NH_2 is transferred to and from GLUTAMINE and ASPARAGINE.

trans effect When substitutions occur in SQUARE-PLANAR complexes, the proportion of cis and trans ISOMERS formed (*see* GEOMETRIC ISOMERISM) depends on the other LIGANDS present in the complex. Certain ligands are able to direct the incoming ligand into a trans position better than the others. The order of trans-directing power for some ligands is

$$CN^- > NO_2 > I^- > Br^- > Cl^- > NH_3 > H_2O$$

transferase Any one of a group of ENZYMES that transfer a chemical group from one substance to another, for example phosphorylases transfer PHOSPHATE groups.

transition metal An element containing one or two electrons in the outer S-ORBITAL and one to 10 electrons in the outer D-ORBITALS, but no electrons in the next P-ORBITALS. The transition metals show similar chemical properties. They are all metals and often exhibit more than one OXIDATION STATE. They are mostly hard materials and good conductors of heat and electricity. Many form coloured IONS and CO-ORDINATION COMPOUNDS.

transition point The temperature at which a PHASE change takes place; in particular the change from one ALLOTROPE to another.

transition state In any reaction having an ACTIVATION ENERGY, the more energetic state through which the molecules of the reagents must pass as they combine to form the products of the reaction. The energy required to create the transition state in a reaction is a measure of the activation energy of the reaction.

transmission electron microscope A type of ELECTRON MICROSCOPE in which the material to be examined is preserved in a suitable fixative, such as glutaraldehyde or osmium tetroxide, then embedded in plastic EPOXY RESIN such as Araldite so that ultrathin sections can be cut. This is necessary since electrons cannot penetrate materials very well. Sections are then stained by various methods to improve their electron scattering ability, often involving HEAVY METALS, and supported on a metal grid that allows electrons to penetrate. Electrons are absorbed by some regions of the material (electron dense regions) but penetrate other electron transparent regions to hit the viewing screen and fluoresce.

transmutation The changing of one element into another as a result of RADIOACTIVITY.

transport coefficient Any quantity that describes some property of a gas that depends on intermolecular collisions to transfer some quantity from one part of the fluid to another. Examples include thermal conductivity, in which energy is transferred from one molecule to another by collisions; VISCOSITY, which relies on molecular collisions to transfer momentum; and DIFFUSION, where collisions limit the spread of molecules from one place to another.

transport number In ELECTROLYSIS, the fraction of the total current that is carried by an ion of a specified type.

transuranic element Any element with atomic number greater than 92 (uranium). All such elements have only radioactive ISOTOPES, with HALF-LIVES that are short compared to the age of the Earth, and so do not occur in nature.

tribromomethane, *bromoform* ($CHBr_3$) A colourless liquid; melting point 8°C; boiling point 150°C. It is prepared by the action of bromine and sodium hydroxide on ethanol or propanone. *See also* HALOFORM.

tricarboxylic acid cycle *See* KREBS CYCLE.

trichloroacetaldehyde *See* TRICHLOROETHANAL.

1,1,1-trichloro-2,2-di(4-chloro-phenyl)ethane Systematic name for DDT.

trichloroethanal, *chloral, trichloroacetaldehyde* (CCl_3CHO) A colourless, oily liquid with a pungent odour; boiling point 98°C. It is manufactured by the action of chlorine on ethanol. Useful compounds are formed by the addition of water, such as chloral hydrate, which is a sleep-inducing (or hypnotic) agent. Trichloroethanal is also used in the manufacture of DDT and is decomposed by alkalis to CHLOROFORM.

trichloromethane *See* CHLOROFORM.

2,4,5-trichlorophenoxyacetic acid *See* 2,4,5-T

triethylamine *See* ETHYLAMINE.

triglyceride A GLYCERIDE in which all three of the HYDROXYL GROUPS on GLYCEROL have combined with a FATTY ACID. The fatty acids can be the same or mixed and the nature of the triglyceride depends on the constituent fatty acids. Triglycerides occur naturally as the main constituents of fats and oils, providing an energy store in living animals. They also provide cooking oils, fats and margarine. The term triglyceride is often used synonymously with FAT. *See also* ESTER, LIPID.

trigonal planar A term describing the arrangement of atoms in a molecule or RADICAL where there is a central atom surrounded by three atoms in a plane, with COVALENT BONDS between the central atom and each of the other atoms, the angle between the bonds being 120°. The carbonate ion, CO_3^{2-}, is an example of a trigonal planar structure.

trigonal pyramidal A term describing the arrangement of atoms in a molecule or RADICAL where an atom at the apex of a triangular-based pyramid makes three COVALENT BONDS to atoms at the corners of the base. Ammonia, NH_3, has this structure rather than a TRIGONAL PLANAR structure, due to the presence of a LONE PAIR of electrons in the nitrogen atom.

trihydric (*adj.*) Describing a compound with three HYDROXYL GROUPS. For example, GLYCEROL is a trihydric ALCOHOL.

triiodomethane, ***iodoform*** (CHI_3) An ANTISEPTIC that forms pale yellow crystals; melting point 119°C. It has a characteristic 'hospital smell'. It is used to test for the $CH_3CO–$ group in carbonyl compounds (*see* CARBONYL GROUP). *See also* TRIIODOMETHANE TEST.

triiodomethane test, ***iodoform test*** A useful method of recognizing the $CH_3CO–$ group in carbonyl compounds (*see* CARBONYL GROUP), or groups in other compounds that may be converted to this. Compounds containing the $CH_3CO–$ group give a precipitate of TRIIODOMETHANE when heated in alkaline solution in an excess of iodine. This has a characteristic yellow colour and 'hospital odour'. If ALCOHOLS containing the $CH_3CH(OH)$ group are subjected to the same test they are first oxidized to $CH_3CO–$ and subsequently give a positive result in the triiodomethane test. So, for example, ethanol, C_2H_5OH, produces triiodomethane in the test but methanol, CH_3OH, does not. Alcohols usually need heating to react whereas carbonyl compounds may react in the cold.

triiodothyronine ($C_{15}H_{12}I_3NO_4$) A HORMONE secreted by the thyroid gland which is concerned with regulation of the body's metabolic rate. Triiodothyronine is derived from the amino acid TYROSINE and contains three molecules of iodine. *See also* THYROXINE.

1,2,3-triketohydrindene hydrate *See* NINHYDRIN.

2,2,4-trimethylpentane, ***iso-octane*** See OCTANE.

2,4,6-trinitrophenol *See* PICRIC ACID.

trinitrotoluene *See* TNT.

triose A MONOSACCHARIDE containing three carbon atoms in the molecule.

triple bond A COVALENT BOND in which three sets of orbitals overlap to form a bond stronger than a DOUBLE BOND. In ethyne, C_2H_2, for example, two sp HYBRID ORBITALS overlap to form a SIGMA-BOND whilst two pairs of P-ORBITALS form two PI-BONDS.

triple point The one combination of temperature and pressure at which the solid, liquid and gas PHASES of a substance can exist together in equilibrium. The triple point of water is at a temperature of 0.01°C and a pressure of 600 Pa.

trisodium phosphate (Na_3PO_4) A white solid occurring as both the DECAHYDRATE and the DODECAHYDRATE. Trisodium phosphate may be formed by neutralizing phosphoric acid with sodium hydroxide:

$$H_3PO_4 + 3NaOH \rightarrow Na_3PO_4 + 3H_2O.$$

Trisodium phosphate is used as a water softener (*see* WATER SOFTENING), precipitating soluble calcium and magnesium salts as insoluble phosphates, for example:

$$3CaCO_3 + 2Na_3PO_4 \rightarrow Ca_3(PO_4)_2 + 3Na_2CO_3$$

tritium The ISOTOPE hydrogen–3. It is widely used in fusion processes and is unstable, with a HALF-LIFE of 12 years.

trivalent, ***tervalent*** (*adj.*) Having a VALENCY of 3.

tropylium ion The ion $C_7H_7^+$, which consists of a ring of seven carbon atoms.

trypsin A PROTEASE enzyme in the vertebrate digestive system that breaks down PROTEINS during digestion. Trypsin does not need an acid environment to function. It has a relative molecular mass of 24,000 and specicifically cleaves PEPTIDE linkages adjacent to ARGININE or LYSINE residues.

trypsinogen *See* TRYPSIN.

tryptophan ($C_{11}H_{12}N_2O_2$) An ESSENTIAL AMINO ACID produced from PROTEINS by the digestive action of TRYPSIN.

tungsten (W) The element with atomic number 74; relative atomic mass 183.9; melting point 3,410°C; boiling point 5,660°C; relative density 19.3. Tungsten has the highest melting point of any metal, which has led to its use in lightbulb filaments. Tungsten carbide, an extremely hard material, is used to tip drills and other cutting tools.

tungsten carbide (WC) A black powder; melting point 2,600°C; boiling point 5,660°C; relative density 15.6. It is formed by a direct reaction between powdered tungsten and AMORPHOUS carbon. It is extremely hard and is used in a sintered form (*see* SINTERING) to tip drills and other cutting tools.

turpentine An oil obtained as exudates or by distillation from pine tress. Various types of turpentine are made, all containing pinene ($C_{10}H_{16}$), which is a TERPENE. Turpentine is used as a solvent and a thinner for paint or varnish.

turquoise A green mineral or gemstone form of the salt copper aluminium phosphate, $CuAl_6(PO_4)_4(OH)_8.4H_2O$.

tuyere A metal pipe through which air is blown into a BLAST FURNACE.

tyrosine ($C_9H_{11}NO_3$) An AMINO ACID that occurs in many proteins. It is the precursor of ADRENALINE.

U

ultracentrifuge A CENTRIFUGE that operates at very high speeds. It can be used in the laboratory to separate COLLOIDS, submicroscopic particles or particles as small as a NUCLEIC ACID or PROTEIN.

ultraviolet ELECTROMAGNETIC WAVES with wavelengths in the range 4×10^{-7} m to about 10^{-9} m. They are produced by the more energetic changes in energy in atomic electrons. Ultraviolet radiation from the Sun is mostly absorbed in the upper layers of the atmosphere (the OZONE LAYER), so relatively little reaches the Earth. That which does reach ground level is responsible for the tanning and burning effect of exposure to sunlight and, with prolonged exposure, is believed to be responsible for skin cancer. Ultraviolet radiation can be detected by PHOTOGRAPHIC FILM and can be made visible by FLUORESCENCE.

unimolecular reaction A reaction involving only a single molecule, for example the decomposition of DINITROGEN TETROXIDE,

$$N_2O_4 \rightarrow 2NO_2$$

unit cell, ***primitive cell*** The smallest part of a CRYSTAL lattice that is needed to describe the structure of a crystal. A crystal can be thought of as being made up of repeated unit cells. For example, the unit cell for sodium chloride, NaCl, can be regarded as a cube containing a central sodium ion with half a chlorine ion in the middle of each face and one eighth of a sodium ion at each corner.

universal gas constant *See* MOLAR GAS CONSTANT.

universal gas equation *See* IDEAL GAS EQUATION.

universal indicator A mixture of INDICATORS designed to produce a continuous variation in colour over the whole scale of pH values. Universal indicator can be used to give a quick, but fairly approximate, indication of the degree of acidity or alkalinity in a solution.

unnil- A prefix used in giving names to elements with ATOMIC NUMBERS greater than 103. The names used are unnilquadium (104), unnilpentium (105), unnilhexium (106), unnilheptium (107), unniloctium (108), unnilennium (109), unnildecium (110).

unsaturated compound Any organic compound in which the carbon atoms are linked by double or triple COVALENT BONDS. ALKENES, ALKYNES and KETONES are all examples of unsaturated compounds.

BROMINE WATER may be used to to determine whether an organic compound is unsaturated. A red-brown solution of bromine in water is decolorized by unsaturated compounds since they react with the bromine.

uracil, ***2,6-dioxytetrahydropyrimidine*** ($C_4H_4N_2O_2$) An organic base that occurs in RNA but not in DNA.

uranium (U) The element with atomic number 92; relative atomic mass 238.0; melting point 1,132°C; boiling point 3,818°C; relative density 19.1. Uranium is the heaviest and most abundant of the naturally radioactive elements, having several ISOTOPES with HALF-LIVES comparable to the age of the Universe.

Uranium is mined commercially at many sites in the world. The most common isotope, uranium–238 has a half-life of 4.5×10^9 years, but the rarer isotope, uranium–235, which makes up just 0.7 per cent of natural uranium, is of greater commercial interest as it undergoes nuclear fission.

uranium(VI) fluoride *See* URANIUM HEXAFLUORIDE.

uranium dioxide *See* URANIUM HEXAFLUORIDE.

uranium hexafluoride, ***uranium(VI) fluoride*** (UF_6) A white solid; sublimes at 65°C; relative density 5.1. Uranium hexafluoride gas is used in the enrichment of uranium by diffusion.

uranium(IV) fluoride, ***uranium dioxide*** (UO_2) A black solid; melting point 3,000°C, decomposes on further heating; relative density 10.9. Uranium oxide is often used as a fuel in nuclear reactors as it has a higher melting point than elemental uranium.

urea ($CO(NH_2)_2$) A white crystalline solid; melting point 135°C. Urea is a waste product formed from the breakdown of ammonia,

NH_3, in the mammalian liver, which is then excreted in the urine. Ammonia is itself a waste product derived from the breakdown of PROTEINS and NUCLEIC ACID, but is very toxic and therefore converted in many vertebrates to urea, which is harmless.

Urea is made by liver cells in a cyclic process called the UREA CYCLE, which is closely linked to the KREBS CYCLE. When purified, urea has some industrial uses, for example in fertilizers and pharmaceuticals.

urea cycle, *ornithine cycle* The series of biochemical reactions occuring in the mammalian liver that convert ammonia to UREA as part of the excretion of metabolic waste products. Ammonia is a waste product from the breakdown of PROTEINS and NUCLEIC ACID, but is very toxic and therefore needs to be converted to the less toxic urea, which can then be excreted in solution as urine. The urea cycle is closely linked to the KREBS CYCLE.

Ammonia enters the urea cycle in two places, firstly in combination with carbon dioxide as carbamyl phosphate, which combines with ornithine to form citrulline. The second molecule of ammonia enters as aspartic acid, which combines with the citrulline to form arginosuccinate. Fumaric acid is removed from the arginosuccinate and can enter the Krebs cycle or be used to regenerate aspartic acid. The removal of fumaric acid produces the AMINO ACID arginine, which is split into urea and ornithine by the ENZYME arginase.

urea-formaldehyde resin Any of a group of syntheic RESINS made by the POLYMERIZATION of UREA and METHANAL (formaldehyde). Urea-formaldehyde resins are used as THERMOSETTING PLASTICS, adhesives and foams.

uric acid ($C_5H_4N_4O_3$) A semi-solid nitrogenous waste produced by most land animals that develop in a shell, including reptiles, insects and birds. Uric acid is produced instead of urine where water is scarce. Humans also produce some uric acid, which if in excess can build up as crystals in joints and tissues, causing gout, or it can form kidney or bladder stones.

V

vacuum A region containing no matter of any kind.

vacuum pump Any device for reducing the pressure of gas in a vessel. The simplest vacuum pumps are rotary pumps driven by an electric motor. Gas from the vessel enters one end of the pump and is forced out by a system of rotating vanes. Once the pressure is so low that the MEAN FREE PATH of a molecule becomes comparable to the size of the vessel, such pumps are no longer effective and diffusion pumps are used.

vulcanite, *ebonite* A hard black plastic made from the VULCANIZATION of RUBBER with a high proportion of sulphur.

valence band In the BAND THEORY of solids, the energy band occupied by the VALENCE ELECTRONS. In metals, this band is either only partially full or overlaps with an empty band. In nonmetals it is completely full, so electrons cannot gain energy to conduct heat or electricity. *See also* CONDUCTION BAND

valence electron An electron in the outer SHELL of electrons of an atom, which may be involved in a bonding process. *See also* BAND THEORY, CHEMICAL BOND.

valence shell The outer electron SHELL of an atom, containing the VALENCE ELECTRONS.

valency A number indicating the number of chemical bonds that can be formed by a given element. Many elements show only one valency, for example the valency of the element magnesium is always 2. Other elements, in particular the TRANSITION METALS, can exist in several different valency states. *See also* COVALENCY, ELECTROVALENCY, OXIDATION STATE.

valeric acid *See* PENTANOIC ACID.

valine ($C_5H_{11}NO_2$) An ESSENTIAL AMINO ACID, present in many PROTEINS.

vanadium (V) The element with atomic number 23; relative atomic mass 50.9; melting point 1,890°C; boiling point 3,380°C; relative density 6.1. Vanadium is a TRANSITION METAL with some useful catalytic properties. It is also used with chromium in the manufacture of some hard steels. In chemistry, vanadium exists in several OXIDATION STATES, each producing ions of distinctive colours.

vanadium(V) oxide, *vanadium pentoxide* (V_2O_5) A white solid that decomposes on heating. Vandium(V) oxide can be formed by burning vanadium metal in oxygen:

$$4V + 5O_2 \rightarrow 2V_2O_5$$

Vandium(V) oxide is widely used for its catalytic properties.

vanadium pentoxide *See* VANADIUM(V) OXIDE.

van der Waals' bond A very weak bond that holds separate molecules together in molecular solids such as solid carbon dioxide. The bond originates from the VAN DER WAALS' FORCE between neutral molecules. The very weak nature of this bond is reflected in the low melting points of most molecular materials.

van der Waals' equation An EQUATION OF STATE for a REAL GAS. For N MOLES of gas at an ABSOLUTE TEMPERATURE T and pressure p, in a volume V, van der Waals' equation is

$$(p + a/V^2)(V - b) = NRT$$

where R is the UNIVERSAL GAS CONSTANT and a and b are constants that represent the attractive INTERMOLECULAR FORCES and the volume of the gas molecules respectively.

van der Waals' force The attractive force between neutral molecules or atoms. The force is strongest if both molecules are POLAR, but a POLAR MOLECULE will create an induced DIPOLE in a non-polar molecule. There is also a weak induced-dipole to induced-dipole force between non-polar molecules. This arises because at any instant the centres of positive and negative charge may not coincide, even though the molecule may be non-polar on average. These forces are relatively weak and give rise to substances that have melting and boiling points below room temperature.

As with all INTERMOLECULAR FORCES, there is a repulsion at shorter distances, due to the effect of the PAULI EXCLUSION PRINCIPLE on the overlapping electron clouds of the molecules. Van der Waals' forces between non-polar molecules can be described by a POTENTIAL ENERGY with a repulsive core that varies like the inverse twelfth power of the separation and an attractive part, dominant at larger separations, but also falling off with distance, which varies like the inverse sixth power of separation. This is called the LENNARD–JONES 6-12 POTENTIAL.

See also REAL GAS.

van't Hoff's isochore An equation giving the rate of change of the EQUILIBRIUM CONSTANT K of a reaction with temperature.

$$d(\ln K)/dT = \Delta H / RT^2$$

where ΔH is the enthalpy change in the reaction, R the MOLAR GAS CONSTANT and T the absolute temperature.

vapour The term sometimes used to describe gaseous state of a material below its boiling point.

vapour density A measure of the density of a gas, found by dividing the mass of a sample of the gas by the mass that the same volume of some reference gas (generally hydrogen) would have at the same temperature and pressure.

vapour pressure The PARTIAL PRESSURE exerted by a VAPOUR, especially the vapour found above the surface of a liquid. If the liquid and its vapour are held in a closed container, the vapour pressure will rise until it reaches the SATURATED VAPOUR PRESSURE. *See also* RAOULT'S LAW.

vat dye *See* DYES.

velocity selector *See* MASS SPECTROMETER.

verdigris A bright green deposit found on copper that has been exposed to the atmosphere. Verdigris is composed of the basic copper carbonate, $CuCO_3.Cu(OH)_2$, and to a lesser extent basic copper sulphate and chloride, $CuSO_4.Cu(OH)_2$ and $CuCl_2.Cu(OH)_2$. It is formed by reaction with acidic pollution or salt water spray in the atmosphere.

vernier A device for fine adjustment or precise measurement. A vernier scale typically consists of a small scale sliding along a main scale. The small scale contains 10 divisions in the space occupied by nine divisions on the main scale. By seeing which pair of scale marks align most accurately, it is possible to read the main scale to one tenth of a division.

Victor Meyer's method A technique for measuring the VAPOUR DENSITY of a volatile liquid. A small tube containing a known mass of liquid is placed in the bottom of a flask and warmed to vaporize the liquid. The air displaced by the vapourized liquid is collected over water, enabling the volume of the vapour produced to be measured.

vinyl chloride, *chloroethene* (CH_2CHCl) A colourless gas; boiling point –14°C. It is used for the manufacture of POLYVINYL CHLORIDE (PVC). Vinyl chloride is made by reacting ethyne with hydrogen chloride over a catalyst. It is CARCINOGENIC.

virial equation An EQUATION OF STATE for a gas, which takes as its starting point the IDEAL GAS EQUATION and than adds a series of pressure dependent terms.

$$pV = n(RT + Ap + Bp^2 + Cp^3 + \ldots)$$

where p and V are the pressure and volume respectively, n the number of moles, R the MOLAR GAS CONSTANT and T the absolute temperature. A, B, C, etc. are constants called virial coefficients.

viscose A type of RAYON, used to make cellophane.

viscosity A measure of the resistance to flow in a fluid. Forces between molecules in a fluid mean that momentum given to one part of the fluid tends to be transferred to nearby regions. This creates a force opposing any tendency for different parts of the fluid to move at different speeds, as happens at a boundary with a solid surface for example.

The viscosity of a fluid is measured as the force F per unit area A between two surfaces in the fluid that are moving at unit velocity v relative to one another and are separated by unit distance x.

$$F = \eta A(dv/dx)$$

where η is a constant of proportionality, called the coefficient of viscosity. *See also* SUPERFLUIDITY, THIXOTROPIC.

visible light ELECTROMAGNETIC WAVES that can be detected by the human eye. Visible light has wavelengths from about 7×10^{-7} m (red) to 4×10^{-7} m (violet). Visible light is produced by very hot objects (such as stars and light-bulb filaments) and also when electrons move from one ENERGY LEVEL to another in an atom. The wavelengths given out depend on which atoms are present and this is an important tool in chemical analysis, particularly in cases where no direct sample can be obtained (for example in looking at the light given out by stars).

vitamin Any one of a group of unrelated organic compounds essential in small amounts for normal body growth and metabolism. Vitamins are classified as water-soluble (B, C, H) or fat-soluble (A, D, E, K). Excess water-soluble vitamins are excreted in the urine; fat-soluble vitamins can be stored (in the liver in humans) but can build up to lethal concentrations if taken in excess. A normal balanced diet usually provides the vitamin requirements but if there are inadequate levels then deficiency disease results. Some vitamins are present in foods as provitamins, which are converted to vitamins inside the body. Many vitamins are destroyed by heat and light, such as by cooking.

vitamin A, *retinol* ($C_{20}H_{30}O$) A fat-soluble VITAMIN that is important in skin structure and to form visual pigments. It is found in dairy foods, fruits, vegetables and liver, in particular fish-liver oils. Lack of vitamin A causes night-blindness (or more severe blindness, xerophthalmia) and dry skin. CAROTENE is converted to vitamin A in the liver, hence this provides an indirect source of vitamin A from green vegetables. RETINAL is a derivative of vitamin A that forms part of the visual pigment RHODOPSIN. This role of vitamin A in vision is thought to be different to its other metabolic roles, which are not fully understood.

vitamin B complex A complex of B_1 (THIAMINE), B_2 (RIBOFLAVIN), B_3 (NIACIN), B_5 (PANTOTHENIC ACID), B_6 (PYRIDOXINE), B_{12} (CYANOCOBALAMIN), BIOTIN and FOLIC ACID. They are water-soluble and serve mostly as COENZYMES in cellular respiration. Most animals obtain the B vitamins in their diet. *See also* CHOLINE.

vitamin C, *ascorbic acid* ($C_6H_8O_6$) A water-soluble VITAMIN that occurs in fresh fruit and vegetables (but is destroyed by soaking or overcooking). It is needed for the synthesis of COLLAGEN, healing of wounds and bone and teeth formation. Most organisms can make vitamin C from GLUCOSE, but humans and certain other primates have to obtain it through their diet. A deficiency of vitamin C leads to the disease scurvy.

vitamin D, *cholecalciferol* A fat-soluble VITAMIN that occurs in fatty fish and margarine and is made in the skin if exposed to enough sunlight. It is needed for the absorption of calcium and phosphorous and is therefore important for the formation of bones and teeth. Lack of vitamin D causes the disease rickets, characterized by weak bones. Vitamin D occurs in two forms, D_2 and D_3, both being steroid derivatives (*see* STEROID HORMONE) and having similar actions. Vitamin D_2 is formed by the action of ultraviolet light on the steroid ergosterol and is found in yeast. Vitamin D_3 (the natural vitamin D or cholecaliferol) is formed by the action of ultraviolet light on a cholesterol derivative in the skin and is found in animals.

vitamin E, *tocopherol* A fat-soluble VITAMIN found in vegetable oil, cereals and green vegetables. Vitamin E prevents the oxidation of fatty acids in cell membranes. It is of unclear function in humans, but causes sterility in rats.

vitamin K, *phytomenadione* A fat-soluble VITAMIN found in leafy vegetables and liver and synthesized by intestinal bacteria. It consists of several similar compounds that are needed as COENZYMES associated with blood clotting proteins. Lack of vitamin K causes haemorrhaging. It is often given to newborn babies to prevent brain haemorrhage, although its routine use has been questioned.

volatile (*adj.*) Easily evaporated.

volt (V) The SI UNIT of POTENTIAL DIFFERENCE or ELECTROMOTIVE FORCE. One volt is an energy of one JOULE per COULOMB of CHARGE.

voltage The measure of the amount of electrical potential energy carried by each unit of charge. The SI UNIT of voltage is the VOLT. *See also* ELECTROMOTIVE FORCE, POTENTIAL DIFFERENCE.

voltaic cell *See* CELL.

voltammeter A container for ELECTROLYSIS experiments with ELECTRODES that can be removed to find their mass. By measuring the change of the mass of the electrodes in an electrolysis experiment, the ELECTROCHEMICAL EQUIVALENT of the element concerned can be found. *See also* HOFMANN VOLTAMMETER.

volume The amount of space occupied by an object. For a cube, the volume is equal to the third power of the length of the sides of the cube. For other shapes, the volume can be calculated by imagining them to be built up from a large number of small cubes. Volume is expressed in SI UNITS in metres cubed.

volumetric analysis A range of techniques of QUANTITATIVE ANALYSIS that involve measuring the volumes of reacting materials. In the gas phase, this involves using graduated vessels attached to MANOMETERS to ensure that the volumes are measured at a known pressure. For liquids, TITRATION is the most common technique. In this, a fixed volume of one reagent has a second reagent added to it from a calibrated vessel, called a BURETTE, until an INDICATOR shows that the reaction is complete.

vulcanization A technique used for treating RUBBER to make it harder, stronger and less affected by external temperatures. The technique was discovered by Charles Goodyear (1800–60) in 1839 and involves heating the rubber and chemically combining it with another element, usually sulphur. The rubber is cross-linked by the sulphur forming bridges between the POLYMER chains. Thus the chains are less easily pulled apart on stretching and the rubber has increased elasticity. For special uses, such as tyres, vulcanizing agents other than sulphur are used along with additives to speed up the process. Objects such as tyres can be shaped and vulcanized at the same time by using heated moulds.

W

Wacker process A method of preparing ETHANAL by the oxidation of ETHENE. An air and ethene mixture is bubbled through a solution containing palladium(II) chloride and copper(II) chloride, which act as catalysts. Ethanal can be further oxidized to ETHANOIC ACID.

Walden inversion A reaction in which one optical isomer (*see* OPTICAL ISOMERISM) of an organic molecule can be converted into the other by the substitution of one FUNCTIONAL GROUP by another, which is then substituted by the original functional group. By the appropriate choice of the reagent that is used to introduce each functional group, it is possible to determine which of the optical isomers will be produced and thus to convert one isomer into the other.

washing soda *See* SODIUM CARBONATE.

water (H_2O) A colourless, odourless, tasteless liquid, an oxide of hydrogen, which is essential to all living organisms and is the most abundant liquid on Earth. Most water is made up of one atom of hydrogen–1) and two atoms of oxygen–16. A small proportion of water contains other ISOTOPES, such as hydrogen–2 and oxygen–18. When hydrogen–2 (deuterium) is incorporated HEAVY WATER results. Water has the highest SPECIFIC HEAT CAPACITY known, due to the presence of HYDROGEN BONDS.

Water freezes at 0°C and boils at 100°C and has its maximum density at 4°C. One cm^3 of water has a mass of approximately 1 g and forms the unit of RELATIVE DENSITY. Ice is less dense than water and so floats on water, allowing organisms to live beneath it. Most water is found in seas, oceans and rivers, which cover 70 per cent of the Earth's surface. The human body contains 60–70 per cent water (about 40 litres) and loss of 8–10 litres can lead to death. *See also* HARD WATER, SOFT WATER, WATER CYCLE.

water cycle The chain of events by which water is re-used in the atmosphere and on the Earth's surface. The condensation of water vapour to produce rain and the evaporation of water from oceans are the key stages in this process, but the absorption of water in soil and its transport to the oceans by rivers are also important, as is the role of respiration in plants and animals, which absorb water from their surroundings and return it to the atmosphere by evaporation.

water gas The gas formed by passing steam over hot COKE. The steam is reduced (*see* REDUCTION) by the coke, producing a mixture of hydrogen and carbon monoxide, which is used as a fuel in some industrial processes:

$$H_2O + C \rightarrow H_2 + CO$$

The reaction cools the coke and is sometimes used in conjunction with the production of PRODUCER GAS. *See also* BOSCH PROCESS, SEMI-WATER GAS.

water of crystallization Water that forms an important part of the crystal structure in certain salts, particularly NITRATES and SULPHATES. This water can be removed by heating the crystals, which then form a powdery, AMORPHOUS solid. Copper(II) sulphate, $CuSO_4.5H_2O$, forms blue crystals, but on heating water vapour is released and the resulting ANHYDROUS copper sulphate is a white powder. On exposure to atmospheric moisture, water is absorbed and the blue colour returns.

water potential (φ) A measure of the tendency of water in a system to go to its surroundings. The water potential is the difference between the CHEMICAL POTENTIAL of pure water under conditions of STANDARD TEMPERATURE AND PRESSURE and the chemical potential of water in the system.

water softening Any process for converting HARD WATER to SOFT WATER. For some purposes, such as washing clothes, chemicals such as 'Calgon' can be added to the water. Calgon contains sodium phosphate complexes, $(NaPO_3)_6$, which react with the dissolved calcium and magnesium ions to form stable soluble complexes such as $[Ca_2P_6O_{18}]^{2-}$, which prevent the metal ions from reacting with soap.

A more traditional way is to use 'washing soda', sodium carbonate. Sodium and magnesium sulphates and hydrogen carbonates react with this to form insoluble carbonates, removing the metal ions from solution, for example:

$$CaSO_4 + Na_2CO_3 \rightarrow CaCO_3 + Na_2SO_4$$

For the softening of water for drinking or washing, ION EXCHANGE methods are used. These involve minerals called PERMUTITES or a modern synthetic equivalent made from fused sand, clay and sodium carbonate.

watt The SI UNIT of POWER. One watt represents work done, or energy converted from one form to another at a rate of one JOULE per second.

wavefunction In QUANTUM MECHANICS, a PROBABILITY AMPLITUDE, the square of which measures the probability of finding the particle at the point specified or the system in the state described by the particular wavefunction. *See also* SCHRÖDINGER'S EQUATION.

wave-particle duality The name given to the dual behaviour of objects originally thought of as either waves or particles. Thus light, traditionally a wave, can behave as a particle (*see* PHOTON, PHOTOELECTRIC EFFECT), whilst electrons, traditionally thought of as particles, can behave in wave-like ways (*see* ELECTRON DIFFRACTION).

wax A natural solid fatty substance made from ESTERS, FATTY ACIDS or ALCOHOLS. In nature, waxes provide a protective, waterproof covering to many animals and plants. Animal waxes such as beeswax, lanolin and wax from sperm-whale oil are used in cosmetics, ointments and polishes. Mineral waxes are obtained from PETROLEUM and provide a number of products including the soft petroleum jelly used in ointments to the hard wax used for making candles.

weak acid An ACID that is not highly IONIZED when dissolved in water, but tends to remain as covalent molecules (*see* COVALENT BOND). Ethanoic acid, CH_3COOH, is an example; only a few ethanoate (CH_3COO^-) ions are formed. Weak acids only react slowly, even with reactive metals. As the supply of ions is used up, the EQUILIBRIUM between ionic and covalent forms is disturbed and more acid DISSOCIATES to the ionic form.

Weston cell A VOLTAIC CELL that produces an accurately reproducible voltage, which can be used in the calibration of voltmeters. The cell traditionally takes the form of an H-shaped glass tube. The anode is composed of mercury covered with a paste of mercury (I) sulphate, the cathode is mercury/cadmium amalgam. The electrolyte is saturated cadmium sulphate solution and the cell usually also contains undissolved cadmium sulphate to ensure the electrolyte remains saturated. The cell gives a voltage of 1.0186 V at 20°C.

white arsenic *See* ARSENIC OXIDE.

white lead *See* LEAD CARBONATE HYDROXIDE.

white spirit A mixture of hydrocarbons obtained from PETROLEUM and used as paint thinners and solvents.

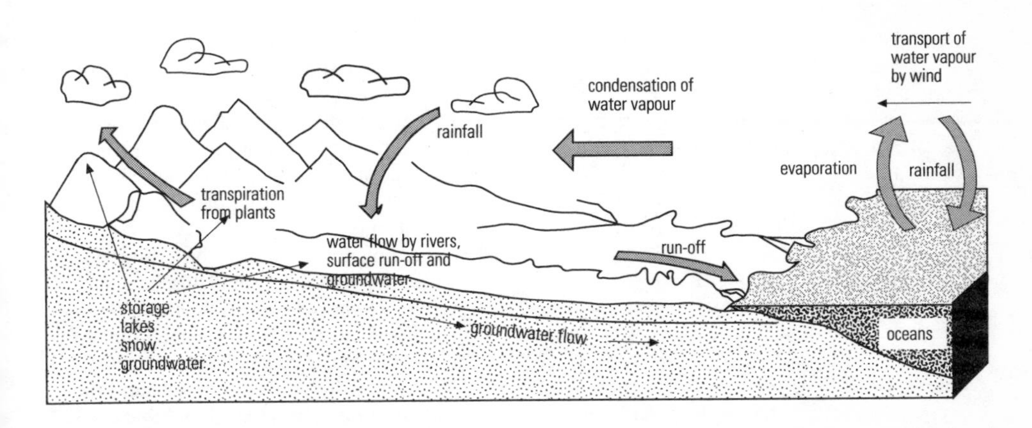

Summary of the water cycle.

Williamson ether synthesis The reaction of HALOGENOALKANES with sodium or potassium ALKOXIDES or PHENOXIDES to give ETHERS. The general reaction is:

$$RHal + NaOR' \rightarrow R\text{-}O\text{-}R' + NaHal$$

where Hal is a HALIDE, and R and R′ are ALKYL GROUPS. *See also* BROMOETHANE.

witherite An ore of BARIUM, a mineral form of BARIUM CARBONATE.

wolfram An old name for TUNGSTEN.

wolframite A dark brown mineral containing a mixture of iron and manganese tungstate, $FeWO_4$, $MnWO_4$. It is mined as a source of MANGANESE.

Woodward–Hoffmann rules Rules put forward by two American chemists that govern the formation of products during certain types of organic reactions. They are concerned with the maintenance of orbital symmetry.

wool A natural PROTEIN (KERATIN) fibre mainly obtained from the fleece of sheep. It consists of coiled protein chains joined by HYDROGEN BONDS, which gives it a springy texture.

work The effect of a force moving through a distance and converting one form of ENERGY to another. The SI UNIT of work is the JOULE.

work hardening The increase in hardness and brittleness of a material (usually a metal) when it undergoes PLASTIC deformation. *See also* ANNEALING.

wrought iron Iron containing less than 0.2 per cent carbon, typically obtained by the PUDDLING process. Wrought iron is easy to work, but too soft to be used in many structural applications. Its use is largely for decorative ironwork.

Wurtz reaction The reaction between HALOGENOALKANES and sodium in dry ether to yield ALKANES.

$$2RX + 2NaX + RR$$

It is named after C. Wurtz (1817–84), who discovered it.

X

xanthates Salts or esters with the formula ROC(S)SH, where R can be an ALKYL or ARYL group. Cellulose xanthate is formed in the viscose process for making RAYON.

xanthone, ***dibenzo-4-pyrone*** ($C_{13}H_8O_2$) A colourless, crystalline compound; melting point 174°C. It is used to make dyes.

xanthophyll A yellow pigment in plants, of the CAROTENOID group. It functions like CHLOROPHYLL in PHOTOSYNTHESIS.

xenon (Xe) The element with atomic number 54; relative atomic mass 131.3; melting point –112°C; boiling point –107°C. Xenon is a NOBLE GAS, extracted by the FRACTIONAL DISTILLATION from liquid air, where it occurs in minute quantities. A gas discharge in xenon has a bright blue-white colour, and is used in flashtubes for photography.

X-ray An ELECTROMAGNETIC WAVE with a wavelength shorter than about 10^{-9} m, produced by the most energetic energy changes of atomic electrons. Such waves are called gamma rays (*see* GAMMA RADIATION) if they are produced by changes within a NUCLEUS. X-rays are a form of IONIZING RADIATION and the shorter wavelengths (hard X-rays) are highly penetrating. They can be detected by photographic film or with a fluorescent screen or by the IONIZATION they produce in a radiation detector. The penetrating quality of X-rays has lead to their use in examining the internal structure of various objects, including the human body.

X-ray crystallography The study of the structure of crystals using X-RAY DIFFRACTION. In a simple experiment, the crystal is placed in the path of an X-RAY beam and surrounded by PHOTOGRAPHIC FILM. When the film is developed, spots appear in the directions of constructive interference. Since a crystal is a three-dimensional structure, it is possible to identify many crystal planes, or layers of atoms. The result is thus complex, but with modern computational techniques it is possible, even with involved crystal structures, to work back to the structure that produced the pattern.

Some samples cannot be obtained as single crystals, but only as a powder or POLYCRYSTALLINE sample, which effectively contains crystals aligned in all possible directions. The result is that the pattern of spots on the X-ray film becomes a pattern of rings. It is the electrons that diffract the X-ray beam and thus information can be obtained about the distribution of the electrons in the atoms or molecules from which the crystal is made. It was in this way that the structure of DNA was first uncovered by Crick and Watson in 1953, and similar techniques have more recently been used to unravel the structure of high-temperature superconductors.

X-ray diffraction The diffraction of an X-RAY beam off the atoms in a crystal. Because the atoms in a crystalline lattice are arranged regularly, they act rather like a diffraction grating for the X-rays, and strong scattering of the X-rays occurs in certain directions according to BRAGG'S LAW. *See also* X-RAY CRYSTALLOGRAPHY.

X-ray fluorescence An analytical technique in which a sample is bombarded with a high energy beam, usually of electrons. Electrons within atoms are excited and the wavelengths of the X-rays given off as these electrons fall back to their GROUND STATES are used to identify the elements present. Because the electron beam can be made very small and scanned over the surface of a sample, X-ray fluorescence is particularly useful for discovering the structure of materials that contain small crystals of varying chemical composition.

xylene *See* DIMETHYLBENZENE.

YZ

yield In any reaction, that proportion of the REAGENTS which react to produce the desired product.

ytterbium (Yb) The element with atomic number 70; relative atomic mass 173.0; melting point 819°C; boiling point 1,196°C; relative density 4.5. Ytterbium is a LANTHANIDE metal used in some special purpose ALLOYS.

yttrium (Yt) The element with atomic number 39; relative atomic mass 88.9; melting point 1,522°C; boiling point 3,338°C; relative density 4.5. Yttrium is a TRANSITION METAL, chemically similar to the LANTHANIDE metals. Yttrium compounds are used in some lasers.

zeolite Any of a class of minerals based on the aluminosilicate ANION that were traditionally used in WATER SOFTENING and other ION EXCHANGE processes. Zeolites can also absorb other small molecules and are used as 'molecular sieves'. They can also be used in SORPTION PUMPS, improving a vacuum by absorbing gas molecules. Natural zeolites have largely been superseded by synthetic materials.

zeroth law of thermodynamics A definition of the concept of temperature, which states that two objects are at the same temperature if there is no net flow of ENERGY between them when they are placed in contact. *See also* THERMODYNAMICS.

Ziegler–Natta catalysts Complex CATALYSTS used in the POLYMERIZATION of ALKENES. The products obtained are denser and tougher than those obtained by traditional high pressure methods of polymerization. Ethene can be polymerized using these catalysts, as can propene, the latter yielding polypropylene. Ziegler–Natta catalysts are prepared by the interaction of an organometallic derivative (*see* ORGANOMETALLIC COMPOUND) and a TRANSITION METAL derivative. For example titanium(IV) chloride and triethylaluminium, $Al(C_2H_5)^3$, are typically combined in a solvent, such as heptane, at low temperatures and pressures.

zinc (Zn) The element with atomic number 30; relative atomic mass 65.4; melting point 290°C; boiling point 732°C; relative density 7.1. Zinc is extracted from its ores (mainly zinc sulphide and zinc carbonate) by REDUCTION with carbon, for example:

$$2ZnCO_3 + C \rightarrow 2Zn + 3CO_2$$

Zinc is a fairly reactive TRANSITION METAL, reacting with acids to give hydrogen and zinc salts, for example:

$$Zn + 2HCl \rightarrow ZnCl_2 + H_2$$

Zinc is used as a protective coating on some steel objects (*see* GALVANIZING) and in the manufacture of ZINC-CARBON CELLS. It is also used in the manufacture of brass, an alloy with copper.

zincate Any compound containing the zincate ion, ZnO_2^{2-}. Zincates are formed by the reaction between zinc oxide and strong alkalis, for example:

$$ZnO + 2NaOH \rightarrow Na_2ZnO_2 + H_2O$$

zinc blende The chief ore of ZINC, comprising mainly zinc sulphide, ZnS.

zinc-carbon cell, *dry cell* A common type of electrochemical CELL in which zinc and carbon electrodes are used with an ELECTROLYTE in the form of a paste.

zinc chloride ($ZnCl_2$) A white solid; melting point 290°C; boiling point 732°C; relative density 2.9. The ANHYDROUS salt is DELIQUESCENT and can be made by the reaction of zinc with hydrogen chloride:

$$Zn + 2HCl \rightarrow ZnCl_2 + H_2$$

Zinc chloride is used as a flux in some soldering operations.

zinc oxide (ZnO) A white solid (yellow when hot); melting point 1,975°C; decomposes on further heating; relative density 5.5. It occurs naturally, and can also be made by heating zinc in air:

$$2Zn + O_2 \rightarrow 2ZnO$$

Zinc oxide is AMPHOTERIC, reacting with acids to form zinc salts, and bases to form ZINCATES, for example:

$$ZnO + 2HCl \rightarrow ZnCl_2 + 2H_2O$$

$$ZnO + 2NaOH \rightarrow Na_2ZnO_2 + 2H_2O$$

Zinc oxide is used as a white pigment and as an antiseptic in some ointments.

zinc sulphate ($ZnSO_4$) A white solid that decomposes on heating. Zinc sulphate usually occurs as the heptahydrate, $ZnSO_4.7H_2O$; relative density 1.9. Zinc sulphate can be made by dissolving zinc in sulphuric acid:

$$Zn + H_2SO_4 \rightarrow ZnSO_4 + H_2$$

It is used as a MORDANT in some dying processes.

zinc sulphide (ZnS) A creamy white solid, which decomposes on heating. Zinc sulphide occurs naturally as zinc blende, and decomposes to zinc on heating:

$$ZnS \rightarrow Zn + S$$

Zinc sulphide is used as a PHOSPHOR in some cathode ray tubes.

zirconium (Zr) The element with atomic number 40; relative atomic mass 91.2; melting point 1,853°C; boiling point 4,376°C; relative density 6.4. Zirconium is a TRANSITION METAL. It is used in the manufacture of fuel-rod casings in nuclear reactors, where its low neutron absorption coupled with good mechanical properties are an advantage.

zwitterion (German, *zwei* = two) An ion with both a positive and a negative charge. For example AMINO ACIDS in water form zwitterions by the loss of a proton from COOH, making it negative. The proton then goes to the NH_2 group, making it positive. In acidic conditions, the positive ion is formed; under basic conditions, the negative ion predominates. Under neutral conditions, the zwitterion exists.

zone refining A technique for obtaining high purity samples of metals and SEMICONDUCTORS, particularly silicon for the electronics industry. A bar of the material is heated in a narrow band, or zone, so this region melts. The heating is arranged so that the molten zone travels along the bar. Impurities tend to dissolve preferentially in either the molten or the solid material, and so can be swept along to one end of the bar leaving a bar of high purity material with impurities concentrated at one end, which is discarded.

Appendix I: SI units

Base units

Physical quantity	*Name*	*Symbol*	*Definition*
length	metre	m	the length equal to the length of path travelled by light in a vacuum in 1/(299,792,458) seconds
mass	kilogram	kg	the mass equal to that of the international prototype kilogram kept at Sèvres, France
time	second	s	the duration of 9,192,631,770 oscillations of the electromagnetic radiation corresponding to the electron transition between two hyperfine levels of the ground state of the caesium–133 atom
electric current	ampere	A	the constant electric current which, if maintained in two straight parallel conductors of infinite length and negligible cross-section, placed 1 metre apart in a vacuum, would produce a force between these conductors equal to 2×10^{-7} metres
thermodynamic temperature	kelvin	K	the fraction 1/273.16 of the thermodynamic temperature of the triple point of water
luminous intensity	candela	cd	the luminous intensity, in a given direction, of a source of monochromatic radiation of frequency 5.4×10^{14} Hz and has a radiant intensity in that direction of 1/683 watt per steradian
amount of substance	mole	mol	the amount of substance containing as many atoms (or molecules or ions or electrons) as there are carbon atoms in 12 g of carbon–12.

Supplementary units

plane angle	radian	rad	the plane angle subtended at the centre of a circle by an arc of equal length to the circle radius
solid angle	steradian	sr	the solid angle that encloses a surface on a sphere equal in area to the square of the radius of the sphere

Derived SI units

Physical quantity	*Name*	*Symbol*	*SI equivalent*
activity	becquerel	Bq	
dose	gray	Gy	
dose equivalent	sievert	Sv	
electric capacitance	farad	F	$A\,sV^{-1}$
electric charge	coulomb	C	As
electric conductance	siemens	S	
electric potential difference	volt	V	WA^{-1}
electric resistance	ohm	Ω	VA^{-1}
energy	joule	J	Nm
force	newton	N	$kg\,ms^{-2}$
frequency	hertz	Hz	s^{-1}
illuminance	lux	lx	
inductance	henry	H	$V\,sA^{-1}$
luminous flux	lumen	lm	
magnetic flux	weber	Wb	
magnetic flux density	tesla	T	
power	watt	W	$J\,s^{-1}$
pressure	pascal	Pa	Nm^{-2}

Appendix II: SI prefixes

Submultiple	*Prefix*	*Symbol*	*Multiple*	*Prefix*	*Symbol*
10^{-1}	deci	d	10^{1}	deca	da
10^{-2}	centi	c	10^{2}	hecto	h
10^{-3}	milli	m	10^{3}	kilo	k
10^{-6}	micro	μ	10^{6}	mega	M
10^{-9}	nano	n	10^{9}	giga	G
10^{-12}	pico	p	10^{12}	tera	T
10^{-15}	femto	f	10^{15}	peta	P
10^{-18}	atto	a	10^{18}	exa	E

Appendix III: Periodic table of the elements

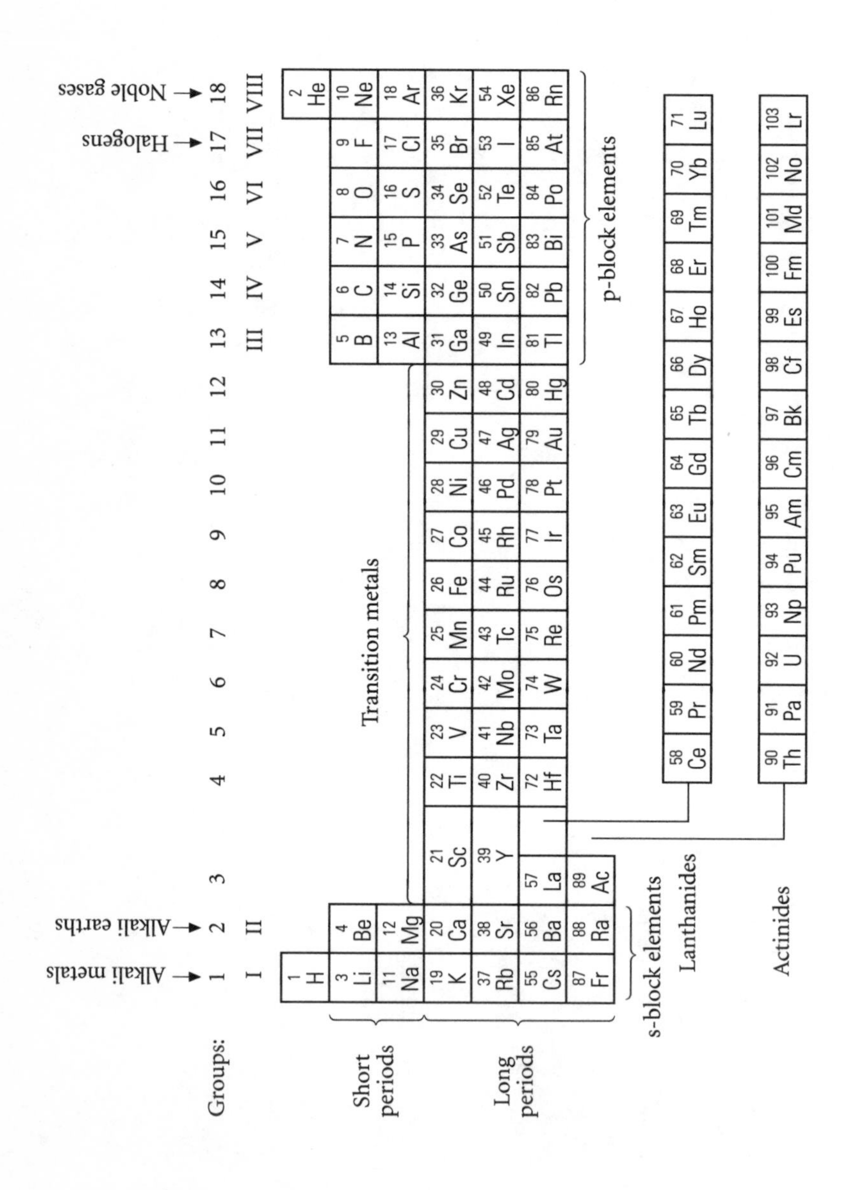

Appendix IV: The chemical elements

Name	*Symbol*	*Atomic number*	*Relative atomic mass*	*Melting point °C*	*Boiling point °C*	*Relative density*
Actinium	Ac	89	227*	1,050	3473**	–
Aluminium	Al	13	26.98	660	1,800	2.7
Americium	Am	95	243*	994	2,607	13.7
Antimony	Sb	51	121.8	630	1,750	6.7
Argon	Ar	18	39.95	–189	–185	1.7×10^{-3}
Arsenic	As	33	74.9	–	613	3.9
Astatine	At	85	(210)	302	377	–
Barium	Ba	56	137.3	725	1,640	3.5
Berkelium	Bk	97	247*	–	–	–
Beryllium	Be	4	9.0	1,285	2,970	1.9
Bismuth	Bi	83	208.98	271	1,560	9.8
Boron	B	5	10.8	2030	2,550	2.4
Bromine	Br	35	79.9	–7	58	3.1 (liquid)
Cadmium	Cd	48	112.4	321	765	8.7
Caesium	Cs	55	132.9	28	690	1.9
Calcium	Ca	20	40.1	840	1,484	1.6
Californium	Cf	98	251*	–	–	–
Carbon	C	6	12.0	–	3,500	2.3
Cerium	Ce	58	140.1	798	3,433	6.8
Chlorine	Cl	17	35.5	–101	–35	3.2×10^{-3} (0°C)

The chemical elements (continued)

Name	*Symbol*	*Atomic number*	*Relative atomic mass*	*Melting point °C*	*Boiling point °C*	*Relative density*
Chromium	Cr	24	52.0	1,900	2,640	7.2
Cobalt	Co	27	58.9	1,495	2,870	8.9
Copper	Cu	29	63.5	1,083	2,582	8.9
Curium	Cm	96	247*	–	–	–
Dysprosium	Dy	66	162.5	1,412	2,567	8.6
Einsteinium	Es	99	254*	–	–	–
Erbium	Er	68	167.3	1,529	2,863	9.0
Europium	Eu	63	152.0	852	1,529	5.2
Fermium	Fm	100	(237)	–	–	–
Fluorine	F	9	19.0	–220	–188	1.7×10^{-3} (0°C)
Francium	Fr	87	223*	27	–	–
Gadolinium	Gd	64	157.3	1,312	3,273	7.9
Gallium	Ga	31	69.7	30	2,403	5.9
Germanium	Ge	32	72.6	937	2,830	5.4
Gold	Au	79	197.0	1,064	2,807	19.3
Hafnium	Hf	72	178.5	2,230	4,602	13.3
Helium	He	2	4.0	–269	–268.9	0.2×10^{-3}
Holmium	Ho	67	164.9	1,472	2,700	8.8
Hydrogen	H	1	1.0	–259	–253	0.1×10^{-3}
Indium	In	49	114.8	157	2,080	7.3
Iodine	I	53	126.9	–	183	4.9
Iridium	Ir	77	192.2	2,410	4,130	22.5

The chemical elements (continued)

Name	*Symbol*	*Atomic number*	*Relative atomic mass*	*Melting point °C*	*Boiling point °C*	*Relative density*
Iron	Fe	26	55.8	1,535	2,750	7.9
Krypton	Kr	36	83.8	–157.4	–152	3.5×10^{-3}
Lanthanum	La	57	138.9	918	3464	6.1
Lawrencium	Lw	103	257*	–	–	–
Lead	Pb	82	207.2	328	1,740	11.4
Lithium	Li	3	6.9	180	1,340	0.5
Lutetium	Lu	71	175.0	1,663	3,402	9.8
Magnesium	Mg	12	24.3	651	1,107	1.7
Manganese	Mn	25	54.9	1,244	2,040	7.4
Mendelevium	Md	101	256*	–	–	–
Mercury	Hg	80	200.6	–39	357	13.6
Molybdenum	Mo	42	94.9	2,610	5,560	10.2
Neodymium	Nd	60	144.2	1,016	3,068	7.0
Neon	Ne	10	20.2	–249	–246	0.8×10^{-3}
Neptunium	Np	93	237*	–	–	–
Nickel	Ni	28	58.7	1,450	2,840	8.9
Niobium	Nb	41	92.9	2,468	4,742	8.6
Nitrogen	N	7	14.0	–210	–196	1.2×10^{-3}
Nobelium	No	102	259*	–	–	–
Osmium	Os	76	190.2	3,045	5,027	22.6
Oxygen	O	8	16.0	–214	–183	1.3×10^{-3}
Palladium	Pd	46	106.4	1,551	3,140	12.3

The chemical elements (continued)

Name	*Symbol*	*Atomic number*	*Relative atomic mass*	*Melting point °C*	*Boiling point °C*	*Relative density*
Phosphorus	P	15	31.0	44	280	1.8
Platinum	Pt	78	195.1	1,772	3,800	21.3
Plutonium	Pu	94	244*	641	3,232	–
Polonium	Po	84	210*	254	–	9.2
Potassium	K	19	39.1	64	774	0.9
Praseodymium	Pr	59	140.9	931	3,512	6.8
Promethium	Pm	61	145*	1,042	3,000	7.3
Protactinium	Pa	91	231*	–	–	–
Radium	Ra	88	226*	700	1,140	5.1
Radon	Rn	86	222*	–72	–62	10.0×10^{-3} (0°C)
Rhenium	Re	75	186.2	3,180	5,620	22.0
Rhodium	Rh	45	102.9	1,966	3,727	12.4
Rubidium	Rb	37	85.5	38	688	1.5
Ruthenium	Ru	44	101.1	2,310	3,900	12.24
Samarium	Sm	62	150.3	1,075	1,791	7.5
Scandium	Sc	21	45.0	1,540	2,850	3.0
Selenium	Se	34	79.0	217	685	4.8
Silicon	Si	14	28.1	1,410	2,335	2.3
Silver	Ag	47	107.9	962	2,212	10.5
Sodium	Na	11	23.0	98	892	1.0
Strontium	Sr	38	87.62	768	1,300	2.5
Sulphur	S	16	32.1	112	445	2.1

The chemical elements (continued)

Name	*Symbol*	*Atomic number*	*Relative atomic mass*	*Melting point °C*	*Boiling point °C*	*Relative density*
Tantalum	Ta	73	180.9	2,996	5,427	16.7
Technetium	Tc	43	99*	2,171	4,876	11.5
Tellurium	Te	52	127.6	451	1,390	6.2
Terbium	Tb	65	158.9	1,365	3,230	8.2
Thallium	Tl	81	204.4	303	1,460	11.9
Thorium	Th	90	232*	1,750	4,790	11.7
Thulium	Tm	69	168.9	1,545	1,950	9.3
Tin	Sn	50	118.7	232	2,270	7.3
Titanium	Ti	22	47.9	1,660	3,280	4.5
Tungsten	W	74	183.9	3,410	5,660	19.3
Uranium	U	92	238*	1,132	3,818	19.1
Vanadium	V	23	50.9	1,890	3,380	6.1
Xenon	Xe	54	131.3	–112	–107	5.5×10^{-3}
Ytterbium	Yb	70	173.0	819	1,196	4.5
Yttrium	Y	39	88.9	1,522	3,338	4.5
Zinc	Zn	30	65.4	290	732	7.1
Zirconium	Zr	40	91.2	1,853	4,376	6.4

*Relative atomic mass of longest-lived isotope

Appendix V: Greek alphabet

Α	α	alpha
Β	β	beta
Γ	γ	gamma
Δ	δ	delta
Ε	ε	epsilon
Ζ	ζ	zeta
Η	η	eta
Θ	θ	theta
Ι	ι	iota
Κ	κ	kappa
Λ	λ	lambda
Μ	μ	mu
Ν	ν	nu
Ξ	ξ	xi
Ο	ο	omicron
Π	π	pi
Ρ	ρ	ro
Σ	σ	sigma
Τ	τ	tau
Υ	υ	upsilon
Φ	φ	phi
Χ	χ	chi
Ψ	ψ	psi
Ω	ω	omega